Protein function

a practical approach

TITLES PUBLISHED IN THE PRACTICAL APPROACH SERIES

Series editors:
Dr D Rickwood
Department of Biology, University of Essex
Wivenhoe Park, Colchester, Essex CO4 3SQ, UK
Dr B D Hames
Department of Biochemistry, University of Leeds
Leeds LS2 9JT, UK

Affinity chromatography
Animal cell culture
Animal virus pathogenesis
Antibodies I & II
Biochemical toxicology
Biological membranes
Biosensors
Carbohydrate analysis
Cell growth and division
Centrifugation (2nd edition)
Clinical immunology
Computers in microbiology
DNA cloning I, II & III
Drosophila
Electron microscopy in molecular biology
Fermentation
Flow cytometry
Gel electrophoresis of nucleic acids (2nd edition)
Gel electrophoresis of proteins (2nd edition)
Genome analysis
HPLC of small molecules
HPLC of macromolecules
Human cytogenetics
Human genetic diseases
Immobilised cells and enzymes
Iodinated density gradient media
Light microscopy in biology
Liposomes
Lymphocytes
Lymphokines and interferons
Mammalian development
Medical bacteriology
Medical mycology
Microcomputers in biology
Microcomputers in physiology
Mitochondria
Mutagenicity testing
Neurochemistry
Nucleic acid and protein sequence analysis
Nucleic acid hybridisation
Nucleic acid sequencing
Oligonucleotide synthesis
Photosynthesis: energy transduction
Plant cell culture
Plant molecular biology
Plasmids
Post-implantation mammalian development
Prostaglandins and related substances
Protein function
Protein purification applications
Protein purification methods
Protein sequencing
Protein structure
Proteolytic enzymes
Radioisotopes in biology
Receptor biochemistry
Receptor – effector coupling
Ribosomes and protein synthesis
Solid phase peptide synthesis
Spectrophotometry and spectrofluorimetry
Steroid hormones
Teratocarcinomas and embryonic stem cells
Transcription and translation
Virology
Yeast

Protein function

a practical approach

Edited by

T E Creighton

MRC, Laboratory of Molecular Biology,
Hills Road, Cambridge CB2 2QH, UK

IRL PRESS
at
OXFORD UNIVERSITY PRESS
Oxford New York Tokyo

IRL Press
Eynsham
Oxford
England

First published 1989
Reprinted (with corrections) 1990

British Library Cataloguing in Publication Data

Protein function.
1. Proteins
I. Creighton, Thomas E. *1940–* II. Series
547.7′5

Library of Congress Cataloging in Publication Data

Protein function.
(Practical approach series)
Includes bibliographies and index.
1. Proteins—Analysis. 2. Proteins—Structure-activity relationships.
3. Proteins—Affinity labeling. 4. Biochemistry—Technique.
I. Creighton, Thomas E., 1940– . II. Series. [DNLM 1. Proteins—physiology. QU 55 P96645]
QP551.P69583 1988 574.19′245 88-31963
ISBN 0 19 963006 2 (hardback)
ISBN 0 19 963007 0 (softback)

Previously announced as:
ISBN 1 85221 139 3 (hardback)
ISBN 1 85221 140 7 (softback)

Printed by Information Press Ltd, Oxford, England.

Preface

The functional properties of proteins vary enormously. Some are simply structural, some catalyse chemical reactions; others transmit information, or interconvert the various forms of energy: chemical, light, movement. Consequently, a comprehensive guide to studying protein function would encompass all of biochemistry.

This volume does not attempt that, but concentrates on those aspects of protein function that are common to most, if not all, proteins. The first of these is the dependence of the functional properties upon the covalent structure and the conformation of each protein. The first priority in any study of protein function should be to preserve the structural and conformational integrity of the protein; to study a 'dying' protein is a hopeless task. Therefore, the very first chapter provides guidelines as to how to maintain protein function, both by minimizing covalent alterations of its structure and by maintaining its folded conformation. Further procedures for studying protein structure and conformation are given in the companion volume *Protein Structure: A Practical Approach*.

The other common theme of protein function is that it invariably involves the protein interacting physically with other molecules; a protein never acts in isolation, but always acts upon something. Therefore, a primary concern is to characterize the interaction of the protein with these other molecules. General procedures for measuring the most basic parameters, the number of ligand molecules bound by a protein and their relative affinities, are described in Chapter 2. Such studies have a very long history, but there are a surprising number of misconceptions about the intepretation of binding data when multiple ligands are bound to the same protein. Consequently, a major part of such a chapter is pointing out how *not* to proceed, and this chapter also provides numerous examples of the errors commonly made in such studies.

Electrophoresis is a major technique in studying protein structure (see *Protein Structure: A Practical Approach* and *Gel Electrophoresis of Proteins: A Practical Approach*), and is also becoming a very useful technique in studying protein function, with the recent advances in blotting techniques described in Chapter 3. The proteins in an electrophoretic gel can be transferred to a membrane, to which they stick tightly. Although I personally find it difficult to understand how it happens, it is an undisputed fact that a sufficient fraction of the protein molecules refold to a sufficient extent to exhibit ligand-binding ability, in spite of having been denatured and stuck to the membrane. With this simple technique, the abilities of numerous proteins to bind virtually any ligand can be tested very simply.

Biologically relevant ligand binding invariably occurs at specific sites on proteins, so it is important to identify and characterize all such binding sites. One of the most direct methods for doing so is by affinity labelling, described in Chapter 4. A reactive group is incorporated into a ligand and reacts with the protein much more rapidly when bound than when free in solution, due to the very high 'effective concentrations' that can occur in ligand–protein complexes. With larger ligands, such as other protein molecules, the interacting molecules can be identified by cross-linking them covalently in the complex. Techniques for doing so are described in Chapter 5. The procedures described in these two chapters are illustrated for a specific class of ligands, but should be readily adapted to other complexes.

Very many proteins consist of multiple polypeptide chains, usually as relatively autonomous structural subunits. This oligomeric structure often has profound implications for the function of the protein, but in many cases the functional implications are not at all obvious. Very many simple, but ingenious, techniques have been devised to examine the roles of subunits in protein function, and these are described in Chapter 6.

One of the most biologically important areas of protein function is in the control of gene expression, which invariably involves protein binding to DNA and RNA. Most of the regulatory proteins occur in very small quantities within cells and have consequently been very difficult to study. Most of the techniques used have relied upon the properties of the nucleic acids, rather than the protein. These involve the identification of DNA–protein complexes by the change in electrophoretic mobility of a small fragment of DNA produced by a protein binding to it; such complexes are extremely tight and consequently dissociate so slowly that the complex can survive an electrophoretic separation, as in 'bandshift gels'. The specific sites on the DNA occupied by the protein can be identified by the aptly-named 'footprinting' technique. Chapter 7 describes these techniques using purified proteins, while Chapter 8 describes how to use them to identify sequence-specific DNA-binding proteins in crude mixtures, then to purify them with the use of DNA affinity chromatography. (The general techniques of protein purification and affinity chromatography with other ligands are described in the volumes *Protein Purification: A Practical Approach* and *Affinity Chromatography: A Practical Approach*.)

The functional groups involved in protein function often have somewhat unusual physical properties, and these may be characterized by the relatively simple technique of competitive labelling described in Chapter 9. This technique can also be used to identify binding sites on proteins for ligands by comparing the protein with the protein–ligand complex, since interaction with another molecules usually causes changes in the reactivities of the functional groups involved. The classical technique for identifying functional groups involved in protein function is to examine the functional effects of chemically modifying the various classes of reactive groups. A number of new approaches and reagents have been developed in recent years, and some of these are described in Chapter 10.

Finally, the most specific modifications of protein structure are those produced by the recently-developed techniques of site-directed mutagenesis, and this technique has become so widely used that no volume on protein function would be complete without it, even though the procedures described do not actually involve the protein. A gene for the protein is required, but this can now be obtained almost routinely by the procedures described in *DNA Cloning: A Practical Approach*, volumes 1–3) or by gene synthesis (*Oligonucleotide Synthesis: A Practical*

Approach). The procedures described in Chapter 11 are some of the most recent and most efficient yet devised.

Use of the techniques described in this volume should provide much information about the functional properties of any protein, but the procedures must be used appropriately, taking into account all the relevant properties of proteins. For a comprehensive description, the reader is referred to my volume (*Proteins: Structures and Molecular Properties*. W. H. Freeman, New York, 1983). For relatively simple techniques to characterize a protein's structure, the companion volume *Protein Structure: A Practical Approach* is highly recommended.

T. E. Creighton

Contributors

G. Ammerer
MRC Laboratory of Molecular Biology, Hills Road, Cambridge CB2 2QH, UK. Present address: Institute of Molecular Pathology, Dr. Bohr Gasse, A1030 Vienna and Institut für Allgemeine Biochemie, Universität Wien, A1090 Vienna, Austria

C. Casiano
Department of Biological Chemistry, University of California School of Medicine, Davis, CA 95616, USA

R. F. Colman
Department of Chemistry and Biochemistry, University of Delaware, Newark, DE 19716, USA

F. Eckstein
Max-Planck-Institut für Experimentelle Medizin, Abteilung Chemie, Herman-Rein-Strasse 3, D-3400 Göttingen, FRG

E. Eisenstein
Molecular Biology-Virus Laboratory, 229 W. M. Stanley Hall, University of California, Berkeley, CA 94720, USA

T. Imoto
Faculty of Pharmaceutical Sciences, Kyushu University 67, Maidashi, Higashi-ku, Fukuoka 812, Japan

H. Kaplan
Department of Biochemistry, University of Ottawa, Ottawa, Canada K1N 9B4

A. M. Klibanov
Department of Chemistry, Massachusetts Institute of Technology, Cambridge, MA 02139, USA

I. M. Klotz
Department of Chemistry, Northwestern University, Evanston, IL 60201, USA

D. Rhodes
MRC Laboratory of Molecular Biology, Hills Road, Cambridge CB2 2QH, UK

J. R. Sayers
Max-Planck-Institut für Experimentelle Medizin, Abteilung Chemie, Herman-Rein-Strasse 3, D-3400 Göttingen, FRG

H. K. Schachman
Molecular Biology-Virus Laboratory, 229 W. M. Stanley Hall, University of California, Berkeley, CA 94720, USA

D. Shore
MRC Laboratory of Molecular Biology, Hills Road, Cambridge CB2 2QH, UK. Present address: Department of Microbiology, Columbia University, New York, NY 10032, USA

P. K. Sorger
MRC Laboratory of Molecular Biology, Hills Road, Cambridge CB2 2QH, UK

A. K. Soutar
MRC Lipid Research Unit, Hammersmith Hospital, DuCane Road, London W12 0HS, UK

R. R. Traut
Department of Biological Chemistry, University of California School of Medicine, Davis, CA 95616, USA

D. B. Volkin
Department of Chemistry, Massachusetts Institute of Technology, Cambridge, MA 02139, USA

D. P. Wade
MRC Lipid Research Unit, Hammersmith Hospital, DuCane Road, London W12 0HS, UK

H. Yamada
Faculty of Pharmaceutical Sciences, Kyushu University 67, Maidashi, Higashi-ku, Fukuoka 812, Japan

N. M. Young
Division of Biological Sciences, National Research Council of Canada, 100 Sussex Drive, Ottawa, Canada K1A 0R6

N. Zecherle
Department of Biological Chemistry, University of California School of Medicine, Davis, CA 95616, USA

Contents

Abbreviations

AP1	yeast homologue of mammalian activator protein 1
APOP	*N*-[4-(*p*-azidosalicylamido)butyl]-3′-(2′-pyridyldithio)propionamide
APTP	*N*(4-azidophenyl)phthalimide
ATCase	aspartate transcarbamoylase
X-BDB-TA 5′-DP	X-[4-bromo-2,3-dioxobutylthio]-adenosine 5′-diphosphate
2-BDB-TAMP	2-[4-bromo-2,3-dioxobutylthio]adenosine 5′-monophosphate
BSA	bovine serum albumin
CBS-Lys	N^{ε}-(4-carboxybenzenesulphonyl) lysine
CBS-Tyr	*O*-(4-carboxybenzenesulphonyl) tyrosine
Chaps	3-[(3-cholamidopropyl)dimethyl ammonio]-1-propane sulphonate
CP1	centromere-binding protein
DMF	dimethylformamide
DMS	dimethylsulphate
DMSO	dimethyl sulphoxide
DNP	dinitrophenyl
DNP-F	1-fluoro-2,4-dinitrobenzene
dNTPαS	deoxynucleoside 5′-*O*-(1-thio) triphosphate
DTBP	dithiobispropionimidate
DTT	dithiothreitol
EDC	1-ethyl-3-[3-(dimethylamino)propyl]carbodiimide
EDTA	ethylenediamine tetraacetic acid
5′-FSBA	5′-*p*-fluorosulphonylbenzoyl adenosine
5′-FSBεA	5′-*p*-fluorosulphonylbenzoyl-1,N^6-enthenoadenosine
5′-FSBG	5′-*p*-fluorosulphonylbenzoyl guanosine
GdmCl	guanidinium hydrochloride
HSE	heat shock element segment of DNA
HSF	heat shock transcription factor protein from yeast
Hepes	*N*-hydroxyethyl-piperazine-*N*′-2-ethanesulphonic acid
HPLC	high-performance liquid chromatography
IPTG	isopropyl-β-D-thiolgalactopyranoside
2-IT	2-iminothiolane
LDL	low-density lipoprotein
Mes	2-(N-morpholino) ethane sulphonic acid
NBS	*N*-bromosuccinimide
NMR	nuclear magnetic resonance
PADR	2′-phosphoadenosine 5′-diphosphoribose
PAGE	polyacrylamide gel electrophoresis
PALA	*N*-phosphonacetyl-L-aspartate
PEG	polyethylene glycol
Pipes	piperazine-*N,N*′-bis-(2-ethanesulphonic acid)

PMSF	phenylmethylsulphonyl fluoride
PRTF	pheromone receptor transcription factor protein
PTH	phenylthiohydantoin
RAP1	repressor activator protein 1
RF IV	double-stranded closed-circular M13 DNA
SBF-B	silencer binding factor-B/ARS binding factor 1 protein
SDS	sodium dodecyl sulphate
SPDP	*N*-succinimidyl 3-(2-pyridyldithio)propionate
2-TA 2′,5′-DP	2-thioadenosine 2′,5′-bisphosphate
TB	90 mM Tris–borate buffer, pH 8.3
TBE	90 mM Tris–borate buffer, pH 8.3, 2 mM EDTA
TCA	trichloroacetic acid
TEMED	*N,N,N′,N′*-tetramethylethylene diamine
TFIIIA	transcription factor IIIA protein from *Xenopus*
THPA	3,4,5,6-tetrahydrophthalic anhydride
TLC	thin-layer chromatography
TLCK	N_{α}-*p*-tosyl-L-lysine chloromethyl ketone
TNBS	trinitrobenzene sulphonic acid
TNM	tetranitromethane
TPCK	*N*-tosyl-L-phenylalanine chloromethyl ketone
Tris	tris (hydroxymethyl) amino methane
VLDL	very low-density lipoprotein
X-gal	5-bromo-4-chloro-3-indolyl-β-galactoside

CHAPTER 1

Minimizing protein inactivation

DAVID B. VOLKIN and ALEXANDER M. KLIBANOV

1. INTRODUCTION

The stabilization of enzymes and proteins against irreversible inactivation is a concern to biochemists and biochemical engineers alike. Whether attempting the purification of a recombinant protein in the laboratory or loading an industrial bio-reactor for the large-scale enzymatic production of a food ingredient, minimizing protein inactivation is a crucial part of any successful isolation, storage or application procedure. Historically, enzyme stabilization has been considered an 'artsy-craftsy' subject and therefore has been largely ignored in general biochemistry textbooks. There is an element of truth to this accusation, because irreversible inactivation is not only a function of external agents such as heat, detergents or pH, but it is also dependent on the nature of the protein. Consequently, as much as we would like to join our colleagues in this book and give the readers straightforward recipes to follow in order to stabilize their particular enzyme or protein, it is simply not possible.

However, one should not quickly admit defeat and blindly seek traditional empirical approaches to stabilize enzymes (stabilization of enzymes is used here to mean either a decrease in the rate constants of inactivation or an increase in the degree of denaturing action required to reach a certain extent of inactivation), such as adding sugars and glycerol or lyophilizing enzyme solutions (although in certain cases these approaches are quite valid). In this chapter we would like to develop a rational strategy toward enzyme stabilization; in other words, a general game plan that can be adapted to any particular situation. We will begin this pursuit by asking the following three questions.

(i) What is the cause (external agent) of inactivation?
(ii) What is the mechanism of inactivation?
(iii) What approaches can we take to prevent, bypass, or at least minimize this mechanism?

In the following sections, we will consider various causes of inactivation and attempt to understand their mechanisms. Suggestions will be offered for both general strategies to prevent reversible unfolding and specific stabilization approaches for specific enzymes. These particular examples will address the most frequently encountered modes of irreversible protein inactivation: aggregation, autolysis, extremes of pH, oxidation and heat. We will use examples from the literature to illustrate some successful strategies and tactics. There has been no attempt to review the literature exhaustively; we simply looked for interesting stories to tell. For a more complete review of the enzyme stabilization literature, review articles can be consulted (1–5).

2. REVERSIBLE VERSUS IRREVERSIBLE INACTIVATION

In order to develop a rationale for enzyme stabilization, we must first consider the differences between reversible and irreversible protein inactivation. The native, catalytically active form of an enzyme molecule is held together by a delicate balance of non-covalent forces: hydrophobic, ionic, and van der Waals interactions and hydrogen bonds. Upon exposure to a certain concentration of denaturing agents or to adverse environmental conditions, these non-covalent forces are first weakened and then broken apart, thereby causing the enzyme molecule to unfold at least partially. This transition to a less ordered conformation disassembles the active site of an enzyme, thus causing inactivation (*Figure 1*). This process of unfolding a protein molecule is a cooperative, two-step phenomenon (between 100% N or 100% U). It must be stressed that this unfolding event usually is completely *reversible*. Upon removal of the antagonistic agent, the enzyme molecule should refold to its catalytically active form, because this conformation is thermodynamically favoured (6). The native, folded form of a protein molecule is only marginally more stable than the unfolded one. In fact, the net difference between the free energies of the folded and unfolded form is 5–20 kcal/mol or, in other words, represents just a few extra hydrogen bonds or ionic interactions. The *reversible* unfolding of proteins has been examined extensively, and its origin and mechanisms are well understood (7–11). This reversible unfolding of a protein molecule is usually the initial stage in the inactivation process. The subsequent *irreversible* events, both covalent and conformational, are specific for individual proteins and individual causes of inactivation (*Figure 2*).

3. CAUSES AND MECHANISMS OF IRREVERSIBLE INACTIVATION

In this section, we will address the first two questions presented in Section 1: what are the causes and mechanisms of protein modification and inactivation? Through a mechanistic understanding of these events, we will be able to put together and organize tactics to circumvent the inactivation reaction(s).

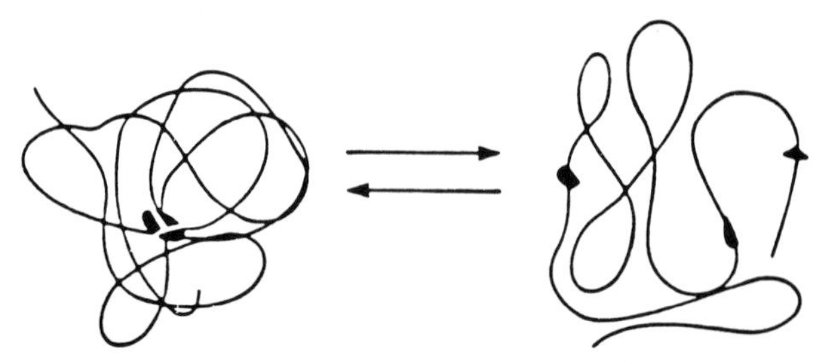

Figure 1. Schematic representation of reversible unfolding (the initial stage of irreversible inactivation) of an enzyme molecule. Filled region is an active center (5). Courtesy of Academic Press.

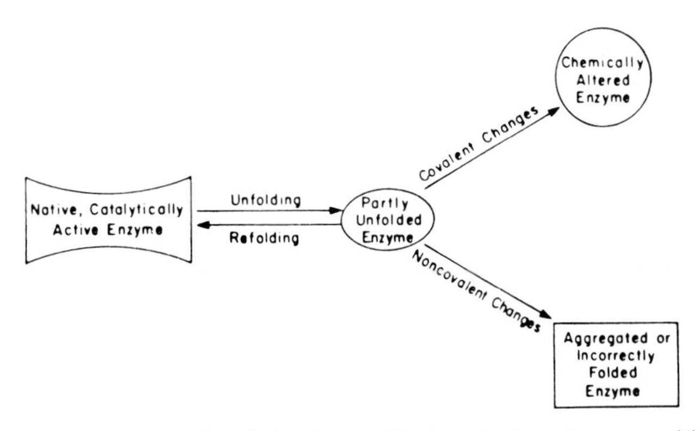

Figure 2. Diagram of events occurring during irreversible inactivation of enzymes (5). Courtesy of Academic Press.

The rates of some well-characterized chemical reactions in proteins are compared in *Table 1*.

3.1 Proteases and autolysis

Operational and storage inactivation of enzymes is often due to the action of microorganisms and exogenous proteases (2). This *in vitro* proteolysis is a common problem that reduces the purification yield of eukaryotic polypeptides from genetically-engineered microbes, especially *Escherichia coli* (20). Proteases presumably can catalyse the hydrolysis of peptide bonds within a protein molecule

Table 1. Examples of deteriorative chemical reactions frequently encountered in proteins and peptides.

Examples	*Conditions*	*Time for 50% of residue modified (min)*
1. Deamidation of an Asn–Gly containing peptide (12)	37°C, pH 7.4	2040
	100°C, pH 7.4	12
2. β-elimination of cystine residues in:		
12 different proteins (13)	100°C, pH 6	540–950
	100°C, pH 8	36–84
19 different proteins and peptides (14)	25°C in 0.2 M NaOH	16–330
3. Peptide bond hydrolysis of:		
Asp–Pro dipeptide (15)	100°C in 0.015 M HCl	11
$Asp_{121}-Ala_{122}$ in RNase A (16)	90°C, pH 4	1560
4. Oxidation of a cysteine residue:		
papain (17)	23°C, pH 6 in 50 μM H_2O_2	10
Cys mutant of subtilisin (18)	25°C, pH 9.5 in 1 M H_2O_2	12.5
α-amylase from *B.stearothermophilus* (19)	90°C, pH 8	30
	(+10 μM $CuCl_2$)	(2)
5. Oxidation of a methionine residue in subtilisin (18)	25°C, pH 9.5 in 0.1 M H_2O_2	2.5

only when that portion of the molecule is in the denatured state; in fact, susceptibility to proteolysis is a common experimental 'probe' for denatured proteins (21). When the protein substrate is also a proteolytic enzyme, the process is called autolysis (self-digestion).

3.2 **Aggregation**

Aggregation has long been recognized as a mechanism of protein inactivation (4). Nearly a century ago, scientists first observed that the heating of aqueous albumin solutions led to precipitate formation. Current interest in the protein folding problem has led to a renewed awareness of aggregation as a cause of inactivation of proteins during their refolding from the denatured state (22).

Aggregation can be thought of as a three step process:

$$N \rightleftharpoons U \rightarrow A \rightarrow A_s$$

where $N \rightleftharpoons U$ represents reversible unfolding, A is aggregated protein and A_s are protein aggregates having undergone thiol–disulphide interchange. First, monomolecular conformational changes must occur leading to reversible protein denaturation. This process exposes the buried hydrophobic amino acid residues to the aqueous solvent. Second, protein molecules with this altered tertiary structure associate with one another to minimize the unfavourable exposure of hydrophobic amino acid residues. Finally, if the protein molecules contain cysteine and cystine residues, inter-molecular thiol–disulphide interchange can occur. Unlike many other causes of protein inactivation, aggregation is not necessarily irreversible. Reactivation may be possible by breaking up the inter-molecular non-covalent forces (hydrogen-bonds or hydrophobic interactions) via the use of denaturants and regenerating native disulphide bonds via reduction and re-oxidation in the absence of denaturant.

On a final note, a distinction should be made between aggregation and simple precipitation. The latter implies that no appreciable conformational changes took place in the macromolecule before dropping out of solution. Therefore, precipitates are easily re-dissolved in aqueous solutions to recover native properties, as is often the case with crystallized or lyophilized proteins.

3.3 **Extremes of pH**

Inactivation and modification of enzymes due to extremes of pH is a common, well-documented event, yet the mechanism of this inactivation can vary with individual enzymes and specific environmental conditions. Furthermore, the degree of inactivation can range from minor conformational changes to irreversible inactivation, depending on incubation conditions. For example, a change in pH can cause the ionization of a catalytically essential group that may result in inactivation without any severe effect on the structure; a simple readjustment of pH will restore activity. Acid and alkaline denaturation of proteins is a well-documented phenomenon in protein denaturation/reactivation studies (9). The important factor at either pH extreme is that, once far away from a protein's isoelectric point, electrostatic interactions between like charges within the protein molecule result in a tendency to unfold. Also, residues that are buried in a protein interior in non-ionized form can ionize only if the protein unfolds; histidine residues are largely responsible for the unfolding of proteins at acid pH. In principle, this process is fully reversible, but these conformational changes can often lead to irreversible aggregation or, in the case of proteases, to autolysis. Finally, extremes of pH can initiate chemical reactions that alter, cross-link or destroy amino acid residues and consequently cause irreversible inactivation.

Peptide bond hydrolysis occurs readily under strongly acidic conditions or by a combination of milder pH and elevated temperatures. While complete acid hydrolysis of protein into its amino acids is obtained under extreme conditions (6 M HCl, 24 h, 110°C), shorter exposures under less acidic conditions show preferred peptide hydrolysis at aspartic acid residues; aspartyl–prolyl linkages are especially vulnerable (23). In addition, the deamidation of asparagine and glutamine residues, which introduces negative charges into the hydrophobic interior of the protein resulting in inactivation, readily takes place under strongly acidic conditions (24), as well as at neutral and basic pH values (12, 16).

Proteins are commonly exposed to alkaline conditions during food processing, and many adverse side reactions have been described (25). These include partial peptide bond hydrolysis, deamidation, hydrolysis of arginine to ornithine, β-elimination and racemization, double bond formation, destruction of amino acid residues and formation of new amino acids. An interesting example of a deteriorative change is the effect of alkali on disulphide bonds in a protein molecule. Base-catalysed β-elimination destroys disulphide bonds with concomitant formation of dehydroalanine and thiocysteine residues:

B: → H
—NH—C(H)(CH$_2$–S–S–CH$_2$–CH(—CO—)(—NH—))—CO— ⟶ —NH—C(=CH$_2$)—CO— + S^-—S—CH$_2$—CH(—CO—)(—NH—)

Dehydroalanine is susceptible to addition reactions, especially from a lysine ε-amino group, thus creating a new intra-molecular cross-link (lysinoalanine) in place of the natural one. This reaction also occurs at neutral pH values and elevated temperatures (13):

H^+

—NH—C—CO—   —NH—CH—CO—

‖ |

CH_2 CH_2

$:NH_2$ $\longrightarrow$ |

| NH_2^+

$(CH_2)_4$ |

| $(CH_2)_4$

—CO—CH—NH— |

—CO—CH—NH—

3.4 **Oxidation**

Oxidation of amino acids, especially those with aromatic side-chains, as well as methionine, cysteine and cystine residues, can occur with a variety of oxidants. Molecular O_2, hydrogen peroxide and oxygen radicals are all known oxidants of proteins, which give rise to different deteriorative reactions. Oxidation of cysteine to cystine residues occurs at basic pH in the presence of transition metal ions like Cu^{2+}. Depending on the strength of the oxidants, however, cysteine can also be converted to sulphenic, sulphinic or sulphonic (cysteic) acids (26).

Hydrogen peroxide is a relatively non-specific oxidizing agent, but under acidic conditions the primary reaction that occurs is the oxidation of methionine to its sulphoxide (27). Although methionine sulphoxide can be reduced *in vivo* by enzymes and *in vitro* by thiols, industrial use or storage of enzymes may be limited by this reaction. Protein inactivation by oxidation can occur in biological systems via hydroxyl ($OH\cdot$) and superoxide (O_2^-) radicals, hydrogen peroxide and hypochlorite ion (OCl^-). For example, neutrophils can produce high levels of oxidizing agents to destroy bacteria (28).

3.5 **Surfactants and detergents**

Detergents cause protein denaturation in a unique way since they interact strongly with proteins at low concentrations. These interactions may be so strong that denaturation is often irreversible (8, 9, 29–31).

Detergents are categorized as either ionic or non-ionic; both groups contain a long-chain hydrophobic moiety called a tail, but differ in the 'head' region groups, which can be either charged or uncharged. Detergents differ greatly in their relative balance between hydrophilic and hydrophobic moieties, which is an important factor determining their behaviour in aqueous (or non-aqueous) environments. A crucial physical property of detergents is their solubility. As shown in *Figure 3*, when detergent monomers are added to an aqueous solution, some of them are dissolved and some form a monolayer at the air/water interface. As more detergent is added, their number approaches the critical micelle concentration (CMC) at which the monomers begin to associate spontaneously to form stable micelles. The driving force for this spontaneous aggregation of surfactant molecules is hydrophobic interactions: the hydrophilic heads point outward toward the aqueous solvent, while the hydrophobic tails associate with

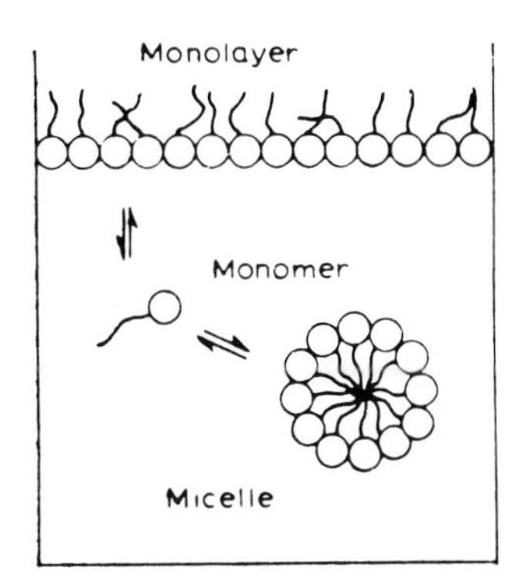

Figure 3. Schematic representation of the equilibrium between monomeric, monolayer and micellar forms of detergent molecules (30). Courtesy of Elsevier Scientific Publishing Company.

each other to shield themselves from thermodynamically unfavourable contact with water molecules.

The binding of anionic detergents [such as sodium dodecyl sulphate (SDS)] to proteins is a function of the free detergent concentration in equilibrium with the protein, and little additional binding of detergent to protein is observed beyond the CMC. When the concentration of monomers is increased above that required to saturate any biological binding sites, binding to other sites begins in a cooperative manner resulting in unfolding of the protein. This causes previously buried hydrophobic amino acid residues to become exposed and available for further detergent binding, until saturation is reached. The maximum amount of SDS bound per gram of protein is similar for nearly all proteins, about 1.4 g/g. The SDS–polypeptide complexes are micellar in nature, with SDS molecules aggregating around exposed hydrophobic areas in the protein. Some proteins have a higher intrinsic stability towards SDS denaturation, but this maximum SDS–protein binding (1.4 g/g) can be obtained for all proteins when heated in the presence of detergent for several minutes.

Cationic detergents such as decyltrimethyl ammonium chloride also show cooperative binding to proteins. However, it usually occurs close to its CMC, which is approximately 10-fold higher than for SDS. Therefore, many proteins are more resistant to the denaturing effect of cationic detergents (compared with SDS) because monomer concentrations cannot be increased much more before the CMC is reached. Non-ionic detergents such as Triton X-100 usually do not denature proteins, presumably because their bulky apolar head groups and over-all rigid structure cannot readily penetrate into the crevices on the surface of the protein molecule to initiate denaturation. Non-ionic detergents presumably could induce protein unfolding if higher monomer concentrations could be reached but, as with cationic surfactants, they are limited by their CMC value. As a result, monomers begin to interact with each other rather than with the protein molecule to form micellar structures.

3.6 Denaturing agents

All of these reagents are known for their protein denaturing capabilities. In principle, the inactivation itself is reversible, but proteins can often be irreversibly denatured due to subsequent aggregation or, in the case of proteases, autolysis. Some of these reagents can also be very useful in stabilizing proteins under the proper conditions. This apparent contradiction is easily clarified once the mechanism of action of each of these agents is understood.

3.6.1 *Denaturants*

High concentrations of urea (8–10 M) and guanidinium chloride (GdmCl, 6 M) are commonly used to denature proteins yet, despite their effectiveness, there is no generally accepted mechanism for their mode of action. Although their denaturation powers result from a combination of effects, two general features are well established. First, these reagents diminish the hydrophobic interactions that play a crucial role in holding together the protein tertiary structure. Second, they interact directly with the protein molecule (11). Regardless of the exact mechanism, urea and GdmCl are widely used as 'probes' of protein conformational stability towards reversible unfolding.

It is important to mention that cyanate is formed spontaneously from urea. An 8 M urea solution contains approximately 0.02 M cyanate at equilibrium (9). The cyanate ion can react with both the amino and sulphydryl groups of amino acids in proteins causing irreversible inactivation. Therefore, urea solutions should be prepared fresh with high quality solid urea, or cyanate ions removed just prior to use.

3.6.2 *High salt concentrations*

At high concentrations, salts exert effects on proteins that depend on both the concentration and the nature of the salt. These effects can be either stabilizing or denaturing; they correlate with the Hofmeister lyotropic series:

$$(CH_3)_4N^+ > NH_4^+ > K^+, \quad Na^+ > Mg^{2+} > Ca^{2+} > Ba^{2+}$$

$$SO_4^{2-} > Cl^- > Br^- > NO_3^- > ClO_4^- > SCN^-$$

Anions and cations to the left of the series are the most stabilizing—they reduce the solubility of hydrophobic groups on the protein molecule by increasing the ionic strength of the solution. In addition, these salts enhance water clusters around the protein which causes a loss in the total free energy of the system (entropy of H_2O decreases). The combination of these two effects stabilizes proteins by 'salting-out' hydrophobic residues causing the molecule to become more compact. Anions and cations to the right of the series are destabilizing and are known to denature proteins. These ions bind extensively to the charged groups of proteins or to the dipoles (peptide bonds), which reduces the number of water clusters around the protein. Therefore, it is less important to the protein whether it is folded or unfolded. This interaction causes the destabilization or 'salting-in' of proteins (32, 33). For example, $(NH_4)_2SO_4$ is a well known stabilizer of enzymes and is routinely used during their storage (1). Conversely, NaSCN destablizes proteins and is a known chaotropic agent.

3.6.3 *Chelating agents*

Metal ion-binding reagents such as EDTA can inactivate metalloenzymes, either by removing the metal ion from the enzyme via the formation of a coordination complex or by binding inside the enzyme as a ligand (3). This type of inactivation is often reversible, but loss of a metal co-factor can also cause a large conformational change that can lead to irreversible loss of activity (4).

On the other hand, proteins that do not require metal ions are generally stabilized by chelating agents, because they remove harmful metal ions.

3.6.4 *Organic solvents*

Enzymes function both in aqueous solution and when suspended in anhydrous organic solvents (34). However, enzyme inactivation is observed when mixtures of the two are formed via addition of water-miscible organic solvents to aqueous protein solutions. When added to an aqueous protein solution, organic solvents can bind directly to the protein via hydrophobic interactions and/or alter the dielectric constant of the solution. The latter affects the balance of non-covalent forces that maintain the protein in its native conformation. Therefore, organic solvents may tend to 'turn proteins inside out' by increasing the solubility of the hydrophobic core while decreasing the solubility of the charged surface.

Ironically, some organic solvents such as polyhydric alcohols and some polar, aprotic solvents like dimethylsulphoxide (DMSO) and dimethylformamide (DMF) can *stabilize* proteins at low concentrations, although at higher concentrations they can inactivate enzymes by denaturation. Why do these solvents stabilize enzymes? There are many examples of the use of glycerol, sugars and polyethylene glycol to stabilize enzyme solutions (1). Substances such as these have a high content of hydroxyl groups; therefore, they have the ability to form multiple hydrogen bonds and act like the solvent water. These reagents cause 'preferential hydration', a thermodynamic phenomenon causing a microscopic phase separation between protein and solvent as well as an increase in the surface tension of water (35–37) that is described in Chapter 14 of *Protein Structure: A Practical Approach.*

3.7 **Heavy metals and thiol reagents**

Inactivation of enzymes via heavy metal poisoning is well documented (38). Heavy metal cations such as mercury (Hg^{2+}), cadmium (Cd^{2+}) and lead (Pb^{2+}) are known to react with protein sulphydryl groups (converting them to mercaptides), as well as with histidine and tryptophan residues. In addition, disulphide bonds can be hydrolytically degraded by the catalytic action of silver or mercury (26).

Thiol-containing reagents can also inactivate enzymes by reducing disulphide bonds, although their action is usually reversible. Low molecular weight disulphide-containing reagents react with protein thiol groups to form mixed disulphides, or even intramolecular protein disulphides between two cysteine residues. The same thiol–disulphide interchange occurs between disulphide reagents like oxidized glutathione and enzymes containing catalytically essential SH-groups (26).

3.8 Heat

Thermal inactivation is perhaps the most frequently encountered and most thoroughly investigated mode of protein inactivation. The behaviour of thermophilic microorganisms and enzymes (39) continues to interest both biologists and biochemical engineers alike. From the biotechnological viewpoint, thermal inactivation is by far the most frequently encountered cause of enzyme inactivation in industry (25). Due to enhanced solubility and reaction rates, as well as reduced microbial contamination and solution viscosity, most industrial enzymatic processes are carried out at elevated temperatures.

As with other denaturing agents, inactivation due to heat is usually a two-step process: the reversible thermal unfolding of an enzyme exposes reactive groups and hydrophobic areas that can subsequently react to result in irreversible inactivation. Two conformational processes can cause irreversible thermal inactivation. First, protein aggregation can occur once the buried, hydrophobic areas are exposed to solvent due to the thermal unfolding. Second, monomolecular conformational scrambling can bring about the inactivation of enzymes. At high temperatures, an enzyme loses its regular non-covalent interactions, but upon return to ambient conditions, non-native, non-covalent interactions 'trap' the enzyme molecule in a scrambled structure, even though the latter is thermodynamically less stable than the native conformation (40).

At high temperatures, such as 90–100°C, pH-dependent covalent reactions limit an enzyme's thermostability: deamidation of asparagine and glutamine residues, hydrolysis of the peptide bonds at aspartic acid residues, oxidation of cysteine residues, thiol–disulphide interchange, and destruction of disulphide bonds. These deteriorative reactions account for the irreversible loss of activity observed at extreme temperatures in ribonuclease (16), lysozyme (41), triose phosphate isomerase (42), bacterial α-amylases (19) and glucose isomerase (43). The chemical processes involving cystine residues (thiol–disulphide interchange and disulphide destruction) have been shown to be general in nature and to occur in many other proteins at high temperatures (13). In addition, if reducing sugars such as glucose are present in the system, they can readily react with the ε-amino groups of lysine. This 'Maillard reaction' is a known cause of thermal inactivation in the food industry (44).

3.9 Mechanical forces

Proteins are susceptible to denaturation by mechanical forces such as pressure, shearing, shaking and ultrasound. Although, in principle, denaturation should be reversible, this is difficult to test, because it is often accompanied by either aggregation or covalent reactions that cause irreversible loss of activity.

3.9.1 *Shaking*

Shaking action can cause protein inactivation. The mechanism of this process is believed to involve an increase in the area of the gas–liquid interface. Proteins align themselves at this interface and unfold to maximize exposure of hydrophobic amino acid residues to the air. The proteins then aggregate because of these exposed hydrophobic areas.

3.9.2 *Shearing*

When enzyme solutions are pumped through tubes or membranes at a high flux, the enzyme may be exposed to a gradient of shearing forces due to the boundary layer at or near the wall. This gradient can cause conformational changes in a protein, resulting in exposure of hydrophobic areas that were previously buried, followed by aggregation. Inactivation is a function of both the shear rate and the exposure time (45).

3.9.3 *Ultrasound*

Ultrasonic pressure waves create microbubbles of dissolved gases that rapidly expand and then violently collapse (cavitation). This results in both mechanical forces and chemical agents (free radicals created in thermal reactions in the microbubbles) that can damage and inactivate proteins (46).

3.9.4 *Pressure*

Inactivation of enzymes by high pressure (0.1–6 kbars) has been investigated in order to understand why high external pressures strongly influence cellular processes. Enzymes may also encounter high pressures in the food or pharmaceutical industry, since high pressure was at one time considered an alternative to heat for sterilization. Pressure-induced inactivation is usually thought to arise from protein denaturation and subsequent aggregation, but two other mechanisms may be responsible for inactivation. First, several multimeric enzymes have been shown to dissociate into monomers under high pressures; upon pressure release, varying amounts of activity could be restored but many of the inactivations were irreversible (47). Second, work on lactate dehydrogenase showed that the oxidation of a cysteine residue was involved in the inactivation mechanism, and that this reaction was connected with structural changes in the enzyme (48).

3.10 **Cold, freezing and thawing, dehydration**

One of the 'gospels' of biochemistry dictates that in order to maximize the storage stability of an enzyme solution one should use low temperatures—the lower the better. However, there are many exceptions to this rule and often refrigeration is preferable to freezing in terms of stability (49). Protein inactivation by lowering the temperature can be both reversible and irreversible.

The reversible cold lability of enzymes is well documented (50). In many cases, it is observed in allosteric enzymes where a decrease in the temperature produces a conformational change. For many years it was assumed that the weakening of hydrophobic interactions was the mechanism causing protein dissociation or denaturation at lowered temperatures. Although this is certainly involved in the reversible inactivation, there is also an observable pH-dependence to this process. Lowering the temperature at a given pH is equivalent to changing the pH at a given temperature for both the rate and extent of loss of enzymatic activity (50). Frequently, this reversible denaturation process also leads to irreversible loss of activity due to subsequent aggregation.

Irreversible loss of enzymatic activity often accompanies the freezing and thawing of enzyme solutions. During the freezing process, solutes such as enzymes and salts are concentrated as the water molecules crystallize, thus causing dramatic changes in pH and ionic strength in the micro-environment of the enzyme. For example, the pH of phosphate buffer can change from 7 to 3.5 upon freezing, which could easily cause acid denaturation of a protein. In addition, concentrating the salts will raise the ionic strength, which could dissociate oligomeric proteins.

Another factor causing irreversible inactivation involves thiol–disulphide interchange or oxidation of sulphydryl groups. As freezing proceeds, the enzyme concentration increases, thereby increasing the concentration of cysteine and cystine residues. When this concentrating effect is accompanied by conformational changes, intra- or inter-molecular thiol–disulphide exchange can readily occur. Furthermore, sulphydryl groups can be readily oxidized at reduced temperatures since the oxygen concentration in a 'partially' frozen system at −3°C is 1150 times that in a solution at 0°C (51). This concentrating effect may also increase the concentration of oxygen radicals.

There are many similarities between freezing and dehydrating enzymes. In fact, many of the arguments made above to explain protein inactivation during freezing also apply to dehydration, since the concentration of liquid water is being reduced in both cases.

3.11 **Radiation**

The effects of both ionizing and non-ionizing radiation on protein inactivation have been extensively investigated, especially the latter due to its potential use as a sterilization technique in the food industry (51).

Although the different ionizing radiations are physically quite heterogeneous (γ-rays, X-rays, electrons, α-particles), the types of chemical changes that they cause in both the protein molecule and the surrounding water molecules are quite similar. Protein inactivation is caused by either direct action (effect of radiation on the protein molecule) or indirect action (effect of by-products of radiolysis of water on the protein molecule) (52). The direct action of ionizing radiation causes covalent changes in the primary structure due to radical formation, followed by cross-linking or amino acid destruction. This results in loss of native conformation or aggregation (53). The indirect action of ionizing radiation is due to the creation of reactive products formed in the aqueous solution. The principal reactive species in pure water are OH · radicals, solvated electrons and H_2O_2, but the list expands if other solutes are present (i.e. pH of solution, salts, buffer ions).

Proteins also can be inactivated by non-ionizing radiation such as visible light or UV radiation. Photochemical oxidation by visible light requires photosensitive dyes that absorb light energy and then oxidize susceptible groups in the protein molecule (cysteine, tryptophan, histidine) (52). UV radiation can directly inactivate proteins via amino acid damage; cystine and tryptophan residues are particularly labile (52).

4. APPROACHES TO MINIMIZE INACTIVATION

It is now apparent that question three of our protein stabilization rationale, that is how to minimize protein inactivation, often can be broken down into two parts:

(i) how do we prevent reversible unfolding?;
(ii) how do we prevent irreversible reactions from occurring once unfolding has taken place?

One can develop a general approach to enzyme stabilization by preventing reversible unfolding, whereas specific approaches will be needed to minimize the reactions that cause irreversible loss of activity. We will now briefly discuss methods and physicochemical rationales to prevent reversible unfolding for the general stabilization of enzymes (the 'rigidification strategy'). By stabilizing the folded form of a protein, we can decrease the unfolding effects of an antagonistic agent, thereby preventing, or at least postponing, the subsequent irreversible reactions. Four approaches can be taken to 'rigidify' an enzyme: an increase in intrinsic stability, additives, immobilization and chemical modification. They are summarized in *Table 2*.

4.1 **Increasing the intrinsic stability**

For certain enzyme applications, such as reactions at elevated temperatures, a more stable form of an enzyme may be available from microorganisms that live in hostile environments such as high temperature or ionic strength. Enzymes from thermophilic sources have been extensively studied (39), and have been found to be similar in structure and function to their mesophilic counterparts; the reasons for their greater stability are usually additional salt bridges or slightly different amino acid compositions.

The ultimate goal would be to engineer stable enzymes ourselves, but this dream requires a complete understanding of how the primary structure of a protein dictates its tertiary structure. Substantial progress has been made recently using both random and site-specific mutagenesis as tools to probe protein structure–function relationships (54).

4.2 **Additives**

Additives can be either specific or non-specific in relation to a particular protein or enzyme. A specific ligand, co-factor or substrate can shift the $N \rightleftharpoons U$

Table 2. Examples of approaches to minimize irreversible inactivation of proteins.

Effectors	*Comments*
Intrinsic stability	
1. Mesophilic versus thermophilic enzymes	Rigidification of enzyme conformation
2. Site-specific mutagenesis	Replacement of labile amino acid residues
Additives	
1. Specific	Shift N $\rightleftharpoons$ U equilibrium toward native form
2. Non-specific	Neutral salts and polyhydric compounds
3. Competitors	Outcompete enzyme for inactivating agent; remove catalysts of deteriorative chemical reactions
Immobilization	
1. Multi-point attachment of enzyme to support	Rigidification of enzyme conformation; steric hindrances prevent interaction with macromolecules, e.g. degradation by proteases
2. Partitioning effects and diffusion restrictions	Chemical and physical properties of support influence the micro-environment around the enzyme molecule
Chemical modification	
1. Cross-linking reagents	Rigidification of enzyme conformation
2. Reagents that alter ionic state or introduce steric hindrances	Modification adds, neutralizes or alters charged residues on enzyme molecule; attachment of soluble macromolecules inhibits interactions with other solutes, e.g. proteases

equilibrium (*Figure 1*) towards the native form. (However, ligand binding can sometimes also cause conformational changes that destabilize the native form.) Examples of non-specific additives are polyhydric compounds such as glycols and alcohols as well as neutral salts. They are known to have both stabilizing and destabilizing effects, as discussed in Sections 3.6.4 and 3.6.2, respectively.

Additives can also be used to 'outcompete' an enzyme for the inactivating agent. For example, bulk proteins can be used to protect enzymes from the gas–liquid interface, thiol reagents can be used to react with free radicals and chelating agents can bind trace metal ions.

4.3 **Immobilization**

Enzyme immobilization is the technology that allowed enzymes to become industrial catalysts. In the form of heterogenous catalysts, enzymes can be readily used in chemical reactors (55). In addition, the characteristics and properties of enzymes (especially stability) may be dramatically altered by the micro-environment of the support. There are numerous methodologies to attach enzymes covalently or non-covalently to inorganic supports, organic polymers or biologically derived materials. These processes have been extensively investigated and reviewed (56, 57).

Immobilization can influence the stability of enzymes by one of the following effects: steric hindrance, partitioning or diffusion restrictions. First, the spatial fixation of enzymes on a support creates steric hindrance; consequently, other macromolecules may have difficulties interacting with the enzyme. This effect can stabilize immobilized enzymes against proteolysis. Second, the concentration of substrates, ligands and hydrogen ions can be unevenly distributed between the bulk solution and the close vicinity of the support. Depending on the chemical nature of the substrate and the support (charge, hydrophobicity, etc.), the 'local concentration' around the support creates a micro-environment for the enzyme that differs from the bulk solution. Therefore, it is common to observe shifts in the pH optimum, the substrate specificity and the kinetic constants of an immobilized enzyme with respect to the free form. In the case of the pH optimum, for example, intrinsic enzymatic properties do not actually change, but the *partitioning* of H^+ ions between the water and the carrier changes the pH around the immobilized enzyme. The optimum pH for enzymatic activity will shift to higher and lower pH values with polyanionic (cation-exchange resin) and polycationic (anion-exchange resin) supports, respectively.

Finally, when an enzyme is embedded within a porous particle, substrates must find their way first to the surface of the particle (external mass transfer) and then into the particle (internal mass transfer) before the enzyme can react with them. These diffusional limitations may cause an apparent 'stabilization' of immobilized enzymes. When the rate of diffusion of substrates and products to and from the immobilized enzyme becomes the limiting step in the reaction, the rate of the reaction is no longer directly proportional to the enzyme concentration as in Michaelis–Menten kinetics (the dependence is now weaker). Let us assume that the inactivation of an enzyme occurs under the action of external factors (heat, pH or denaturing agents) and that the intrinsic stabilities of the free and immobilized enzymes are equal. If we express the inactivation process as 'activity versus inactivation time' or 'activity versus degree of inactivation' (see definition of stabilization, Section 1), an apparent 'stabilization' of immobilized enzyme seems to have been achieved whereas, in fact, the effect is due only to substrate diffusional limitations (2, 56).

Immobilization may rigidify an enzyme and increase its resistance toward unfolding if multi-point attachment can be achieved (5), since covalent attachment or entrapment of an enzyme should restrict its freedom to unfold.

4.4 **Chemical modification**

The chemical modification of proteins is an extensive field (58), but for the purposes of our 'rigidification strategy', we will consider only bifunctional reagents such as glutaraldehyde. There are many reviews on this topic available for detailed methodologies (2, 56). Bifunctional reagents react with functional groups on the enzyme to create artificial cross-links. In addition, the enzyme can be enriched with additional functional groups before cross-linking. Actual stabilization of the native conformation via cross-linking is difficult in practice, however, due to steric problems and inactivation due to modification. Chemical modification may cause either a decrease or increase of a protein's susceptibility to reversible unfolding (59). However, there are some successful examples of enhancement of conformational stability via intra-molecular cross-linking (5).

Attachment of carbohydrates (water-soluble polymers) can stabilize enzymes against inactivation (60). Both natural and chemically synthesized glycoproteins are known to exhibit frequently greater stability toward heat and denaturants than their apo-proteins, presumably due to the multi-point attachment mechanism discussed in the immobilization section.

5. CASE STUDIES IN MINIMIZING PROTEIN INACTIVATION

This section presents case studies from the literature to demonstrate the process of stabilizing a particular enzyme against irreversible inactivation. The first step in the methodology is to identify the cause of inactivation. Then by elucidating the mechanism of how activity is lost, approaches to minimize enzyme inactivation can be developed. The following stories were chosen because they represent problems that are frequently encountered when working with enzymes: proteolytic digestion, aggregation, pH changes, oxidation and thermal inactivation.

5.1 **Proteolytic digestion**

Proteolytic enzymes are widely used in the food industry and for detergent applications. These enzymes are notorious for their operational instability, because they catalyse hydrolytic self-degradation in aqueous solutions. Stabilization against autolysis is achieved in many cases by chemical modification, yet the rationales behind these success stories are quite diverse. For example, trypsin shows a reduced rate of autolytic degradation when conjugated with dextran (61), since large macromolecules such as polysaccharides create steric hindrances that retard the interactions between protease molecules.

Another approach to prevent autolysis is to modify chemically the NH_2-groups of proteases using anhydrides of dicarboxylic acids. The rationale for this procedure consists of two parts: first, transformation of positively charged amino groups into carboxyl groups will increase the number of negative charges on the molecules. Therefore, increased electrostatic repulsion between acidic proteins will minimize inter-molecular interactions. Second, many proteases cleave peptide bonds that are adjacent to positively charged amino groups. Therefore, acylation of lysine residues greatly reduces the number of cleavable bonds. As shown in *Figure 4*, the microbial protease from *Streptomyces caespitosus* was stabilized against autolysis by succinylation of surface amino groups of lysine residues. Furthermore, this stabilization was similar to that achieved via immobilization, where autolysis is prevented by the spatial fixation of the enzyme (62).

Immobilization can also minimize the effects of proteolytic digestion by rigidifying an enzyme. D-Amino acid oxidase was protected by immobilization to agarose. Even though parts of this enzyme were cleaved by trypsin, the immobilized enzyme retained its active conformation, whereas the free enzyme

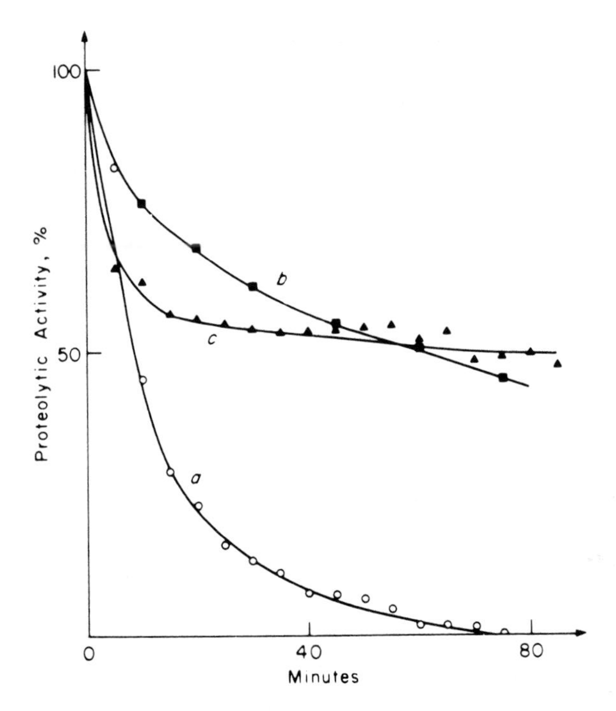

Figure 4. Time course of inactivation of *Streptomyces caespitosus* protease at 45°C: (**a**) the free enzyme; (**b**) the enzyme covalently attached to Sepharose; (**c**) succinylated protease (62). Courtesy of John Wiley and Sons.

was completely inactivated under the same conditions. Moreover, when the partially-digested immobilized enzyme was removed from its support, all enzymatic activity was lost (63).

5.2 Aggregation

The degree of aggregation depends on the extent of denaturation, which in turn is determined by external environmental factors (pH, ionic strength, ligands). For instance, inactivation brought about by aggregation usually increases with temperature. During denaturation/reactivation studies with phosphoglycerate kinase, aggregation occurred at 23°C, but was avoided by lowering the temperature to 4°C (64).

Aggregation is a second (or higher)-order kinetic process and therefore is dependent on the protein concentration. A decrease in protein concentration will lower the rate and degree of aggregation. Under denaturing conditions, most proteins will aggregate at concentrations of the order of 0.01–10 mg/ml (4). Since protein solubility is a function of pH and ionic strength, aggregation can be minimized when the pH of the solution is shifted away from the protein's isoelectric point. For example, thermally-induced aggregation of chymotrypsinogen (pI = 9) was significantly reduced at pH 2–3 due to electrostatic repulsion of similarly charged groups on the surface (65).

Studies on the refolding of denatured bovine trypsinogen illustrate the advantages of immobilization and chemical modification techniques to minimize aggregation. The fully reduced and denatured protein free in solution was shown to undergo rapid inactivation via aggregation upon removal of denaturant. However, when the enzyme was immobilized to an agarose support, the molecules could not interact, and thus aggregation was avoided. The denatured enzyme could now successfully refold to its native conformation upon removal of the denaturant (66). Aggregation was also minimized by synthesizing a mixed disulphide between glutathione and fully reduced and denatured trypsinogen. The newly-added charged groups of glutathione improved the solubility of the unfolded polypeptide chain, which allowed the denatured molecule to proceed with refolding upon removal of the denaturant.

5.3 Extremes of pH

Most enzymes are denatured rapidly at low pH values. In the case of chicken lactate dehydrogenase, lowering the pH to 3.2 results in complete loss of enzymatic activity in less than 1 h. However, as shown in *Figure 5*, immobilization of this enzyme to glass beads caused a dramatic resistance to pH inactivation (67). This remarkable pH stability is probably due to the partitioning effect of immobilization. Since the micro-environment of this carrier possesses buffering capacity, pH changes in the bulk solution will not appreciably affect the micro-environment of the immobilized enzyme.

With the recent advent of protein engineering, more refined procedures such as site-specific mutagenesis are emerging to replace specifically labile amino acid residues. For instance, recombinant-derived interleukin-1α has been stabilized against a pH-dependent covalent reaction, deamidation, by replacing a particularly labile asparagine residue with a serine (68).

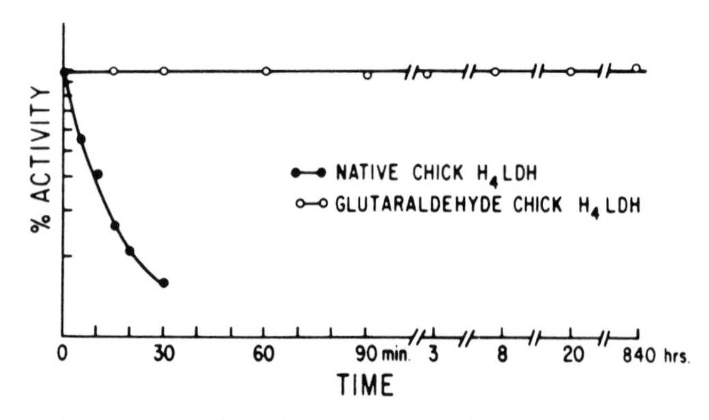

Figure 5. Comparison of the stabilities of soluble and glutaraldehyde glass-bound chick heart lactate dehydrogenase in 0.1 M acetate at pH 3.2 (67). Courtesy of Academic Press.

5.4 **Oxidation**

The stabilization of an oxygen-labile hydrogenase from *Clostridium pasteurianum* was achieved by immobilizing the enzyme to a charged support (69). The rationale was based on the 'salting-out' phenomenon, in which the dissolved oxygen concentration in aqueous solution decreases with increasing salt concentration. Therefore, immobilization to a charged support will bury the enzyme in a micro-environment equivalent to a concentrated salt solution.

In the presence of air, the free enzyme inactivates with a half-life of approximately 4 min. Adsorption of the enzyme to DEAE–cellulose led to a 25-fold stabilization. The positive charge density of the micro-environment could be further increased by using polyethyleneimine–cellulose, which has a longer, more charged group than DEAE: another increase in enzyme stability to 400 times the original level was observed. Finally, by changing the counter-ion on the support from chloride to phosphate, a final stabilization of 3000-fold was achieved, that is the half-life of the hydrogenase in the presence of air increased from 4 min to over 2 weeks.

Another common example of oxidative inactivation of enzymes is the oxidation of cysteine residues to cystine residues or sulphenic acids. Rhodanese is a cysteine-rich enzyme. It retains full enzymatic activity when stored under nitrogen or in the presence of high concentrations (>200 mM) of the reducing agent dithiothreitol (DTT); however, lower concentrations of DTT (2–100 mM) inactivate the enzyme (70). Why does an enzyme stabilizer become an enzyme inactivator? Upon exposure to air, DTT is oxidized (in the presence of trace metal ions) producing hydrogen peroxide, which in turn can oxidize protein SH groups. This peroxide formation explains the observed protein inactivation, because when catalase (an enzyme that breaks down hydrogen peroxide) is added to the system, rhodanese retains all of its catalytic activity despite the presence of inactivating concentrations of DTT.

5.5 **Heat**

To examine the applicability of the rigidification strategy toward thermal inactivation of enzymes, chymotrypsin and trypsin were chosen as model systems. Several immobilization and chemical modification techniques were tested to determine their effects on stabilization against irreversible heat inactivation (autolysis was eliminated) (71–73).

Multi-point attachment of these enzymes to a support was achieved by modifying them with an analogue of a polymeric gel monomer, followed by co-polymerization with the monomer itself. This technique resulted in an enzyme covalently attached to the three-dimensional lattice of a polymeric gel. As shown in *Figure 6*, the number of attachment sites to the gel could be manipulated by varying the degree of chemically modified residues on the enzyme molecule before co-polymerization. It was found that the number of multi-point attachments directly influences the rate of thermal inactivation of chymotrypsin. In other words, increasing the number of attachment sites caused increased rigidification and greater thermal stabilization.

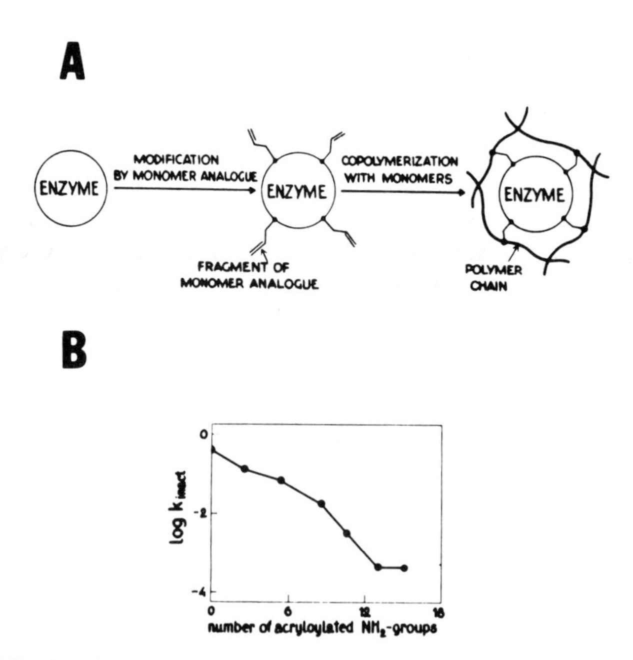

Figure 6. Stabilization of chymotrypsin against thermal inactivation (**A**) Schematic representation of co-polymerization of the enzyme modified by a monomer analogue with the monomer proper. (**B**) The dependence of the effective first-order rate constant (k_{inact}, min^{-1}) for mono-molecular thermal inactivation of acryloylated chymotrypsin chemically entrapped in polymethacrylate gel on the degree of modification of the enzyme at the first step in immobilization (71). Courtesy of Elsevier/North Holland Biomedical Press.

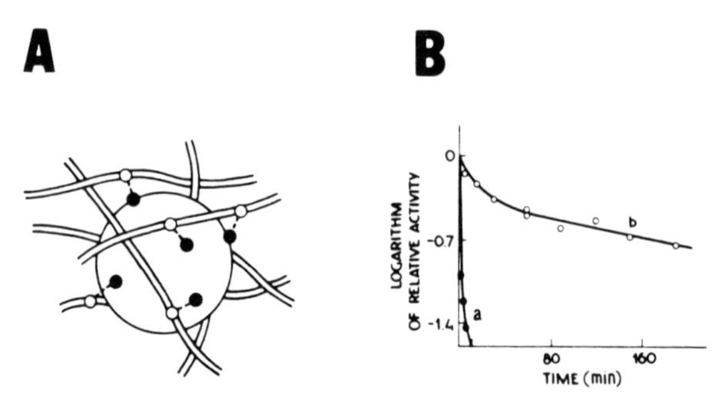

Figure 7. Stabilization of chymotrypsin against thermal inactivation. (**A**) Schematic representation of multi-point non-covalent attachment of enzyme within a three-dimensional lattice of a polymeric gel. (**B**) Kinetic curves of mono-molecular thermal inactivation of chymotrypsin: (**a**) in an aqueous solution in the absence of a monomer; (**b**) in 44% (w/w) polymethacrylate gel (72). Courtesy of Elsevier/North Holland Biomedical Press.

Another approach to rigidify these enzymes was to entrap them non-covalently in various polymeric gels. The effect of the gel concentration on their thermal stability could then be examined. The rationale behind this experiment is that by increasing the charge on the gel, the non-covalent contacts (electrostatic and hydrogen bonding) between the enzyme and the gel matrix will also increase, which will result in a rigidified enzyme and enhanced thermostability. As shown in *Figure 7*, whether free in solution [or entrapped in a neutral polyacrylamide gel over the range of 0–50% (w/w)], the rate of thermal inactivation of chymotrypsin was quite rapid. However, when entrapped in a charged polymethacrylate gel [concentration >30% (w/w)], the thermal stability of this enzyme dramatically increased by several orders of magnitude.

As a final example, the effect of intra-molecular cross-links of different lengths on the thermal stability of soluble chymotrypsin was studied. To synthesize these cross-links, the carboxyl groups of the enzyme were first activated with carbodiimide, and then the treated protein was derivatized with diamines of the type $H_2N(CH_2)_nNH_2$, with n varying from 0 to 12. Bifunctional reagents, such as diamines, introduce synthetic cross-links into the protein molecule which, in theory, should rigidify it. However, the optimal size of the cross-linking reagent will depend upon the specific enzyme since the distance between activated residues will differ from molecule to molecule. As shown in *Figure 8*, the dependence of the rate constant for thermal inactivation on the chain length of the cross-linking agent showed maximum stabilization at an intermediate length. This suggested that cross-linking with either longer or shorter length chains introduced steric constraints into the enzyme molecule. Although not as dramatic an effect as the polymeric gel examples discussed previously, a 3-fold stabilization of chymotrypsin against thermal inactivation was achieved.

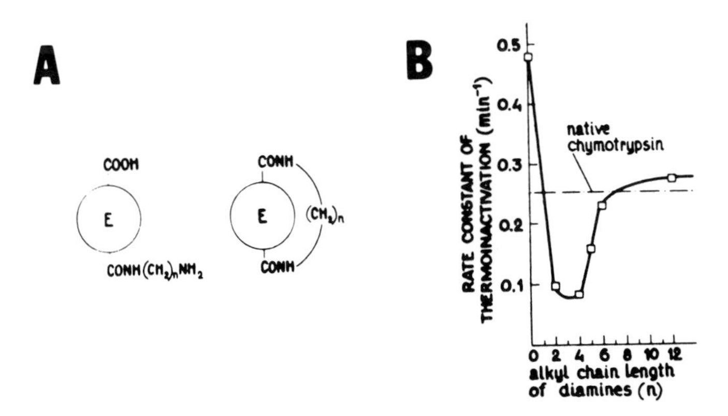

Figure 8. Stabilization of chymotrypsin against thermal inactivation. (**A**) Schematic representation of enzyme activated by carbodiimide and then cross-linked with diamines of chain length n. (**B**) First-order rate constant for mono-molecular thermal inactivation of α-chymotrypsin cross-linked intra-molecularly with diamines of varying chain lengths, $H_2N(CH_2)_nNH_2$ (73). Courtesy of Elsevier/North Holland Biomedical Press.

6. CONCLUDING REMARKS

For the convenience of the reader, we have separated the causes of inactivation of enzymes into many different categories, but we hope that the relationships between these processes were made clear. For example, we have seen how the different mechanisms of thermal inactivation are not only dependent on other inactivating agents, such as pH and salts, but also on the presence or absence of aggregation and autolysis. The stabilization of enzymes is also a process not easily separated into categories, and, in many cases, cumulative effects can be achieved by combining the different strategies presented. For example, an enzyme can be stabilized against heat inactivation by:

(i) adding a ligand to shift the equilibrium toward native protein;
(ii) immobilizing to prevent aggregation;
(iii) adjusting the pH;
(iv) removing dissolved oxygen to prevent oxidative covalent reactions.

In this chapter we discussed the many potential causes of irreversible inactivation and their modes of action. We then presented the most common approaches used to minimize these deteriorative events and demonstrated their advantages by examining some case studies. We made no attempt to complete the matrix of all potential stabilization techniques versus all causes of inactivation, but hopefully the rationale presented here will enable the reader to tackle any particular case of enzyme inactivation.

7. ACKNOWLEDGEMENT

This research was supported, in part, by the National Science Foundation, grant number DMB-852072.

8. REFERENCES

1. Wiseman, A. (1978) In *Topics in Enzyme and Fermentation Biotechnology*. Wiseman, A. (ed.), Ellis Horword, Chichester, Vol. 2, p. 280.
2. Klibanov, A. M. (1979) *Anal. Biochem.,* **93**, 1.
3. Schmid, R. D. (1979) *Adv. Biochem. Eng.,* **12**, 41.
4. Mozhaev, V. V. and Martinek, K. (1982) *Enzyme Microb. Technol.,* **4**, 299.
5. Klibanov, A. M. (1983) *Adv. Appl. Microbiol.,* **29**, 1.
6. Anfinsen, C. B. and Scheraga, H. A. (1975) *Adv. Protein Chem.,* **29**, 205.
7. Kauzmann, W. (1959) *Adv. Protein Chem.,* **14**, 1.
8. Tanford, C. (1968) *Adv. Protein Chem.,* **23**, 121.
9. Lapanje, S. (1978) *Physicochemical Aspects of Protein Denaturation*. J. Wiley, New York.
10. Privalov, P. L. (1979) *Adv. Protein Chem.,* **33**, 167.
11. Creighton, T. E. (1984) *Proteins: Structures and Molecular Properties*. W.H. Freeman, New York.
12. Geiger, T. and Clarke, S. (1987) *J. Biol. Chem.,* **262**, 785.
13. Volkin, D. B. and Klibanov, A. M. (1987) *J. Biol. Chem.,* **262**, 2945.
14. Florence, T. M. (1980) *Biochem. J.,* **189**, 507.
15. Marcus, F. (1985) *Int. J. Peptide Protein Res.,* **25**, 542.
16. Zale, S. E. and Klibanov, A. M. (1986) *Biochemistry,* **25**, 5432.
17. Lin, W. S., Armstrong, D. A. and Gaucher, G. M. (1975) *Can. J. Biochem.,* **53**, 298.
18. Estell, D. A., Graycar, T. P. and Wells, J. A. (1985) *J. Biol. Chem.,* **260**, 6518.
19. Tomazic, S. J. and Klibanov, A. M. (1988) *J. Biol. Chem.,* **263**, 3086.

20. Marston, A. O. (1986) *Biochem J.*, **240**, 1.
21. Okunuki, K. (1961) *Adv. Enzymol.*, **23**, 29.
22. Jaenicke, R. (1984) *Angew. Chem. Intl. Ed. Engl*, **23**, 395.
23. Inglis, A. S. (1983) In *Methods in Enzymology*. Hirs, C. H. W. and Timasheff, S. N. (eds), Academic Press, New York, Vol. 91, p. 324.
24. Manjula, B. N., Acharya, A. S. and Vithayayhil, P. J. (1977) *Biochem. J.*, **165**, 337.
25. Whitaker, J. R. and Fujimaki, M. (ed.) (1980) *Chemical Deterioration of Proteins*. American Chemical Society Symposium Series 123, Washington, D.C.
26. Torchinsky, Y. M. (1981) *Sulfur in Proteins*. Pergamon Press, Oxford.
27. Neuman, N. P. (1972) In *Methods in Enzymology*. Hirs, C. H. W. and Timasheff, S. N. (eds), Academic Press, New York, Vol. 25, p. 393.
28. Brot, N. and Weissbach, H. (1983) *Arch. Biochem. Biophys.*, **223**, 271.
29. Makino, S. (1979) *Adv. Biophys.*, **12**, 131.
30. Helenius, A. and Simons, K. (1975) *Biochim. Biophys. Acta*, **415**, 29.
31. Steinhardt, J. and Reynolds, J. A. (1969) *Multiple Equilibrium in Proteins*. Academic Press, New York.
32. Von Hippel, P. H. and Schleich, T. (1969) In *Structure and Stability of Biological Macromolecules*. Timasheff, S. N. and Fasman, G. D. (eds), Marcel Dekker, New York, p. 417.
33. Arakawa, T. and Timasheff, S. N. (1984) *Biochemistry*, **23**, 5912.
34. Klibanov, A. M. (1986) *CHEMTECH*, **16**, 254.
35. Lee, L. L. Y. and Lee, J. C. (1987) *Biochemistry*, **26**, 7813.
36. Gekko, K. and Timasheff, S. N. (1981) *Biochemistry*, **20**, 4667.
37. Gekko, K. and Timasheff, S. N. (1981) *Biochemistry*, **20**, 4677.
38. Valle, B. L. and Ulmer, D. D. (1972) *Annu. Rev. Biochem.*, **41**, 91.
39. Brock, T. D. (1986) *Thermophiles: General, Molecular and Applied Microbiology*. J. Wiley, New York.
40. Klibanov, A. M. and Mozhaev, V. V. (1978) *Biochem. Biophys. Res. Commun.*, **83**. 1012.
41. Ahern, T. J. and Klibanov, A. M. (1985) *Science*, **228**, 1280.
42. Ahern, T. J., Casal, J. E., Petsko, G. A. and Klibanov, A. M. (1986) *Proc. Natl. Acad. Sci. USA*, **84**, 675.
43. Volkin, D. B. and Klibanov, A. M. (1989) *Biotechnol. Bioeng.*, in press.
44. Whitaker, J. R. (1972) *Principles of Enzymology for the Food Sciences*. Marcel Dekker, New York.
45. Charm, S. E. and Wong, B. L. (1978) *Biotechnol. Bioeng.*, **20**, 451.
46. Coakley, W. T., Brown, R. C., James, C. J. and Gould, R. K. (1973) *Arch. Biochem. Biophys.*, **159**, 722.
47. Penniston, J. T. (1971) *Arch. Biochem. Biophys.*, **142**, 322.
48. Schmid, G., Ludemann, H. D. and Jaenicke, R. (1978) *Eur. J. Biochem.*, **86**, 219.
49. Bergmeyer, H. U., Bernt, E., Gawehn, K. and Michal, G. (1974) In *Methods of Enzymatic Analysis* 2nd Edition, Bergmeyer, H. U. (ed.), Academic Press, New York, Vol. 1, p. 169.
50. Bock, P. E. and Frieden, C. (1978) *Trends Biochem. Sci.*, **3**, 100.
51. Schwimmer, S. (1981) *Source Book of Food Enzymology*. Avi Publishing, Westport.
52. Florkin, M. and Stolz, E. H. (eds) (1967) *Comprehensive Biochemistry*. Vol. 27, Elsevier, Amsterdam.
53. Yamamoto, O. (1977) *Adv. Exp. Med. Biol.*, **86A**, 509.
54. Mathews, B. W. (1987) *Biochemistry*, **26**, 6685.
55. Klibanov, A. M. (1983) *Science*, **219**, 722.
56. Mosbach, K. (ed.) (1976) *Methods in Enzymology*. Academic Press, New York, Vol. 44.
57. Zarbosky, O. R. (1973) *Immobilized Enzymes*. CRC Press, Cleveland.
58. Means, G. E. and Feeney, R. E. (1971) *Chemical Modification of Proteins*. Holden-Day, San Francisco.
59. Habeeb, A. F. S. A. (1971) In *Chemistry of Cell Intersurfaces. Part B*. Brown, H. D. (ed.), Academic Press, New York, p. 259.
60. Marshall, J. J. (1978) *Trends Biochem. Sci.*, **3**, 79.
61. O'Neil, S. P., Wykes, J. R., Dunnill, P. and Lilly, M. D. (1971) *Biotechnol. Bioeng.*, **13**, 319.
62. Maneepun, S. and Klibanov, A. M. (1982) *Biotechnol. Bioeng.*, **24**, 483.
63. Naoi, M., Naoi, M. and Yagi, K. (1978) *FEBS Lett.*, **88**, 231.
64. Mitraki, A., Betton, J. M., Pesmadril, M. and Yon, J. M. (1987) *Eur. J. Biochem.*, **163**, 29.
65. Eisenberg, M. A. and Schwert, G. W. (1951) *J. Gen. Physiol.*, **34**, 583.
66. Light, A. (1985) *BioTechniques*, **3**, 298.

67. Dixon, J. E., Stolzenbach, F. E., Berenson, J. A. and Kaplan, N. O. (1973) *Biochem. Biophys. Res. Commun.*, **52**, 905.
68. Wingfield, P. T., Mattaliano, R. J., MacDonald, H. R., Craig, S., Clove, G. M., Gronenborn, A. M. and Schmiesner, U. (1987) *Protein Eng.*, **1**, 413.
69. Klibanov, A. M., Naplan, N. O. and Kamen, M. D. (1978) *Proc. Natl. Acad. Sci. USA*, **75**, 640.
70. Costa, M., Pecci, L., Pensa, B. and Canella, C. (1977) *Biochem. Biophys. Res. Commun.*, **78**, 596.
71. Martinek, K., Klibanov, A. M., Goldmacher, V. S. and Berezin, I. V. (1977) *Biochim. Biophys. Acta*, **485**, 1.
72. Martinek, K., Klibanov, A. M., Goldmacher, V. S., Tchernysheva, A. V., Mozhaev, V. V., Berezin, I. V. and Glotov, B. O. (1977) *Biochim. Biophys. Acta*, **485**, 13.
73. Torchilin, V. P., Maksimenko, A. V., Smirnov, V. N., Berezin, I. V., Klibanov, A. M. and Martinek, K. (1978) *Biochim. Biophys. Acta*, **522**, 277.

CHAPTER 2

Ligand–protein binding affinities

IRVING M. KLOTZ

1. INTRODUCTION

Binding of an effector molecule by a biomacromolecule is the primary step in most biological functions. An enzyme combines with a substrate to launch the catalytic conversion. A neurotransmitter must be bound to its synaptic receptor to initiate the neurobiological process. Carrier proteins bind small molecules to transport them in tissues or vascular fluids. An antibody combines with an antigen to initiate an immunological response.

In this chapter, the treatment of such interactions will be confined to situations where one of the combinants is a small molecule and the other a homogeneous biomacromolecule. The former is usually designated the 'ligand', the latter the 'receptor', for this volume, a protein. Many experimental methods have been devised for measuring ligand–receptor binding, but for soluble systems the simplest procedure is classical equilibrium dialysis (1). Detailed instructions will be given, therefore, for this technique in its simplest form. Variations will be described sufficiently to draw out their special features. Brief descriptions will also be given of methods based on other principles, such as spectroscopy.

Binding measurements provide the essential data for the evaluation of ligand–receptor affinities. However, the path for going from the former to the latter in all but the simplest systems is very slippery, and one can easily slide into illusions. To steer clear of self-deception, one must understand the fundamental conceptions of different equilibrium constants and recognize the inherent characteristics of various graphical and analytical representations of binding data.

2. PRINCIPLES OF EQUILIBRIUM DIALYSIS

2.1 **Basic premise**

In the equilibrium dialysis method, a solution of protein, or other biomacromolecule, on one side of a semi-permeable membrane is allowed to reach equilibrium with a ligand in the same solvent on the other side (*Figure 1*). At equilibrium, the chemical potential, and hence the thermodynamic activity, of the free ligand must be the same in both compartments. The protein concentration generally is low so that its effect on the activity coefficient of the ligand is negligible. Consequently, the concentration of free ligand should be the same on both sides of the membrane. In principle, then, if one determines the total amount

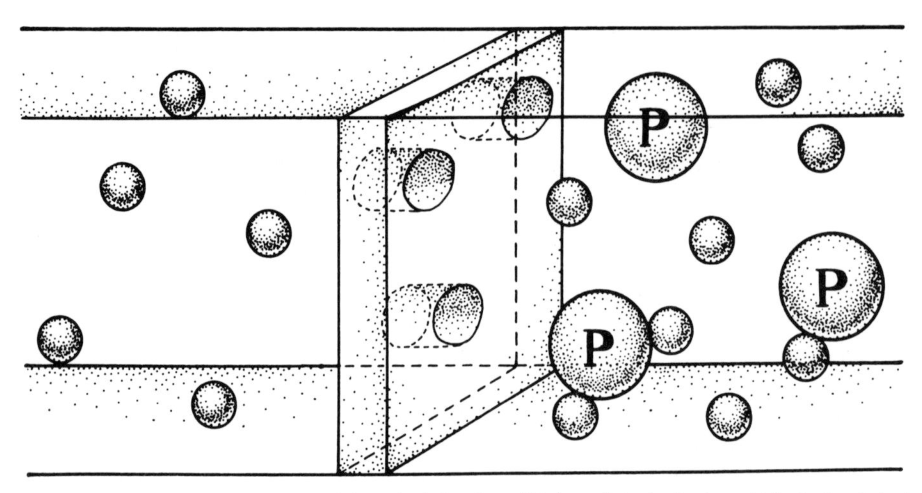

Figure 1. Graphic representation of the principle of equilibrium dialysis. In the middle is the dialysis membrane that is permeable to the ligand (small spheres) but not to the protein (large spheres).

of ligand in the entire container and combines that with a measurement of the free ligand concentration in the protein-free compartment, one can compute the amount of bound ligand.

2.2 Calculation of extent of binding

The extent of binding, B, is normally expressed as

$$B = \frac{L_P}{P_T} \tag{1}$$

where L_P is the number of moles of ligand on the receptor and P_T is the total moles of receptor. If the total moles of ligand in the container, L_T, is known and the concentration of free ligand, $[L]$, has been determined, then

$$L_P = L_T - V_T[L] \tag{2}$$

where V_T is the total volume of solution, the sum of both compartments. In practice, most ligands are also bound by the membrane of the dialysis apparatus (*Figure 1*) and sometimes also by the walls of the container. If L_M represents the moles of this peripheral binding, then the correct expression for L_P becomes

$$L_P = L_T - V_T[L] - L_M \tag{3}$$

Two methods have been used to obtain L_M. In the first, for each individual binding experiment, a companion is arranged with a corresponding dialysis apparatus containing identical solutions, except that protein is omitted. The amount of ligand bound by this blank control apparatus, L'_M, is

$$L'_M = L_T - V_T[L'] \tag{4}$$

where $[L']$ is the concentration of free ligand in this control. Using Equation 4 to replace L_T in Equation 3, we find that

$$L_P = V_T([L'] - [L]) + (L'_M - L_M) \quad (5)$$

If the amount of ligand bound by the membrane in the blank, L'_M, were equal to that picked up when protein is present, L_M, then

$$L_P = V_T([L'] - [L]) \quad (6)$$

This is probably an adequate correction when $[L']$ is near $[L]$, that is when the extent of binding by receptor is not large. (It is also an adequate correction in the rare case in which very little non-specific binding occurs so that L'_M and L_M are both small numbers.) Yet it is very evident that if binding by the membrane and apparatus is reversible, as it would be for a simple adsorption phenomenon, the uptake by the apparatus with the blank would be larger than that with the protein; for when $[L']$ is greater than $[L]$, as it must be when the receptor binds ligand, then the extent of non-specific adsorption, L'_M, will be higher than L_M.

In the second method for making the non-specific adsorption correction, one arranges a *series* of blanks covering a range of free ligand concentrations $[L']$ that spans the equilibrium concentrations $[L]$ in the presence of protein receptor. With this series of L'_M values calculated from the battery of controls, one can draw a graph (or prepare a table) showing the variation of non-specific adsorption with free ligand concentration. Then the appropriate value for L_M to be used in Equation 3 is that read from the calibration curve for the experimentally determined value of $[L]$, the free ligand concentration in equilibrium with the receptor.

It is also possible to circumvent a correction for non-specific binding by measuring the *total* ligand concentration in solution in the protein compartment (*Figure 1*) and in the solution in the protein-free compartment. Then L_P is given by the equation

$$L_P = V_P([L_P] - [L]) \quad (7)$$

where $[L_P]$ is the total concentration of ligand, bound and free in the protein compartment of volume V_P. Since $[L]$, the concentration of free, non-bound ligand, is the same in both compartments, its value can be obtained from a measurement in the protein-free compartment. However, there are some drawbacks to this procedure. In general, the attainment of osmotic equilibrium between the two compartments, only one of which has protein, involves transfer of some solvent. Hence V_P is not the initial volume of the protein-containing solution. If one also measures the concentration of protein, $[P]$, at equilibrium, then

$$[P]V_P = P_T \quad (8)$$

so, knowing P_T initially added, one can calculate V_P. In practice, analysing for concentration of protein $[P]$ in the presence of ligand, and in fact for $[L_P]$ in the presence of protein, may be experimentally difficult.

2.3 **Donnan effects**

If the receptor protein carries a net charge and the ligand is also electrically charged, then, even in the absence of binding, the concentration of free ligand will not be the same in both compartments (*Figure 1*) because of the Donnan effect (2). This difference in concentration can be made negligibly small by having an adequate quantity of buffer salt or neutral salt in the solution. For example, for the binding of a monovalent ion ligand by 0.05% serum albumin at pH 6.1, about 1 unit above its isoelectric point, 0.025 M phosphate buffer renders the Donnan effect negligible. At pH values further from the isoelectric point, or at higher protein concentrations, higher concentrations of supporting electrolyte may be necessary.

3. EXPERIMENTAL PROCEDURES IN EQUILIBRIUM DIALYSIS

3.1 **Components of apparatus**

3.1.1 *Semi-permeable membrane*

The membrane must be freely permeable to water and to small molecules and ions, but at the same time must not permit passage of the protein. Three sources for appropriate cellulose tubing membranes are the Union Carbide Corp., Spectrum Medical Industries, Inc., and Sartorius GmbH. A range of sizes is available. That of 8/32 inch (inflated diameter) is convenient for a volume of 1 ml for the protein solution. For 5 ml volumes, 18/32 inch is suitable, for 10–20 ml samples, 23/32 inch is generally used. For flat membranes, cut out pieces from large tubing, for example 36/32 inch, in the geometric shape appropriate for the dialysis chambers. Some suppliers specify the molecular weight cut-off for permeability (generally 10 000–30 000). If there is any uncertainty with regard to a specific protein, place a portion of protein solution in one compartment of the dialysis apparatus, and an equal volume of protein-free solution in the other; test the latter (for 24 h) for any leakage of protein into it.

3.1.2 *Containers*

Pyrex test tubes of appropriate size serve as dialysis vessels when dialysis bags are used (3). For volumes of 5–20 ml of protein solution, tubes of 25 × 200 mm are convenient (*Figure 2a*). Use number four rubber stoppers to close the tubes to prevent evaporation. They should be covered with thin sheets of polyethylene to minimize any risk of contamination of the solution by the rubber stopper. Alternatively, close the dialysis tube with a rubber serum cap. Attach a fibreglass thread (No. E-181) to the dialysis bag to facilitate stirring and to permit easy removal of the bag from the tube. Sever the desired length of thread from the spool by melting it in the flame of a small burner; cutting the threads leads to unravelling of the fibres.

Alternative suitable containers are glass vials with rubber-lined screw caps (4), sample vials with polyethylene caps or glass tubes constructed from a pair of male

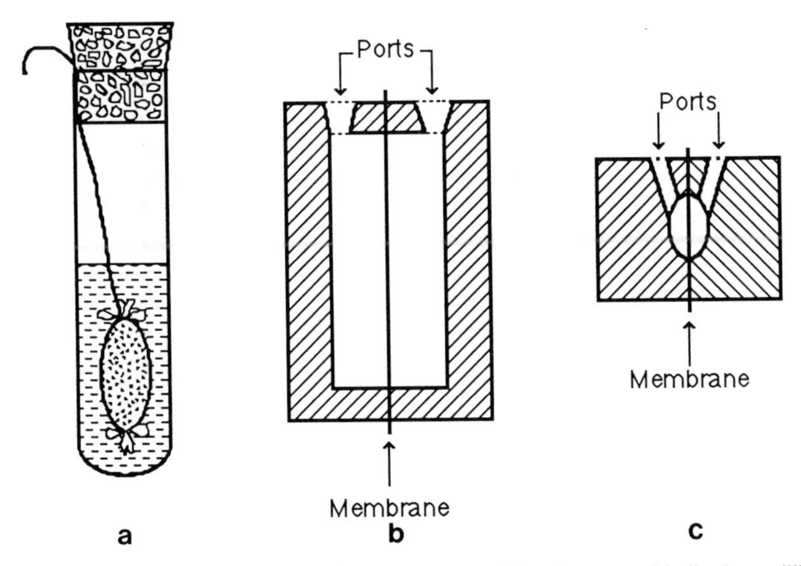

Figure 2. Different forms of dialysis vessels for measurement of ligand–receptor binding by equilibrium dialysis. In **a**, a protein-containing dialysis bag is immersed in a protein-free solution contained in a test tube. This configuration is convenient for volumes of several or many millilitres. In **b**, two chambers are machined precisely to clamp a flat sheet of cellulose membrane between them. The volume of each chamber is about 1 ml. The principle of construction in **c** is similar to that in **b**, but the scale is much smaller and the volume in each chamber can be as small as 20 μl.

and female standard taper joints (5). To deal with quantities of solution of the order of 1 ml, use a cell in which a flat sheet of cellulose membrane is held between two flat, cylindrical, 1-ml compartments (6). The chambers can be fabricated from glass, metal or plastics, such as Teflon, plexiglas or polyacrylamide (*Figure 2b*). Each cell compartment also has a port so that solution can be inserted or removed. Samples even smaller than 1 ml can be accommodated by a dialysis cell such as that shown in *Figure 2c*, which holds volumes of 20 μl in each chamber (7).

3.1.3 *Thermostat and shaker*

Adequate agitation is necessary to reach equilibrium in a matter of hours. Without agitation, the attainment of equilibrium may take days, especially with large volumes of sample. Efficient mixing is facilitated if an air bubble is enclosed within the dialysis bag or at the top of the sample compartments. All dialysis experiments should be carried out in a thermostat at a precise, fixed temperature. Although in many cases ligand binding is not strongly dependent on temperature, any calculations of binding affinities and related thermodynamic quantities require specification of the precise temperature.

3.2 **Operations and manipulations**

3.2.1 *Preparation of the membrane*

Cellulose tubing is generally manufactured by the xanthate process, which leaves substantial quantities of impurities in the product. These must be removed by extensive washing.

(i) Take a roll of cellulose tubing and cut off lengths appropriate to the volume of protein solution to be used in the dialysis experiment. For 23/32-inch diameter tubing, a 10-inch length is suitable for 10–20 ml of solution.
(ii) Place the cut tubing in a beaker of glass-distilled water and heat on a steam bath for 1 h.
(iii) Transfer the tubing to fresh water and repeat the heating. Then soak the tubing in fresh samples of water at room temperature for several hours; repeat this step twice.
(iv) Finally, transfer the tubing to a beaker containing the buffer solution or supporting electrolyte that will be used in the binding measurements, and allow it to soak for 12–24 h. If the supporting electrolyte is a neutral salt instead of a buffer, adjust the pH of the solution to the value desired for the dialysis experiment by adding dilute HCl; the tubing itself tends to lower the pH of an unbuffered solution. The tubing must be kept wet until it is used to form the dialysis bag.

Harsher treatment of the tubing than that described may impair the properties of the dialysis membrane. Heating tubing on a steam bath for 72 h weakens the bags and makes some permeable to proteins of molecular weight as high as 70 000. Trace metal impurities may be diminished by extraction with chelators or dilute acid (pH 3).

3.2.2 *Solutions*

In general a buffer must be used to maintain the pH at some defined value. For proteins with isoelectric points slightly below 7, a constant self pH can be maintained, but electrolyte is still essential to swamp out the Donnan effect. Thus for serum albumin with an isoelectric point of 5, an appropriate neutral solvent is 0.1 M NaCl. The compositions of a few suitable buffers to establish pH values from 5 to 9 are listed in *Table 1*. An ionic strength of 0.1 will reduce the Donnan effect to a negligible value for most proteins at concentrations of 1% or less. Solutions of lower ionic strength at the same pH can be obtained by diluting the buffers of *Table 1*.

Prepare the protein solution in supporting electrolyte by dissolving crystalline or lyophilized material, if available. For a receptor with a ligand-binding equilibrium constant of 10^4–10^5 M^{-1}, the protein concentration should be about 0.2% by weight. The moisture content of the original solid protein should be determined to calculate the actual protein concentration in solution. For this purpose, weigh a sample of about 10 mg into a small dish and place in a drying pistol *in vacuo* over a good desiccant, such as P_2O_5, at a temperature of 100°C until constant weight is reached. If some spectrophotometric or other assay for the protein is well established, it can be used to determine the protein concentration. For example, if the molar absorption coefficient is known, an absorbance

Table 1. Composition of some buffers for equilibrium dialysis.

Electrolyte	*pH*	*Molarity*	*Ionic strength*
Sodium acetate	5.0	0.100	0.100
Acetic acid		0.056	
Potassium acid phthalate	5.0	0.050	0.100
NaOH		0.025	
NaCl	5.0[a]	0.100	0.100
Citric acid	5.7	0.0234	0.100
NaOH		0.0567	
Mes	6.2	0.100	0.100
HCl		0.100	
Na_2HPO_4	6.8	0.035	0.132
KH_2PO_4		0.028	
Na_2HPO_4	6.9	0.0139	0.053
KH_2PO_4		0.0110	
Hepes	7.6	0.100	0.100
HCl		0.100	
Tris	8.3	0.100	0.100
HCl		0.100	
$NaHCO_3$	8.6	0.100	0.100
Glycine	9.1	0.400	0.100
NaOH		0.100	
Na_2CO_3	10.5	0.0345	0.100
HCl		0.0035	

[a] pH obtained by addition of 4.65 ml of 0.0025 M HCl to 500 ml of solution.

measurement provides directly the protein concentration, provided that other absorbing substances are absent or can be compensated for.

Dissolve the pure ligand in the same supporting electrolyte as that for the protein. For a ligand–receptor binding constant of 10^4–10^5 M^{-1}, the ligand concentration should cover a range of 10^{-4} to 10^{-5} M. Prepare a stock solution of 10^{-4} M and dilute aliquots with supporting electrolyte to obtain lower concentrations.

3.2.3 *Filling the dialysis tubes*

(i) Take a length of dialysis tubing from the final buffer solution in which it has been soaking. Remove adhering liquid promptly by placing the tubing between sheets of white absorbent paper and press out the liquid.
(ii) Make two tight knots at one end of the tubing.
(iii) Pipette in a precise quantity of protein solution (e.g. 10 ml), and force out most of the air above the liquid by manipulation with your fingers.
(iv) Close the top end of the dialysis bag by making two tight knots, the first of which should leave a bubble of air above the liquid.

(v) Cut off loose open ends of the tubing close to the knots and discard these remnants.
(vi) Tie a fibreglass thread between the double knots at the top of the bag.
(vii) Pipette a precise quantity of ligand solution into the test tube (*Figure 2a*).
(viii) Suspend the dialysis bag in the ligand solution with the free end of the thread hanging over the lip of the test tube, place the polyethylene-lined stopper securely in the tube and mount the test tube in the shaker in the thermostat.

3.2.4 *Determination of time for equilibration*

(i) Set up a series of five dialysis tubes each containing the same concentration of protein within the bag and the same concentration of ligand external to the bag.
(ii) Place each of the tubes in the thermostated shaker.
(iii) For the types of equipment illustrated in *Figure 2*, equilibrium is generally attained in 4–6 h. Therefore, remove one of the tubes after 3, 4, 6, 12 and 18 h, respectively.
(iv) Analyse the external solution for ligand concentration to determine the time required for equilibrium.

It may still be convenient, however, to plan experiments so that the assembly of the dialysis tubes is completed near the end of a working day and equilibration is carried out overnight.

3.2.5 *Confirmation of impermeability of membrane to protein*

Even if the manufacturer specifies a molecular weight above which the protein will not pass through the membrane, it is prudent to confirm that the membrane will retain the receptor in the chamber to which it was originally added.

(i) Set up a series of three dialysis tubes each containing the same concentration of protein within the bag and the same concentration of (ligand-free) buffer external to the bag.
(ii) Place each of the tubes in the thermostatted shaker.
(iii) After 18 h, remove the tubes and analyse for protein by some spectrophotometric or sensitive colorimetric assay.

Ideally, no detectable amount of protein should be present. In practice, if less than 0.1% of that originally present in the bag has leaked through, the membrane is acceptable. This experiment is carried out in triplicate (at least) because occasionally a single sample of cellulose membrane can be defective.

3.2.6 *Analysis of solutions at equilibrium*

At the end of the equilibration period, remove the dialysis bags from the container, take out an aliquot of external solution with a pipette and carry out an analysis for the ligand. In general, each ligand has a well-established specific mode of analysis for the concentration range being studied. By analysing the external solution one avoids complications arising from the presence of receptor.

4. COMPUTATION OF BOUND LIGAND

4.1 **Correction for binding to membrane**

To illustrate these calculations, a set of actual experimental data (8) have been assembled in *Table 2*.

Six protein-free containers were used to obtain data for membrane binding. Let us examine the numerical entries for container 1_o (*Table 2, part A*). A precise quantity (10 ml, footnote a) of solution of ligand of known concentration $[L]_o$ (6.474×10^{-5} M, first entry under 1_o) was added to the test tube container, to constitute the external compartment. Thus the total moles of ligand added, L_T, must be 0.01 litre $\times$ 6.474×10^{-5} M $= 6.474 \times 10^{-7}$ mol, which is the second entry in the column under 1_o. Then the dialysis bag containing a precise quantity (10 ml,

Table 2. Analyses and computations for binding of 2-azo-*p*-dimethylaniline by bovine serum albumin.

	A. Protein-free tubes					
	1_o	2_o	3_o	4_o	5_o	6_o
$[L]_o$, initial concentration of ligand in external compartment, M $\times$ 10^5	6.474	4.849	3.249	2.782	2.161	1.530
L_T^a, total moles of ligand added, $\times 10^7$	6.474	4.849	3.249	2.782	2.161	1.530
$[L']$, concentration of free ligand at equilibrium, M $\times$ 10^5	2.543	1.827	1.193	1.010	0.801	0.514
$V_T^a[L']$, total moles of free ligand in both compartments, $\times 10^7$	5.086	3.654	2.386	2.020	1.602	1.028
L'_M, moles of ligand bound by membrane, $\times 10^7$	1.388	1.195	0.863	0.762	0.559	0.502
	B. Protein-containing tubes					
	1	2	3	4	5	6
$[L]_o$, initial concentration of ligand in external compartment, M $\times$ 10^5	19.449	13.892	9.725	6.474	4.849	3.249
L_T^a, total moles of ligand added, $\times 10^7$	19.449	13.892	9.725	6.474	4.849	3.259
P_T, total moles receptor, $\times 10^7$	4.425	4.425	4.425	4.425	4.425	4.425
$[L]$, concentration of free ligand at equilibrium, M $\times$ 10^5	2.275	1.655	1.091	0.719	0.524	0.322
$V_T^a[L]$, total moles of free ligand in both compartments, $\times 10^7$	4.550	3.310	2.182	1.438	1.048	0.644
$L_T - V_T[L]$, total moles ligand bound to receptor and membrane, $\times 10^7$	14.899	10.582	7.543	5.036	3.801	2.605
L_M, moles ligand bound to membrane, $\times 10^7$	1.382	1.120	0.847	0.621	0.491	0.310
L_P, moles ligand bound to protein, $\times 10^7$	13.517	9.462	6.696	4.415	3.310	2.295
B, moles bound ligand per mole of receptor	3.055	2.138	1.513	0.998	0.748	0.519

[a] The total volume, V_T, in each container was 20 ml, 10 ml in the dialysis bag and 10 ml in the external compartment.

footnote a) of buffer (protein-free) was placed in the ligand solution in the test tube, and the container was shaken gently in a thermostat until equilibrium was reached. The bag was removed, and the ligand concentration $[L']$, was determined, in this case by measurement of the optical absorbance, and found to be 2.543×10^{-5} M (third entry under 1_o). Therefore, the total moles of free ligand in the entire test tube, dialysis bag plus external compartment, must be $(0.01 + 0.01)$ litre $\times 2.543 \times 10^{-5}$ M $= 5.086 \times 10^{-7}$ mol (fourth entry under 1_o). Hence the amount bound by the membrane and other components of the container is, following Equation 4, $(6.474 - 5.086) \times 10^{-7}$ or 1.388×10^{-7} mol (the fifth entry under column 1_o). Thus this experiment established that at a free ligand concentration of 2.543×10^{-5} M, the correction for membrane binding in this specific type of dialysis apparatus is 1.388×10^{-7} mol.

Similarly, test tube 2_o established that at $[L'] = 1.827 \times 10^{-5}$ M, $L'_M = 1.195 \times 10^{-7}$ mol, and so on in columns 3_o–6_o. From these numbers one prepares a graph (*Figure 3*) of the membrane-binding correction L'_M as a function of free, unbound ligand concentration $[L']$.

4.2 Calculation of ligand bound to receptor

In part B of *Table 2*, protein-containing solution was added to the dialysis bags. The initial loading concentrations of ligand were higher than in part A, since a

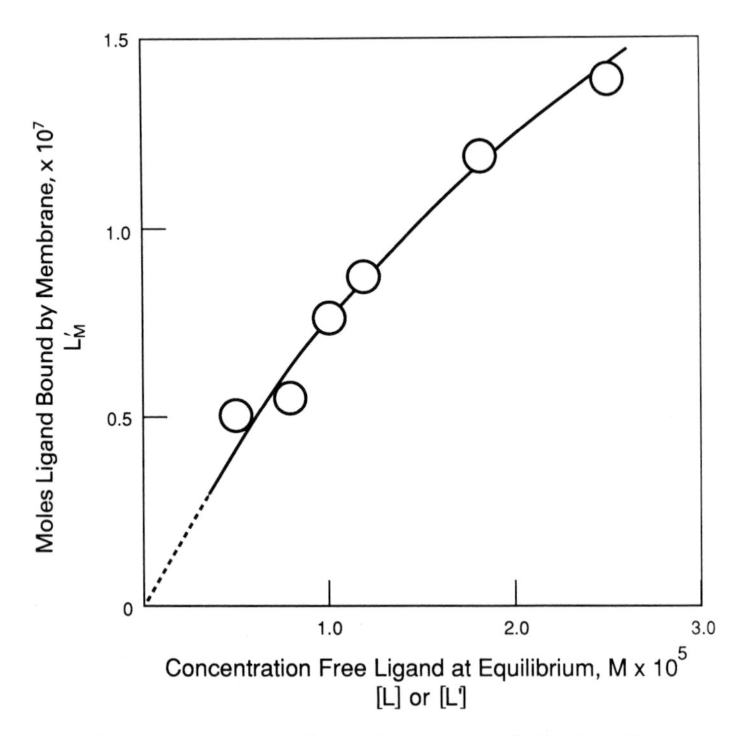

Figure 3. Example of a calibration curve for making corrections for binding of ligand to membrane and container.

large fraction of the ligand would be bound by the receptor protein. Thus in column 1, 19.449×10^{-5} M [first entry under tube 1) was the initial ligand concentration in the 10 ml of protein-free solution added to the test tube, to constitute the external compartment in the dialysis apparatus. In this tube, therefore, the total moles of ligand, L_T, is 0.01 litre $\times$ 19.449×10^{-5} M = 19.449 $\times$ 10^{-7} mol (second entry under 1). The third entry, total receptor protein added to the dialysis bag, is 4.425×10^{-7} mol, and is the same in each column because the same quantity of protein was used in each tube. After equilibration, the concentration of free ligand, $[L]$, was determined by analysis of the external compartment. The same value must hold for both compartments at equilibrium. Therefore, the total moles of unbound, free ligand in container 1 must be $(0.01 + 0.01)$ litre $\times$ 2.275×10^{-5} M = 4.55×10^{-7} mol (the fifth entry under 1). Hence the ligand removed from free solution by binding to the receptor and to the membrane, $L_T - V_T[L]$, is $(19.449 - 4.550) \times 10^{-7} = 14.899 \times 10^{-7}$ mol (sixth entry under 1). Now find the correction for binding by the membrane, L_M (see Equation 3). That quantity is a function of $[L]$ the free ligand concentration, which in tube 1 is 2.275×10^{-5} M. At this concentration, according to *Figure 3*, L_M is 1.382×10^{-7} mol (seventh entry under 1), which is then subtracted from 14.899×10^{-7} mol to give 13.517×10^{-7} mol, the ligand bound by the protein receptor, L_P (eighth entry under 1). Finally to obtain the extent of binding, B, divide 13.517×10^{-7} mol (eighth entry) by 4.425×10^{-7} mol (third entry), as specified by Equation 1, and we obtain the extent of binding, 3.055 mol ligand/mol receptor for the experiment in container 1. A corresponding sequence of computations for tubes 2–6 gives the numerical entries in the bottom row of *Table 2B*.

5. ALTERNATIVE METHODS OF STUDYING LIGAND BINDING

A vast array of physicochemical and biological methods have been developed to measure and study ligand binding (*Table 3*). Each of these techniques depends on one of the following procedures.

(i) Determination of the concentration of free, unbound ligand.
(ii) Detection of a change in a physicochemical or biological property of the bound ligand.
(iii) Detection of a change in a physicochemical or biological property of the receptor.

5.1 Determination of concentration of free ligand

Among available methods, equilibrium dialysis is the most widely used. This is one of a class of procedures that measure the distribution of ligand between two phases, only one of which can accommodate protein (*Figure 1*). In dialysis and in its closely related technique, ultrafiltration, the phases are separated by a semi-permeable membrane. Alternatively, the two phases can be mutually-immiscible liquids, with the interfacial surface providing the barrier. In gel filtration, the

Table 3. Methods of studying binding of small ligands by macromolecules.[a]

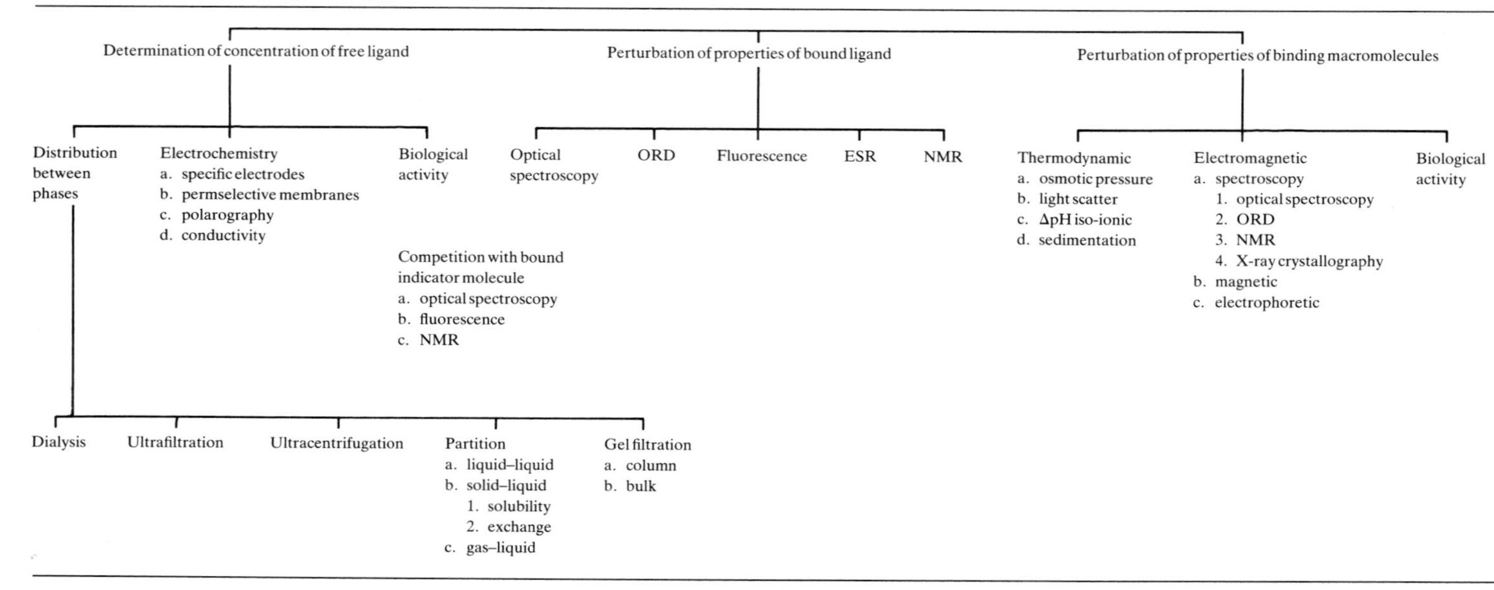

[a] For specific references to use of these methods, see ref. 9.

portals to the microchannels maintain the gel interior separate from the bulk solution.

5.2 Perturbation of properties of bound ligand

On the receptor a bound ligand experiences a different molecular environment than it does in bulk solution. Thus electronic and other spectroscopic energy levels are perturbed, and spectra of all types are often modified, in frequency or amplitude or both. Shifts in optical absorption spectra were used very early to measure ligand binding (10). Analogous changes may appear anywhere from the UV to the IR to the NMR region, as well as in ORD and fluorescence behaviour.

In the simple circumstances that bound ligand shows the same spectroscopic properties no matter how many ligand molecules are bound to the receptor, then the fraction of ligand in the bound state, α, can be calculated from absorbance measurements by the equation

$$A = \alpha A_{\mathrm{b}} + (1 - \alpha)A_{\mathrm{f}} \tag{9}$$

and from NMR shifts by

$$\delta = \alpha\delta_{\mathrm{b}} + (1 - \alpha)\delta_{\mathrm{f}} \tag{10}$$

where A is the absorbance, δ is the chemical shift, and the subscripts b and f refer to the bound and free ligands, respectively. Corresponding equations can be formulated for fluorescence intensities or CD parameters.

When, however, bound ligand molecules can be in different environments, A_{b} (or δ_{b}) need not be a constant but could reflect different molar extinction coefficients for each of the bound ligand molecules. The number of parameters in Equation 9 must be increased markedly, but methods for treating these situations have been formulated (11).

5.3 Perturbation of properties of receptor protein

A whole gamut of physicochemical and biological properties of the receptor protein may be modified by combination with a ligand (*Table 3*). These techniques are not generally applicable, however, since they depend entirely upon the properties of the protein, but they often provide special insights into ligand–receptor complexes. In particular, ligands binding to different sites on a protein will often give different responses, so the different sites can be distinguished. This can be a great advantage over equilibrium dialysis, which does not distinguish between different binding sites on a protein molecule. Methods for detecting binding of ligands to specific sites on proteins are described for some of the most important cases in Chapters 3–9.

6. GRAPHICAL REPRESENTATIONS OF DATA

To probe the biological response to an effector ligand, we need to know the amount bound, B, as a function of the free effector concentration $[L]$, that is

Table 4. Binding of carbamoyl phosphate by aspartate transcarbamoylase.[a]

B	$[L]$	B	$[L]$
0.20	0.14	2.74	10.22
0.40	0.30	2.74	10.30
0.40	0.34	2.65	11.52
0.74	0.72	3.10	15.05
0.37	0.36	3.02	15.10
0.72	0.73	3.44	19.11
0.90	0.96	2.70	15.34
1.54	2.29	3.23	19.00
1.53	2.34	4.10	41.0
1.48	2.51	3.86	41.51
1.54	2.64	3.92	42.15
1.41	2.54	4.77	57.47
1.90	4.26	4.41	60.41
2.28	5.85	4.40	67.69
2.22	6.13	5.32	84.44
2.32	6.04	4.56	76.00
2.25	6.34	4.26	85.20
2.17	6.46	5.11	141.7
2.10	6.36	5.34	161.8

[a] See ref. 20.

$B = g([L])$. When the necessary experimental data have been collected (e.g. *Table 4*), then $B = g([L])$ can be presented graphically, or analytically in terms of an algebraic equation. Among many possible graphical presentations, three are widely used (12, 13).

6.1 Direct linear scales

The simplest plot to prepare is one of B versus $[L]$. In cases where the ligand–receptor affinity is very strong, one obtains a graph such as that in *Figure 4* for the binding of leucine by α-isopropylmalate synthase (14). It is very clear in this example that at higher ligand concentrations B is constant within experimental error over a more than 10-fold range of $[L]$, so one can conclude that this receptor protein can be saturated by ligand. A disadvantage of this presentation is that very many of the experimental points are squeezed into a small region of the figure near the ordinate axis so the variation of B with $[L]$ (and hence the biological response) is not readily discernible.

Figure 4 presents an exceptional situation. In most studies of ligand–receptor binding reported in the literature, saturation values (or total number of binding sites) have been claimed, but not convincingly demonstrated. It is very easy to be misled by a direct graph with linear scales for B and $[L]$. A very telling illustration of the pitfalls is shown in *Figure 5*, which displays the binding of laurate by human

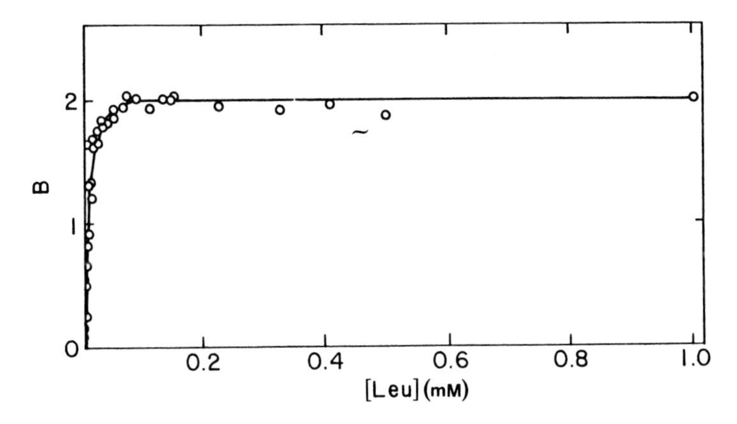

Figure 4. Direct graph for binding of leucine by α-isopropylmalate synthase (14) using linear scales for B and $[L]$.

serum albumin (15). If the experiments had been terminated at $[L] \simeq 1.6 \times 10^{-5}$ M (*Figure 5a*), one would have been inclined to claim saturation of protein at $B_{sat} = 7$. A series of experiments that stopped at about $[L] = 5.0 \times 10^{-5}$ M would have suggested $B_{sat} = 8$ (*Figure 5b*). Further binding studies up to $[L] = 30 \times 10^{-5}$ M (*Figure 5c*) would have led to a claim of saturation at $B_{max} = 10$. It is clear that in fact a saturation level cannot be specified for this system.

6.2 Semi-logarithmic graph

Erroneous conclusions about the total number of binding sites on a receptor can be avoided if B is plotted as a function of log $[L]$. This is illustrated in *Figure 6* which depicts the same experimental data as in *Figure 5*. Now, however, it is overwhelmingly obvious that the total number of binding sites for laurate on albumin cannot be determined from the available data since the curve does *not* approach an asymptotic plateau, the saturation level. Most published ligand–receptor binding data fail to show an asymptotic plateau (at the highest available values of $[L]$) if plotted as B versus log $[L]$.

Another attractive feature of the semi-logarithmic graph is that it spreads out the experimental data uniformly. This is illustrated in *Figure 7* which depicts the same data as in *Figure 4*. No longer are the results at low $[L]$ severely compressed, yet the existence of an asymptotic plateau at high $[L]$ is unequivocally established. One can readily follow the course of B as free effector L increases in concentrations from the very lowest to the very highest range.

This type of plot is analogous to that commonly used to present acid (or base) titration curves. Again binding (this time of H^+ ions) is being measured over a broad expanse of free proton concentrations. The data can be spread out uniformly, and a saturation level, if it exists, can be readily determined.

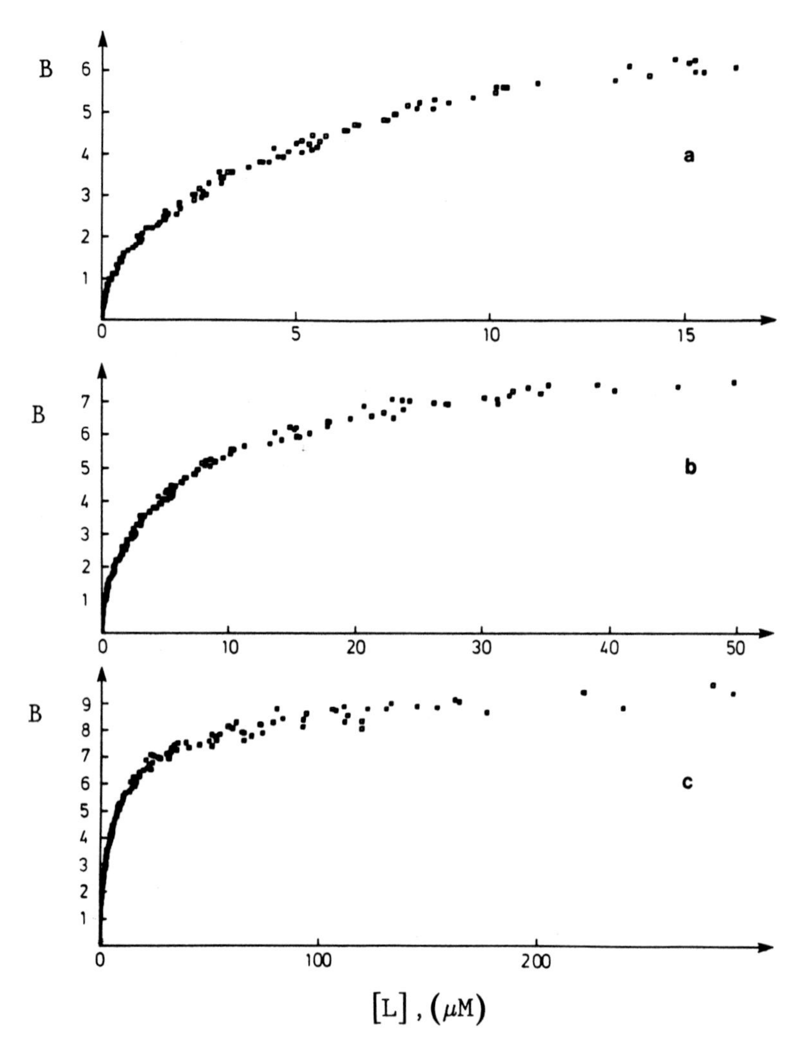

Figure 5. Linear scale plots of the binding of laurate ion by human serum albumin, showing three different ranges of concentration. [Reproduced from (15) with permission of the authors.]

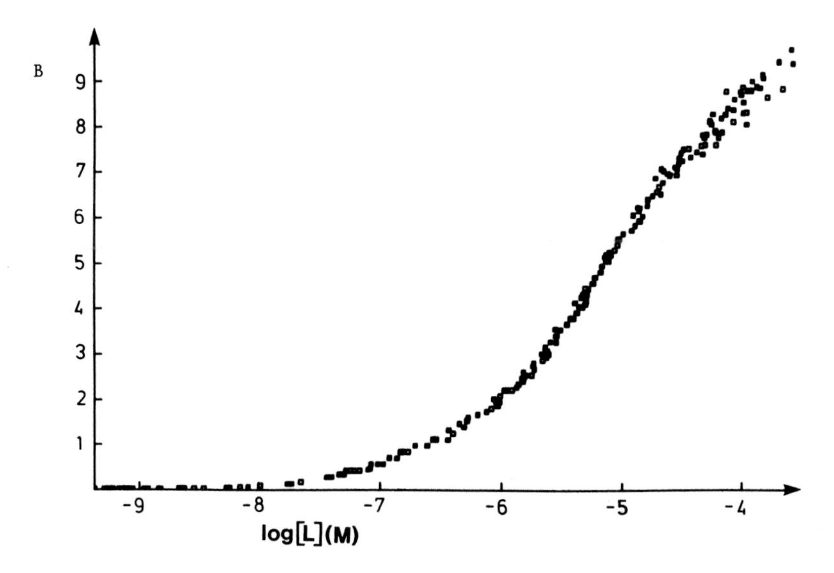

Figure 6. Graph of B versus log $[L]$ for the binding of laurate ion by human serum albumin (15) using the same experimental data as in *Figure 5*.

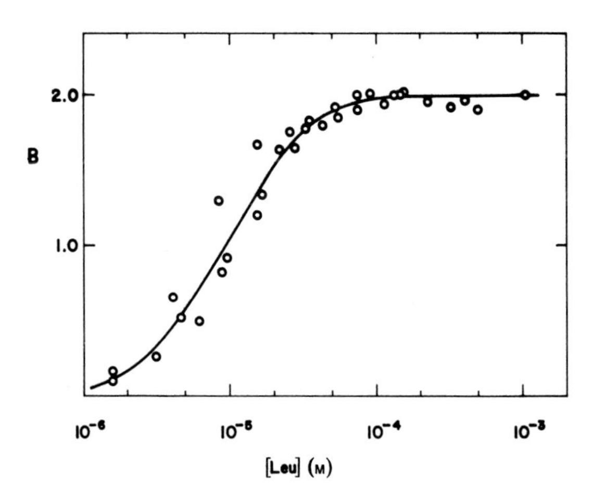

Figure 7. Semi-logarithmic representation of binding of leucine by isopropylmalate synthase (14) using the same experimental data as in *Figure 4*.

6.3 Alternative graphs

A variety of alternative graphs have appeared in the literature, all suggested by various transforms of $B = g[(L)]$ that are linear in simple circumstances, as was pointed out by B. Wolff more than half a century ago (16). Of these plots, the Scatchard graph (17) with $B/[L]$ as ordinate and B as abscissa, is the most popular. It has the attractive property, for a simple system of identical binding sites, of being linear, with the $B/[L]$ intercept giving the value of the association constant and the B intercept the limiting value of B. This form of graphical depiction is not recommended, however, because it has a propensity to be misleading in several respects.

When data are plotted on a $B/[L]$ versus B graph, there is a strong, sometimes irresistible, temptation to fit them to a straight line, either by eye or by least squares methods. In most cases in the literature of ligand–receptor binding, it can be shown that the conclusions derived from a $B/[L]$ versus B graph are untenable. For example, for the binding of diazepam by benzodiazepine receptors, *Figure 8a* has been used to summarize the results (18). Individual points have been fitted to a straight line with a very good correlation coefficient (0.97), and the total number of receptor sites, 830, determined from the intercept on the B axis (18). However, if the same data points are plotted on a semi-logarithmic graph (*Figure 8b*) it becomes obvious that there is no hint of saturation at 830. Since the graph in *Figure 8a* is purported to be a straight line, that in *Figure 8b* should be an ideal, S-shaped curve which at its inflexion point would have a value of B halfway to the saturation plateau (12). Even if we arbitrarily assume that the highest observed experimental value of B, about 700, is at the inflexion point in *Figure 8b*, the saturation plateau would be double this value, 1400, and may be even higher. Consequently in the Scatchard graph, *Figure 8a*, the intercept on the B axis ought to be very far to the right of that shown, and the points clearly should not be fitted to a straight line.

Even when the presence of curvature in a plot of $B/[L]$ versus B cannot be overlooked, the graph may entice one into untenable conclusions. For example, if the binding of laurate by serum albumin is plotted in a Scatchard graph (*Figure 9*), one could accept for the total number of sites any value above 10 that fitted one's predisposition. In this experimental case, the semi-logarithmic graph (*Figure 6*) had already unequivocally demonstrated (15), however, that saturation has not been approached at the highest value of $[L]$ attained. On the other hand, when only a Scatchard graph is used (19) to depict data, with no semi-logarithmic counterpart, as in representations of binding of carbamoyl phosphate by aspartate transcarbamoylase (*Table 4*), it is difficult to appreciate how extremely compressed are the data in the neighbourhood of the B axis (*Figure 10*) (21) and hence how indeterminate is a projected intercept on the abscissa at high B in a Scatchard graph. This point is strikingly reinforced in *Figure 11*, which depicts the same information (*Table 4*) as in *Figure 10* but in a form that makes very visible the very large uncertainties in $B/[L]$ at high binding, which are not apparent in *Figure 10*. Furthermore, in *Figure 11*, there is not even a hint of approach to a saturation

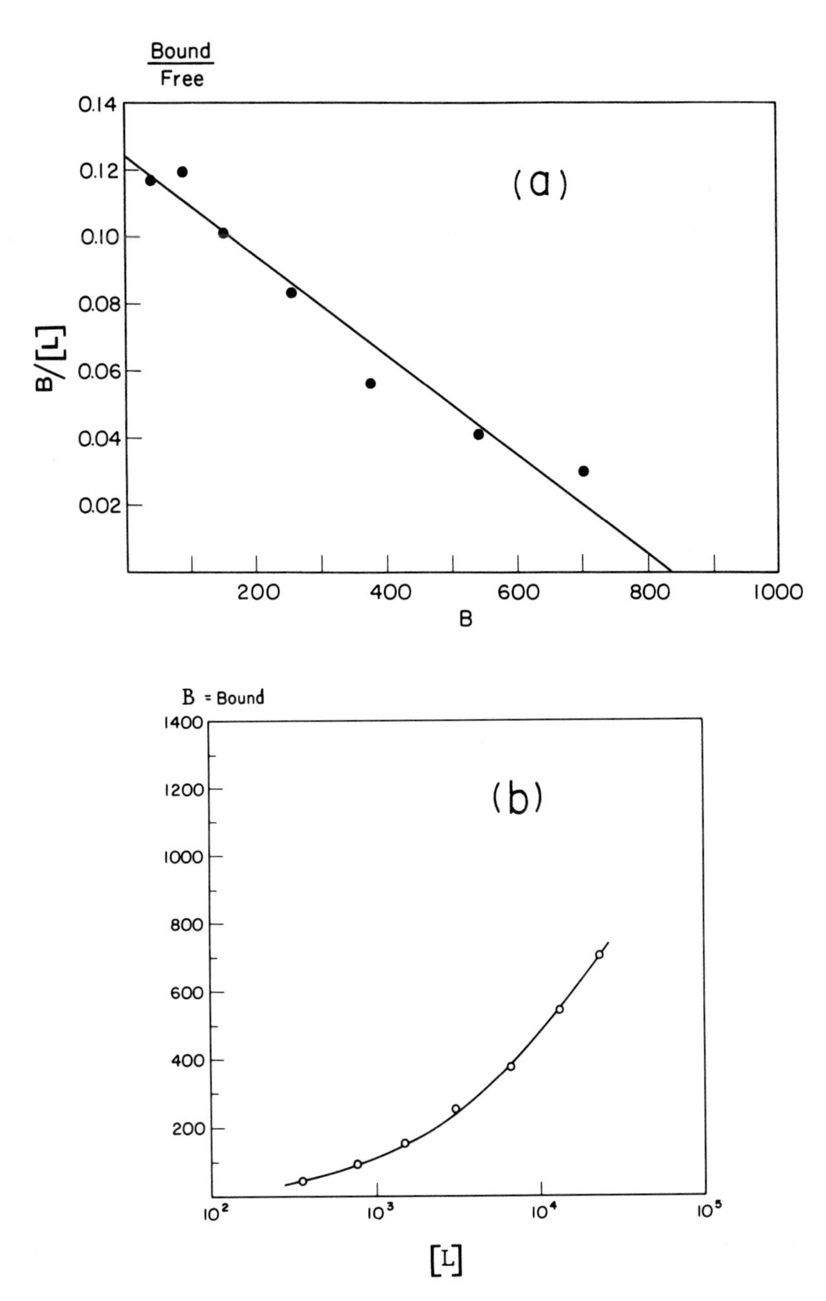

Figure 8. Binding of diazepam by benzodiazepine receptors. (**a**) Scatchard graph; (**b**) semi-logarithmic graph. Original data in ref. 18.

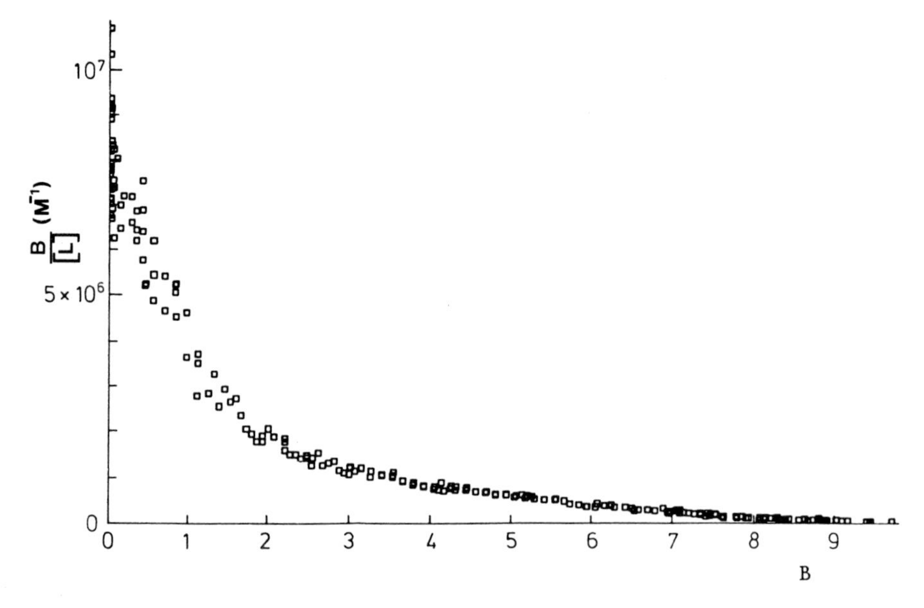

Figure 9. Scatchard graph of binding of laurate ion by human serum albumin (15), using the same data as in *Figure 6*.

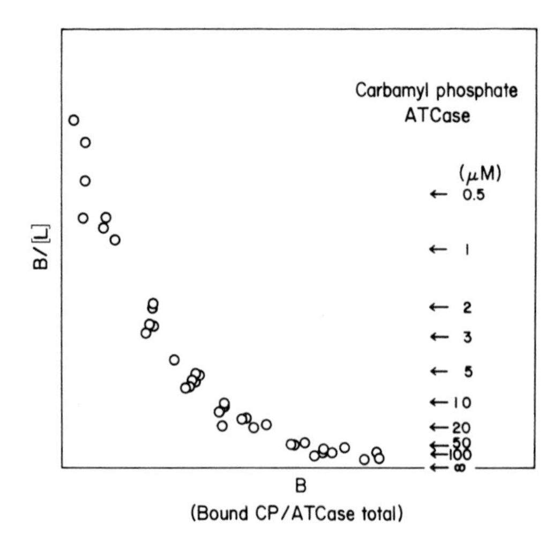

Figure 10. Scatchard graph of data for binding of carbamoyl phosphate by aspartate transcarbamoylase (*Table 4*). To emphasize compression in the ordinate axis at *left*, corresponding values of $[L]$ are shown at the *right*.

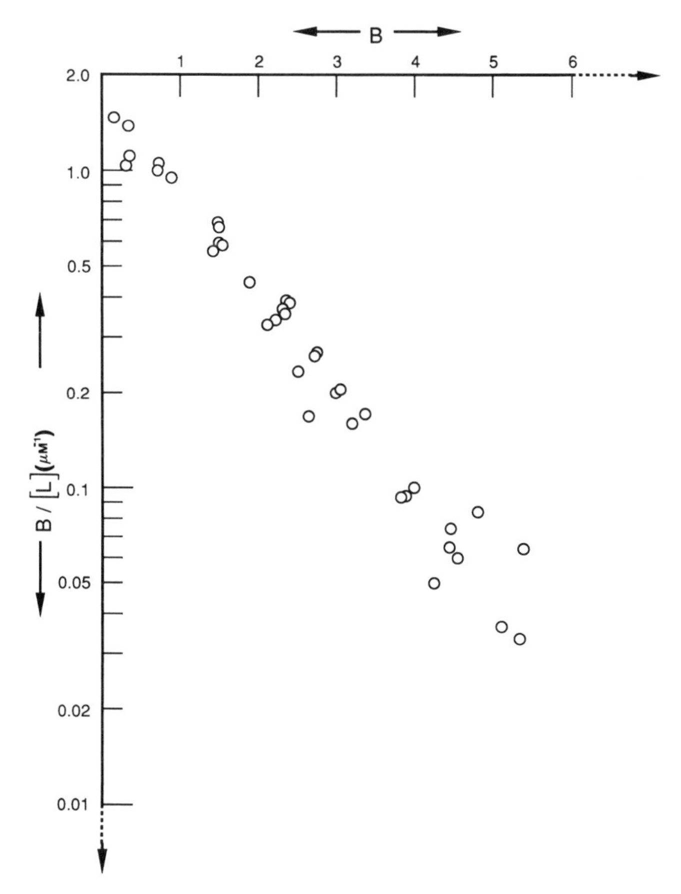

Figure 11. Binding of carbamoyl phosphate by aspartate transcarbamoylase (from *Table 4*) plotted with an ordinate of log $[B/[L]]$ instead of $B/[L]$ as in *Figure 10*. The abscissa axis is the same as in *Figure 10*. The latter figure obscures the very large uncertainties in data at high B and gives the false impression that one is near saturation of the receptor.

value for B, which, if it existed, should be apparent from downward curvature toward the *right*.

Still another common mis-interpretation occurs widely when concave curvature in a Scatchard graph is unequivocal, as in *Figure 10*. Such curvature is due to one, or both, of the following factors:

(i) the population of receptor molecules has two or more classes of sites with different but fixed affinities that do not change with extent of occupancy by ligand;
(ii) the affinities for ligand of at least some of the sites are decreased with increasing occupancy of sites by the ligand.

Concave curvature in a Scatchard graph is often ascribed to 'negative cooperativity', that is to factor (ii), but such a conclusion is totally unwarranted,

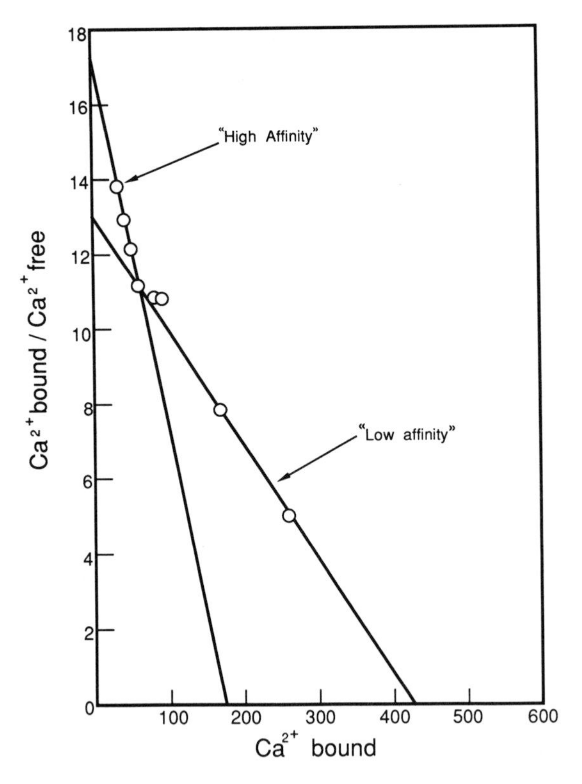

Figure 12. Scatchard graph for binding of Ca^{2+} by a transport protein from mitochondrial membranes (22). The inappropriate assignment of 'high affinity' and 'low affinity' sites was made in the original paper.

since factor (i) may be the reason for the curvature. There is no way of telling from equilibrium binding data alone whether classes of sites of fixed affinities or decreases of affinities with increased occupancy are responsible for the observed concavity in a Scatchard graph.

A complementary error that appears widely in the literature of curved Scatchard graphs is the drawing of two straight lines through the data (e.g. *Figure 12*). A wide collection of such incorrect deconvolutions of curved Scatchard plots has been published by Norby *et al.* (23). All of them ascribe one of the lines, that of higher slope, to 'high affinity' sites and the other to 'low affinity' sites. Even if one knew, from some extra-thermodynamic information, that there were two classes of sites on a particular receptor, it is simply wrong to use two lines through the data to calculate the number and affinity of sites in each class. The slopes and intercepts of such lines are complicated functions of site parameters (24, 25) and generally unsuited for evaluation of binding constants.

7. ALGEBRAIC REPRESENTATIONS OF DATA

If more than one ligand is bound by the receptor, the successive affinity constants in practice have to be evaluated from algebraic expressions. These, in turn, must be formulated in terms of binding equilibrium constants.

7.1 **Stoichiometric binding constants**

There are two distinct approaches to the formulation of binding equations for multiple equilibria (13). The first is a purely thermodynamic one, which does not even have to recognize the molecular concept of sites on the receptor. It takes cognizance only of the multiple stoichiometry observed in binding. Thus if P represents the receptor and L the ligand, we may write a general step in the stoichiometric equilibria as

$$PL_{i-1} + L \leftrightarrow PL_i \tag{11}$$

For each such step we define a *stoichiometric equilibrium constant,* K_i, appropriately indexed to correspond to the stoichiometric step:

$$K_i = \frac{[PL_i]}{[PL_{i-1}][L]} \tag{12}$$

In addition, the number of molecules of L bound (if and) when the receptor is saturated is designated as n, and the corresponding stoichiometric equilibrium constant for the final step is K_n.

The moles of bound L per mole of receptor (in all stages of binding), B, can be related (13) to the concentration $[L]$ of free ligand and the stoichiometric association constants by the equation

$$B = \frac{K_1[L] + 2K_1K_2[L]^2 + \cdots + i(K_1K_2\cdots K_i)[L]^i + \cdots + n(K_1K_2\cdots K_n)[L]^n}{1 + K_1[L] + K_1K_2[L]^2 + \cdots + (K_1K_2\cdots K_i)[L]^i + \cdots + (K_1K_2\cdots K_n)[L]^n} \tag{13}$$

Equation 13 is used to evaluate the stoichiometric equilibrium constants for any specific ligand–receptor system.

7.1.1 *Evaluation of first stoichiometric binding constant*

It is almost always possible to obtain K_1 by a simple graphical procedure. As is evident from Equation 13, if we factor out $[L]$ from the numerator and transfer it to the left-hand side of the equation, then

$$\lim_{[L]\to 0} \frac{B}{[L]} = K_1 \tag{14}$$

An example of such a graphical extrapolation based on the data in *Table 4* is shown in *Figure 13*. As a rule, the first stoichiometric binding constant can be determined within a precision of about 20%.

7.1.2 *Evaluation of successive stoichiometric binding constants by iterative fitting*

The K_i values can be established by iterative fitting of Equation 13 to the experimental data (3). For example, for the binding of laurate ion by human

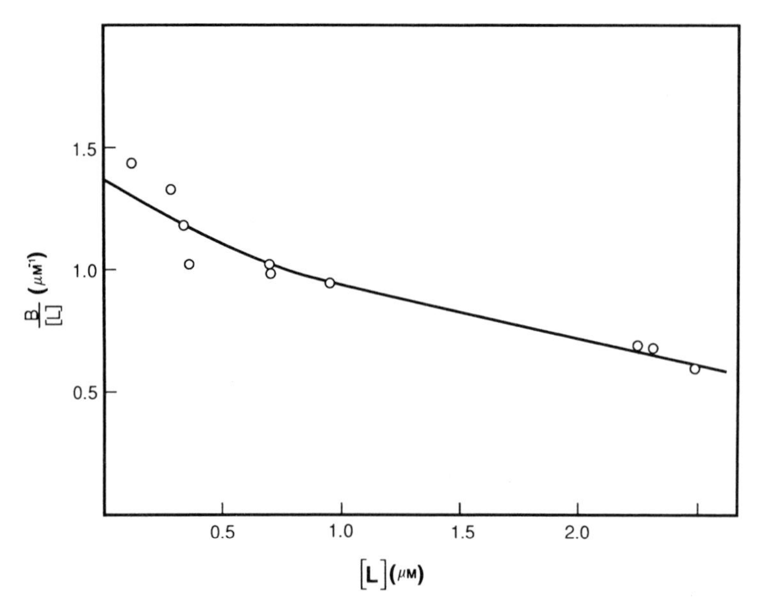

Figure 13. Evaluation of first stoichiometric binding constant for binding of carbamoyl phosphate by aspartate transcarbamoylase (*Table 4*) by extrapolation of $B/[L]$ to the limiting value as free ligand concentration approaches zero.

serum albumin (shown in *Figures 5* and *6*), iterative, least-squares fitting to 220 experimental points (15) provided K_i values up to $i = 10$. The stoichiometric constants giving the best-fit approximation are the following.

Stoichiometric step	*Stoiochiometric binding constant* (M^{-1})
1	8.31×10^6
2	1.67×10^6
3	2.88×10^5
4	6.18×10^5
5	1.07×10^2
6	1.98×10^8
7	9.25×10^1
8	1.43×10^7
9	1.86×10^3
10	1.94×10^4

It should be stressed that the termination of entries at K_{10} does not imply that the receptor, serum albumin, is saturated with ligand at $n = 10$. As is obvious in *Figure 6* there is no indication of a saturation plateau. Nevertheless, since experimental data for B values appreciably above 9 mol/mol receptor are not available, K_i values above $i = 10$ cannot be determined.

7.1.3 *Calculation of stoichiometric binding constants from ghost site binding constants*

An alternative statistical fitting procedure that is very extensively used is based on an equation for B as a function of $[L]$ that is different, but readily derived from Equation 13. Since the significance of the binding parameters in this alternative equation is widely misunderstood, it is essential to examine the algebraic connections before one interprets the numerical results.

Since the denominator of Equation 13 is a polynomial of degree n, it can be replaced by an equivalent algebraic expression in terms of the n roots of the polynomial. Furthermore, it is also known (13) that the numerator is a derivative of the denominator. Consequently, by straightforward manipulation, one can show (13) that

$$B = \frac{K_\alpha[L]}{1 + K_\alpha[L]} + \frac{K_\beta[L]}{1 + K_\beta[L]} + \cdots + \frac{K_\nu[L]}{1 + K_\nu[L]} = \sum_{\alpha}^{\nu} \frac{K_\omega[L]}{1 + K_\omega[L]} \quad (15)$$

in which the number of terms ν equals the degree of the polynomial in the denominator of Equation 13. The coefficients, K_ω, in Equation 15 are constants, related to the roots of the polynomial; however, they do *not* correspond individually to any specific one of the stoichiometric equilibrium constants. (Since each of the equilibrium constants K_α, K_β, etc. respectively, does not correspond to any specific stoichiometric step in the ligand–receptor binding nor to any specific actual site on the receptor system, we can assign each to a 'ghost' site. The ensemble of n imaginary, non-existent ghost sites with corresponding equilibrium constants K_ω faithfully reproduces the experimentally observed dependence of B on L.) Furthermore, they are *not* site-binding constants (13) despite the fact that Equation 15 resembles a site-binding equation for B in special circumstances (non-interacting sites). Equations 13 and 15 are equivalent modes of expression of the stoichiometric formulation of ligand–receptor binding.

If experimental measurements of binding can be made at sufficiently high concentrations of L to saturate the receptor, the number of terms in Equation 15 will be n. If this is not feasible, however, the number of terms in Equation 15 corresponds merely to the degree, i, of the polynomial in Equation 13 that is needed to encompass all of the available experimental binding data.

Algebraic linkages between stoichiometric binding constants and ghost site binding constants can be derived (13, 26) from the equivalence of Equations 13 and 15. Appropriate statistical fitting and allied computer programs have been described in the literature (27–30).

As a concrete illustration of this procedure, consider the data for the binding of leucine by isopropylmalate synthase (*Figure 7*). This is a divalent system, that is, 2 mol of ligand can be bound by the receptor. Equations 13 and 15 for this system can be written as

$$\frac{K_1[L] + 2K_1K_2[L]^2}{1 + K_1[L] + K_1K_2[L]^2} = B = \frac{K_\alpha[L]}{1 + K_\alpha[L]} + \frac{K_\beta[L]}{1 + K_\beta[L]} \quad (16)$$

With some modest algebraic manipulation (13) we find

$$K_1 = K_\alpha + K_\beta \tag{17}$$

$$K_1K_2 = K_\alpha K_\beta \tag{18}$$

Computer programs using the right-hand equality of Equation 16 to fit the data give for leucine binding by isopropylmalate synthase the binding constants

$$K_\alpha = (2.4 \times 10^4 + 1.1\sqrt{-1} \times 10^5)\ \text{M}^{-1}$$

$$K_\beta = (2.4 \times 10^4 - 1.1\sqrt{-1} \times 10^5)\ \text{M}^{-1}$$

from which, using Equations 17 and 18, we obtain the constants

$$K_1 = 0.48 \times 10^5\ \text{M}^{-1}$$

$$K_2 = 2.5 \times 10^5\ \text{M}^{-1}$$

Thus the stoichiometric binding constants are real numbers, with K_2 being greater than K_1; that is the affinity of the receptor is increased with increasing occupancy by ligand. When such 'cooperativity' exists, we find that the ghost site binding constants, K_α and K_β, are complex or imaginary numbers, eminently appropriate for ghost sites. Clearly they cannot be meaningful values for actual site binding constants, and obviously cannot be assigned to actual individual sites. They can be related algebraically to combinations of site constants, as will be illustrated in Section 7.2. If, in addition, one assumes an appropriate molecular model for the receptor, it may be feasible to compute site binding constants from K_ω values. An illustration of this procedure will also be presented in Section 7.2.

As another example, we can list the binding constants for interactions of carbamoyl phosphate with aspartate transcarbamoylase (binding data in *Table 4* and *Figure 10*).

Ghost site binding constant (M^{-1})	*Stoichiometric binding constant* (M^{-1})
$K_\alpha = 0.884 \times 10^6$	$K_1 = 1.45 \times 10^6$
$K_\beta = 0.488 \times 10^6$	$K_2 = 0.403 \times 10^6$
$K_\gamma = 0.552 \times 10^5$	$K_3 = 0.104 \times 10^6$
$K_\sigma = 0.259 \times 10^5$	$K_4 = 0.376 \times 10^5$
$K_\varepsilon = 0.218 \times 10^5$	$K_5 = 0.151 \times 10^5$
$K_\zeta = 0.201 \times 10^5$	$K_6 = 0.821 \times 10^4$

In this case all the ghost site binding constants are real numbers. It is obvious, however, that they have different numerical values than the stoichiometric binding constants. It should be noted that for this (hexavalent) ligand–receptor system, the stoichiometric affinities decrease progressively with increasing occupancy by ligand.

To a rough approximation, one could say that for carbamoyl phosphate–aspartate transcarbamoylase K_α and K_β are the same (K'_ω) and that K_γ, K_σ, K_ε and K_ζ have essentially the same value (K''_ω). For such circumstances, the six terms required by Equation 15 for this system can be reduced to two:

$$B = \frac{2K'_\omega[L]}{1 + K'_\omega[L]} + \frac{4K''_\omega[L]}{1 + K''_\omega[L]} \tag{19}$$

In fact one can obtain a very good fit of Equation 19 to the data in *Table 4* or *Figure 10* with $K'_\omega = 0.44 \times 10^6\ \mathrm{M}^{-1}$ and $K''_\omega = 0.16 \times 10^5\ \mathrm{M}^{-1}$. However, this does *not* mean that there are two classes of sites for carbamoyl phosphate on aspartate transcarbamoylase. K'_ω and K''_ω are *not* site binding constants, nor are they stoichiometric binding constants. In this ligand–receptor system K'_ω and K''_ω are real numbers, but still can only be assigned to non-real ghost sites. The relationships of such binding constants to site binding constants are described in the following Section 7.2.

7.2 Site binding constants

In most cases proteins bind ligands at specific sites. If there are no interactions between the sites, we designate (13) each site on the receptor by a simple index number, 1, 2, . . ., j, and represent the respective equilibrium at any specific site, ${}_{\mathrm{j}}P$, as

$$ {}_{\mathrm{j}}P + L \leftrightarrow {}_{\mathrm{j}}PL \tag{20} $$

Then one can define site equilibrium constants, k_{j}, by

$$ \frac{[{}_{\mathrm{j}}PL]}{[{}_{\mathrm{j}}P][L]} = k_{\mathrm{j}} \tag{21} $$

However, if there are interactions between the sites, we must recognize that our notation ${}_{\mathrm{j}}PL$ is inadequate to distinguish among the different species at equilibrium. For example, suppose one species of ${}_{\mathrm{j}}PL$ has another molecule of ligand filling site 1, as well as one species with site 1 open. Then the individual equilibria must be distinguished, for example, as follows

$$ {}_{1,\mathrm{j}}P_1L + L \leftrightarrow {}_{1,\mathrm{j}}P_{1,\mathrm{j}}L_2 \tag{22} $$

$$ {}_{1,\mathrm{j}}P + L \leftrightarrow {}_{1,\mathrm{j}}P_{\mathrm{j}}L \tag{23} $$

In both cases L is going into site j but Equation 22 represents the situation in which site 1 is already filled (and stays filled) and Equation 23 that in which site 1 is, and remains, open. Of course, if the equilibrium at site 1 has no influence on the affinity of site j, then the detail of Equations 22 and 23 is unnecessary, and the simple notation of Equations 20 and 21 fits the circumstances correctly. To be general, however, we want to include ligand–receptor systems where the sites do influence each other. In that case, we have to define different equilibrium constants for Equations 22 and 23:

$$ \frac{[{}_{1,\mathrm{j}}P_{1,\mathrm{j}}L_2]}{[{}_{1,\mathrm{j}}P_1L][L]} = k_{1,\mathrm{j}} \tag{24} $$

$$ \frac{[{}_{1,\mathrm{j}}P_{\mathrm{j}}L]}{[{}_{1,\mathrm{j}}P][L]} = k_{\mathrm{j}} \tag{25} $$

That of Equation 25 is the same as the one already defined in Equation 21 for in both we are considering the ligation of one ligand molecule by site j when site 1 is

Table 5. Comparison of numbers of different kinds of binding constants for a macromolecule with n binding sites.

No. of binding sites, n	Total no. of site binding constants ($k_{1,2,\ldots,j}$)	No. of independent site binding constants	No. of stoichiometric binding constants (K_i)	No. of ghost site binding constants (K_ω)
2	4	3	2	2
3	12	7	3	3
4	32	15	4	4
6	192	63	6	6
8	1024	255	8	8
12	24 576	4095	12	12

empty and unchanged. However, $k_{1,j}$ is, in general, not equal to k_j because occupancy of site 1 by L makes the receptor a different species whose affinity for L at site j could be markedly different.

In a detailed analysis (13), we find that the number of site binding constants that need to be defined, to treat binding generally, goes up astronomically (see *Table 5*). Even for a hexameric enzyme, such as aspartate transcarbamoylase, the number of site constants is near 200; for a dodecameric enzyme, such as glutamine synthetase, the number approaches 25 000. Clearly site equilibrium constants are not tractable binding parameters, except when molecular structural information permits one to set up some additional constraining relations between them.

When, for example, the individual sites have fixed, unchanging affinities for ligand, (that is, if the value of each k_j is the same regardless of whether other sites are empty or occupied) then the total moles of bound ligand at all sites is given by (13)

$$B = \frac{k_1[L]}{1 + k_1[L]} + \frac{k_2[L]}{1 + k_2[L]} + \cdots + \frac{k_j[L]}{1 + k_j[L]} + \cdots + \frac{k_n[L]}{1 + k_n[L]} \quad (26)$$

Superficially, Equation 26 resembles the algebraic transform Equation 15, but it is *not* the same because, in general, when there are interactions between the binding sites

$$K_\alpha \neq k_1, \ldots, K_\nu \neq k_n.$$

We can illustrate these distinctions by examining a concrete example, a bivalent ligand–receptor system. The full complement of binding constants is shown graphically in *Figure 14*. For a bivalent system, the relationships between K_α and K_β and the site binding constants can be shown (13) to be

$$K_\alpha = \tfrac{1}{2}(k_1 + k_2) \pm \tfrac{1}{2}[(k_1 + k_2)^2 - 4k_1k_{1,2}]^{\frac{1}{2}} \quad (27)$$

$$K_\beta = \tfrac{1}{2}(k_1 + k_2) \mp \tfrac{1}{2}[(k_1 + k_2)^2 - 4k_1k_{1,2}]^{\frac{1}{2}} \quad (28)$$

As we can see, there are three independent site constants in these two equations

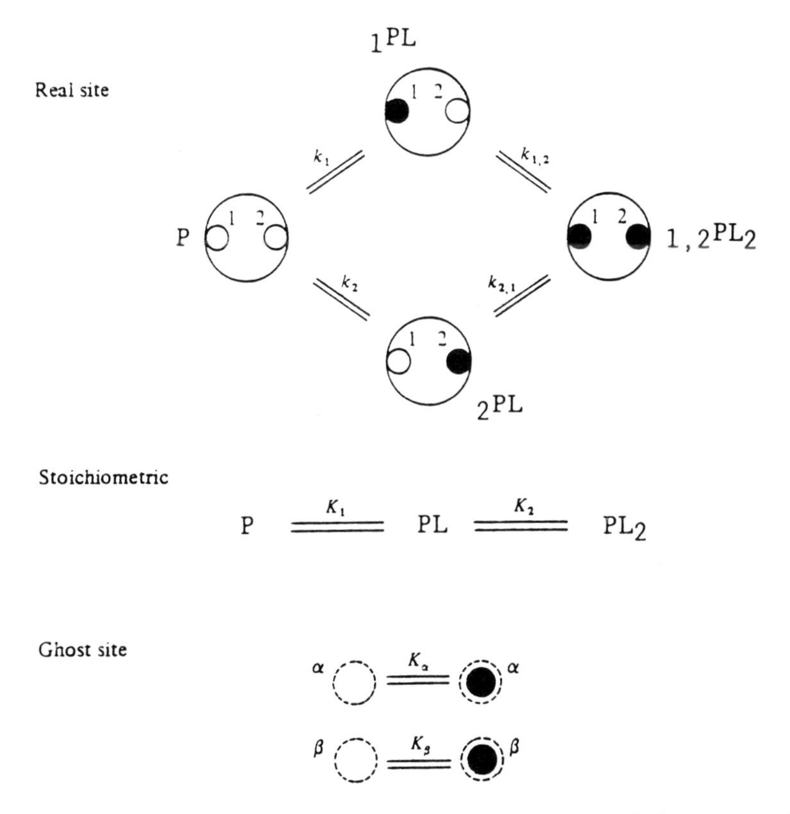

Figure 14. Comparison of meanings of different types of ligand–receptor binding constants for a two-site receptor.

(the fourth is fixed by the other three because of the requirement that ΔG° for a cycle must be zero and hence the product of all four *k*s must equal unity); so all three *cannot* have specified values. On the other hand, if one proposes some form of restraining molecular model, then the site constants may be derivable. For example, the receptor isopropylmalate synthase (*Figure 4*) is known from structural information to be constituted of two identical protomers. Since the protomers are assumed to be identical when no ligand is attached to the receptor, we can state that

$$k_1 = k_2 \equiv k_{\mathrm{I}}.$$

Furthermore (*Figure 14*) one would expect that

$$k_{1,2} = k_{2,1} \equiv k_{\mathrm{II}}.$$

Under these circumstances one can use the known values of K_α and K_β for leucine binding (see Section 7.1.3) together with Equations 27 and 28 to show that

$$k_1 = 0.24 \times 10^5\ \mathrm{M}^{-1} = k_2$$

$$k_{1,2} = 5.0 \times 10^5 \text{ M}^{-1} = k_{2,1}$$

Thus the values of the site binding constants have been numerically determined. We should notice also that even though in our model *both binding sites are identical* at the outset, their binding of ligand fits a two-term Equation 15, that *superficially looks* like an expression for two different classes of sites.

Finally, we should recognize that if for a bivalent system we were to insert all four site constants into Equation 15 to obtain

$$B = \frac{k_1[L]}{1 + k_1[L]} + \frac{k_2[L]}{1 + k_2[L]} + \frac{k_{1,2}[L]}{1 + k_{1,2}[L]} + \frac{k_{2,1}[L]}{1 + k_{2,1}[L]} \tag{29}$$

the result is an incorrect meaningless expression. Each term on the right-hand side in Equation 29 approaches 1 as $[L] \to \infty$; hence the four terms approach 4. However in a divalent system the maximum value of B is 2. Obviously a site binding equation of the format of Equation 29 is not valid. On the other hand, a two-term equation of this form with the ghost site constants K_α and K_β is fully valid.

8. REFERENCES

1. Osborne, W. A. (1906) *J. Physiol.*, **34**, 84.
2. Donnan, F. G. (1924) *Chem. Rev.*, **1**, 73.
3. Klotz, I. M., Walker, F. M. and Pivan, R. B. (1946) *J. Am. Chem. Soc.*, **68**, 1486.
4. Karush, F. (1950) *J. Am. Chem. Soc.*, **72**, 2705.
5. Gurd, F. R. N. and Goodman, D. S. (1952) *J. Am. Chem. Soc.*, **74**, 670.
6. Karush, F. (1956) *J. Am. Chem. Soc.*, **78**, 5519.
7. Brodersen, R., Andersen, S., Jacobsen, C., Sønderskov, O., Ebbesen, F., Cashore, W. J. and Larsen, S. (1982) *Anal. Biochem.*, **121**, 395.
8. Hughes, T. R. (1955) Ph.D. Dissertation, Northwestern University, Evanston, IL, p. 56.
9. Klotz, I. M. (1973) *Ann. N.Y. Acad. Sci.*, **226**, 18.
10. Klotz, I. M. (1946) *J. Am. Chem. Soc.*, **68**, 2299.
11. Knudsen, A., Pedersen, A. O. and Brodersen, R. (1986) *Arch. Biochem. Biophys.*, **244**, 273.
12. Klotz, I. M. (1985) *Qt. Rev. Biophys.*, **18**, 227.
13. Klotz, I. M. (1986) *Introduction to Biomolecular Energetics, Including Ligand–Receptor Interactions.* Academic Press, Inc. Orlando, Florida.
14. Teng-Leary, E. and Kohlhaw, G. B. (1973) *Biochemistry*, **12**, 2980.
15. Pedersen, A. O., Hust, B., Andersen, S., Nielsen, F. and Brodersen, R. (1986) *Eur. J. Biochem.*, **154**, 552.
16. See, for example, Haldane, J. B. S. and Stern, K. G. (1932) *Allgemeine Chemie der Enzyme.* Steinkopf, Dresden, p. 119.
17. Scatchard, G. (1949) *Ann. N.Y. Acad. Sci.*, **51**, 660.
18. Skolnick, P., Moncada, V., Barker, J. L. and Paul, S. M. (1981) *Science*, **211**, 1448
19. Feldman, H. A. (1983) *J. Biol. Chem.*, **258**, 12865.
20. Suter, P. and Rosenbusch, J. (1976) *J. Biol. Chem.*, **251**, 5986.
21. Klotz, I. M. (1983) *Trends Pharmacol. Sci.*, **4**, 253.
22. Jeng, A. C., Ryan, T. E. and Shamoo, A. E. (1978) *Proc. Natl. Acad. Sci. USA*, **75**, 2125.
23. Norby, J. G., Ottolenghi, P. and Jensen, J. (1980) *Anal. Biochem.*, **102**, 318.
24. Klotz, I. M. and Hunston, D. L. (1971) *Biochemistry*, **10**, 3065.
25. Honoré, B. and Brodersen, R. (1984) *Mol. Pharmacol.*, **25**, 137.
26. Simms, H. S. (1926) *J. Am. Chem. Soc.*, **48**, 1239.
27. Fletcher, J. E., Spector, A. A. and Ashbrook, J. D. (1970) *Biochemistry*, **9**, 4580.
28. Feldman, H. A. (1972) *Anal. Biochem.*, **48**, 317.
29. Klotz, I. M. and Hunston, D. L. (1977) *Proc. Natl. Acad. Sci. USA*, **74**, 4959.
30. Munson, P. J. and Rodbard, D. (1980) *Anal. Biochem.*, **107**, 220.

CHAPTER 3

Ligand blotting

ANNE K. SOUTAR and DAVID P. WADE

1. INTRODUCTION

Ligand blotting is a method of detecting or identifying proteins on the basis of their specific biological function. It is similar to immunoblotting ('Western' blotting) in that cell proteins are solubilized, fractionated by polyacrylamide gel electrophoresis and transferred electrophoretically to a nitrocellulose or similar membrane ('blotted'). However, specific proteins can then be identified on the membrane by their ability to bind a particular ligand, rather than with a specific antibody as in immunoblotting. The ligand can be another protein or peptide, a nucleic acid or, indeed, any molecule that binds with high affinity and for which a detection system is available. So far, the method has been particularly useful for studies on membrane proteins, such as cell surface receptors, which can be difficult to assay in cell extracts.

The success of the technique depends on two factors.

(i) At least some of the biological activity of the protein must survive the rigours of solubilization, fractionation and transfer. Although these procedures might be expected to disrupt the biological function of many proteins, in practice sufficient refolding to restore specific binding activity often occurs during the transfer of proteins from the gel to the membrane. Alternatively, as in the case of the low-density lipoprotein receptor (LDL receptor), the native structure of the protein is at least partially maintained during detergent treatment by its numerous intra-chain disulphide bonds. If highly denaturing systems are used, for example a gel containing sodium dodecyl sulphate (SDS), the method is unlikely to be useful in the detection of proteins whose full biological activity is dependent on subunit interactions, especially where the subunits are not identical. Having said this, binding activity itself could reside with just one of the subunits, and might be detectable. It must be emphasized that binding of a particular ligand to a protein on a blot does not in itself constitute sufficient evidence that a specific binding protein of physiological significance has been identified. Some confirmatory observations, for example changes in the amount of the protein under different conditions or differences between tissues, should be sought.

(ii) The second factor is that a sufficiently sensitive means of detecting the bound ligand must be available. This may be by direct identification of a radioactive or visible label, or by means of an indirect identification with, for example, a

Table 1. Examples of proteins that have been detected by ligand blotting.

Protein and source	*Ligand*	*Method of detection*	*Reference*
Epidermal growth factor (EGF) receptor from human epidermoid carcinoma A431 cell membranes	Mouse EGF	Anti-EGF antiserum + [^{125}I] protein A	1
Thyroid stimulating hormone (TSH) receptor from human and pig thyroid plasma membranes	Bovine TSH	Anti-TSH antiserum + peroxidase-linked anti-IgG	2
Acetylcholine receptor from plasma membranes of the *Torpedo californica* electric organ	^{125}I-labelled α-bungarotoxin (neurotoxin from snake venom)	Autoradiography	3
Glycoproteins from rat brain synaptosomal plasma membranes	Concanavalin A	Horseradish peroxidase (binds to Con A)	4
Mitochondrial proteins from whole rat brain	^{32}P-labelled rat brain microtubule-associated proteins	Autoradiography	5
Low-density lipoprotein (LDL) receptor from bovine adrenal cortex membranes and fibroblasts	LDL	Anti-LDL antiserum + ^{125}I-labelled anti-IgG	6
	Biotin-LDL	Streptavidin biotinylated horseradish peroxidase complex	7
High-density lipoprotein (HDL) receptor from bovine aortic endothelial cell membranes	^{125}I-labelled HDL	Autoradiography	8
Nuclear protein from human rhabdomyosarcoma A 673 cells	[^{32}P] DNA (promotor region of human transferrin receptor gene)	Autoradiography	9
Histones from chicken and trout	^{32}P-labelled λ phage or trout DNA	Autoradiography	10

specific antibody or the biotin–streptavidin–peroxidase complex. In our laboratory we have used a variety of methods to detect different lipoprotein ligands bound to the LDL receptor from a number of different tissues and cultured cells. In the detailed laboratory protocols we will describe our methods, but will attempt to point out where it may be necessary to use an alternative approach to optimize the results for other systems. Some examples of successful use of ligand blotting are shown in *Table 1*. The list is not complete!

2. SOLUBILIZATION OF CELLS AND TISSUES

If the subcellular localization of the protein to be blotted is known, then it may be preferable, or essential if the protein is a minor component, to carry out a brief subcellular fractionation procedure before solubilization to increase the 'specific activity' of the protein in the extract. For example, a mitochondrial, nuclear, cytosolic or crude membrane fraction can be prepared by centrifugation. In this way, it is possible to load less total protein onto the gel to obtain the same signal and thus improve the resolution on the gel. Put another way, it will be possible to load sufficient protein onto the gel to detect anything at all! However, there is always a chance that a hitherto unrecognized binding protein could be missed in this way, and we have successfully ligand-blotted the LDL receptor in unfractionated extracts of cultured cells. An alternative approach is to solubilize the whole tissue (or the subcellular fraction) and carry out some form of purification step that will both concentrate and enrich the specific protein in the extract. For example, absorption and elution of the extract from an ion-exchange medium such as DEAE–cellulose or an affinity matrix can be straightforward and rapid procedures when carried out on a small scale.

The precise nature of the solutions used to solubilize the cells or tissue will depend largely on the nature of the protein to be detected. The method we have used for solubilizing both whole cultured cells and crude membrane fractions of tissues is given in *Table 2*, in which we have also include some suggestions for factors that can be varied. A variant of this solution was originally described for solubilizing bacterial cells and it is therefore applicable to a wide variety of cell types.

Soluble extracts of both cells and membrane fractions can be stored as aliquots in liquid nitrogen or at −70°C for several months with no apparent loss of LDL receptor binding activity but this should be checked for other proteins. It is best to avoid repeated freezing and thawing of extracts.

3. FRACTIONATION BY POLYACRYLAMIDE GEL ELECTROPHORESIS

3.1 **Choice of method**

The most straightforward method of fractionating proteins on polyacrylamide gels is that originally described by Laemmli (11), using gels containing SDS and with a

Table 2. Solutions for solubilizing cells and cell subfractions.

Solubilizing solution

5 M urea
1.6% (w/v) Triton X-100
1 mM phenylmethylsulphonyl fluoride } protease inhibitors (Sigma Chem. Co.)
0.3 mM leupeptin }

Protocol

1. Add 0.5 ml of solubilizing solution/100 mg wet weight.
2. Suck up and down 10 times with a syringe and 21-gauge needle to homogenize.
3. Leave on ice for 20 min and then spin at 100 000 *g* for 60 min to remove insoluble material.

Factors that can be varied

Chaotropic agent (e.g. urea)	Should be omitted if it might irreversibly affect binding activity (e.g. by dissociating subunits).
Detergent	Try different types to increase solubility or maintain activity, e.g. ionic—sodium dodecyl sulphate (SDS) —sodium deoxycholate (DOC) zwitterionic—3-[(3-cholamidopropyl)dimethylammonio]-1-propane sulphonate (Chaps) other non-ionic—Tween, octylglucoside Note that detergent concentrations greater than ~2% (w/v) may adversely affect electrophoresis
Buffer	May be necessary to control pH by inclusion of buffer, especially if a chromatographic concentrating step is included in the procedure.
Salt	May be necessary to increase ionic strength for maximum solubilization or stability of some proteins, but note that high salt concentration interferes with electrophoresis.

discontinuous buffer system in which there is a stacking or sharpening of the individual protein bands. Most proteins are completely unfolded and dissociated when treated with SDS and sulphydryl reducing agents, to form single randomly coiled polypeptide chains that have bound sufficient SDS to give them a constant negative charge per unit mass. As a result proteins migrate on reduced SDS gels as a log function of size, so that the apparent molecular weight of any protein band can be estimated by comparing it with standard proteins run on the same gel (*Figure 1*).

Generally, 10% acrylamide gels are used, but the percentage of acrylamide in the gel can be altered to obtain optimum resolution in a given molecular weight range. As a rough guide, if the protein of interest is greater than 300 000 daltons, then a lower percentage, for example 5–6%, should be used, while proteins of less than 30 000 daltons should be fractionated on 15–17% gels. Gels of low acrylamide content are less easy to handle than 10–12% gels. Note that certain glycoproteins behave anomalously on SDS gels and, in the absence of reducing agents, proteins with inter- or intra-chain disulphide bonds may not be fully denatured, which can also result in anomalous behaviour.

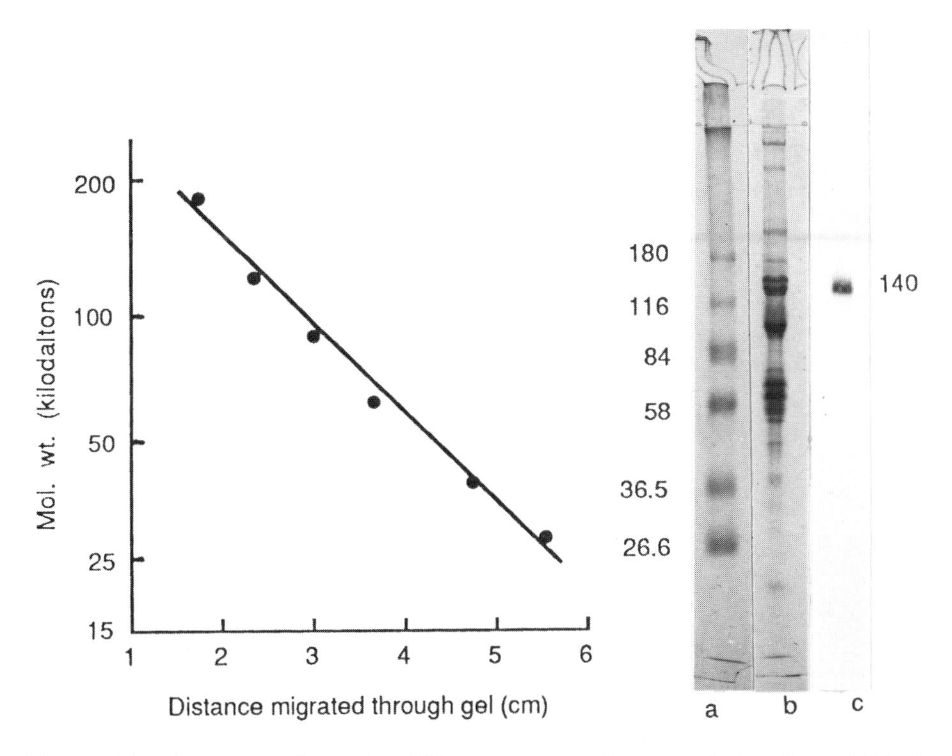

Figure 1. Ligand blotting of the rat liver LDL receptor in crude extracts fractionated by SDS gel electrophoresis. (**a**) A mixture of pre-stained proteins comprising approximately 5 μg each of α-macroglobulin (mol. wt 180 000), β-galactosidase (116 000), fructose-6-phosphate kinase (84 000), pyruvate kinase (58 000), lactic dehydrogenase (36 000) and triosephosphate isomerase (26 600) was fractionated by electrophoresis on a 10% polyacrylamide gel containing SDS. The graph shows the relationship between molecular weight and the distance migrated on the gel. Solubilized rat liver membrane proteins (50 μg protein/lane) were fractionated under non-reducing conditions on the same gel. One lane was stained for protein (**b**), while proteins from a duplicate lane were transferred electrophoretically to nitrocellulose (**c**). LDL receptor protein was detected by incubation of the nitrocellulose strip with ^{125}I-labelled β-very low-density lipoprotein (β-VLDL) (0.5 μg protein/ml, specific radioactivity 800 c.p.m./ng), followed by autoradiography for 16 h with pre-flashed film and intensifying screens. The apparent molecular weight of the LDL receptor was estimated from the standard curve. All details of the methods are given in the text.

Not surprisingly, not all proteins will exhibit biological activity after electrophoresis on reduced SDS gels even when the denaturants have been removed during blotting. For example, treatment of the LDL receptor with sulphydryl reducing agents destroys its binding ability, and thus for successful ligand blotting of this protein, 2-mercaptoethanol must be omitted from the gel system. On the other hand, the oxidation of essential free sulphydryl groups that can occur during the preparation of cell extracts may inhibit the activity of other proteins. If no binding activity can be detected after SDS–gel electrophoresis it may be necessary to use a non-denaturing system to separate the proteins. We have successfully employed a system (12) in which membrane proteins are fractionated on gradient

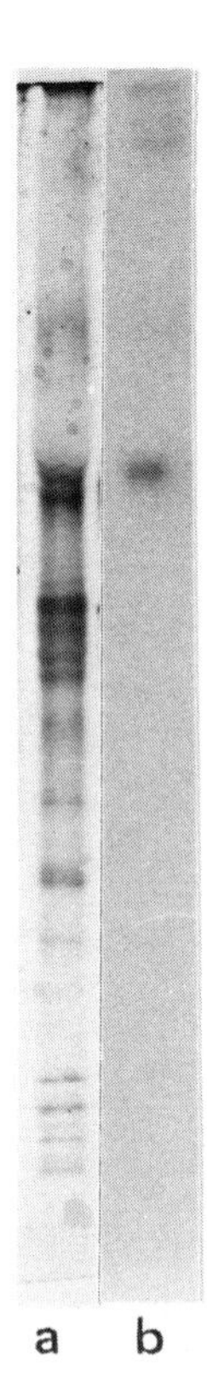

Figure 2. Ligand blotting of the LDL receptor in human liver membranes fractionated on a non-denaturing gel system. Human liver membranes solubilized with Triton X-100 (100 μg protein) were fractionated on a 3–15% polyacrylamide gradient gel containing 0.1% (w/v) Triton X-100. One strip was stained for protein (**a**). Proteins from a duplicate strip were transferred to nitrocellulose (**b**) and the LDL receptor was detected by incubation of the nitrocellulose strip with 0.5 μg/ml ^{125}I-labelled β-VLDL (sp. act. 800 c.p.m./ng) followed by autoradiography for 16 h with pre-flashed film and intensifying screens. All details of the methods are given in the text.

gels containing the non-ionic detergent Triton X-100 (*Figure 2*). On this gel system protein subunits should not be dissociated and proteins will migrate according to a combination of factors, including their net charge, shape and molecular weight. Gradient gels permit separation of proteins of a wide range of molecular weight, but eventually proteins will reach a point at which they are trapped in the gel matrix and cannot readily be transferred to nitrocellulose. For this reason, gradient gels for ligand blotting should not be run to equilibrium. Thus, to determine the molecular weights of the constituents of a band on a Triton gel it is necessary to cut out the band and separate the proteins on an SDS gel.

3.2 **SDS–polyacrylamide gel electrophoresis**

The method described below is for a work-shop produced gel apparatus, and the equipment for protein gel electrophoresis from different manufacturers will vary considerably in their details. When in doubt, always follow the supplier's instructions for your own equipment.

(i) The stock solutions and apparatus required are listed in *Table 3*. Gloves should be worn when handling unpolymerized gel solutions and acrylamide powder, since both are neurotoxic. Polymerized gels are safe to handle.

(ii) Clean the glass plates with detergent, rinse well with water and then finally with ethanol. Assemble the plates with the spacers as shown in *Figure 3a* and clip them together. Seal the edges of the plate with molten 2% agarose; some people prefer to use waterpoof tape. Stand the plate vertically on the clips and make certain the top is horizontal.

(iii) Fill the plate with freshly-prepared running gel solution to within 2 cm of the top of the cut-out plate. Avoid trapping air bubbles. You can tilt the plates *gently*, but be careful not to disturb the clips and agarose, or the gel will leak. Layer water gently on the top of the gel solution to fill the plates. Take care not to disturb the gel solution or the top will be uneven and the resolution less good. Leave the plate at room temperature for 30–60 min. The gel has polymerized when a sharp line re-appears between the gel and the water. (A gel can be stored overnight at this stage, with the top wrapped in plastic film to prevent evaporation. Some further polymerization may occur so that the gel will not give exactly the same R_f values for proteins as one used immediately, but this will not affect molecular weight determination so long as standards are run simultaneously.)

(iv) Pour off the water, and rinse the top with 1–2 ml of spacer gel solution. Fill the space completely with spacer gel solution and insert the comb (*Figure 3b*). Avoid trapping air bubbles. To contain the spilt gel solution it is useful to stand the plate on a tray or at least on paper towels at this stage. Leave the gel for approximately 30 min to polymerize. Check by looking for sharp lines around the top of the comb, or by tilting the gel very gently. (Some liquid will be excluded from the gel during polymerization.)

(v) Remove the comb and gently straighten any of the gel pieces between the sample wells with a hypodermic needle. Remove the lower spacer, leaving the side spacers in place. Rinse the top sample wells and the lower edge of the gel with running buffer by squirting the buffer in with a Pasteur pipette or with a syringe and bent needle. Fill the lower reservoir of the gel apparatus with running buffer and check that the seal is greased. Place the gel assembly in the apparatus with the cut-out plate facing the upper reservoir. The plate should rest on the small plastic supports in the lower buffer chamber. Clip the gel in place near the seal and fill the upper reservoir with buffer. Check for leaks (*Figure 3c* and *d*).

(vi) Apply the samples and standards. Either, layer the samples underneath the buffer by holding a plastic pipette tip just above the sample well, resting against the front plate, and allowing the sample to run down the plate into the well. Or, first remove some of the buffer from the upper tank and all the buffer from the sample wells with a syringe and needle. Apply the samples as just described and then gently layer running buffer on top. Fill the upper

Table 3. Fractionation of proteins on gels containing SDS.

Stock solutions

A. 30% (w/v) acrylamide 0.2% (w/v) bisacrylamide.
B. 30% (w/v) acrylamide 0.8% (w/v) bisacrylamide.
C. 48 ml 1 M HCl + 36.6 g Tris + 230 μl TEMED + water to 100 ml (pH 8.9).
D. 48 ml 1 M HCl + 5.98 g Tris + 460 μl TEMED + water to 100 ml (pH 6.7).
10% (w/v) SDS.
10% (w/v) ammonium persulphate (freshly prepared daily)
×10 running buffer (30 g Tris, 144 g glycine/litre, pH 8.6).
2% (w/v) agarose. (Boil the agarose suspension for ~15 min or until a clear solution is formed. For use, melt the agarose in a boiling water bath.)
Pre-stained standard protein mixture (Sigma SDS-7B).

Equipment

Power pack—capable of producing 40–50 mA, 200 V.
Gel apparatus, Glass plates, Spacers, Comb, Bulldog clips — usually purchased as a kit.

Working solutions

Running gel

For 30 ml of gel solution, mix the components shown below, adding ammonium persulphate last, just before pouring the gel.

% Acrylamide	A	C	Water	SDS	Ammonium persulphate
	(ml)	(ml)	(ml)	(ml)	(μl)
5	5	3.75	20.7	0.3	250
10	10	3.75	15.7	0.3	250
15	15	3.75	10.7	0.3	250

Spacer gel

2.5 ml of B
5.0 ml of D
0.2 ml of SDS
12 ml of water
100 μl of ammonium persulphate solution

Running buffer

100 ml of ×10 running buffer
10 ml of 10% SDS
Dilute to 1 litre with water

Sample buffer

(a) Reduced gels
1.0 ml of buffer D
0.8 ml of 10% SDS
0.8 ml of glycerol
8 μl of 2-mercaptoethanol

(b) Non-reduced gels
1.0 ml of buffer D
0.8 ml of 10% SDS
0.8 ml of glycerol

Samples are prepared by mixing 50 μl of protein extract (containing not more than 200 μg protein) with 25 μl of sample buffer. Heat at 100°C in a boiling water bath for 2.0 min. Cool briefly, and apply to the gel. Standard protein mix is treated in the same way.

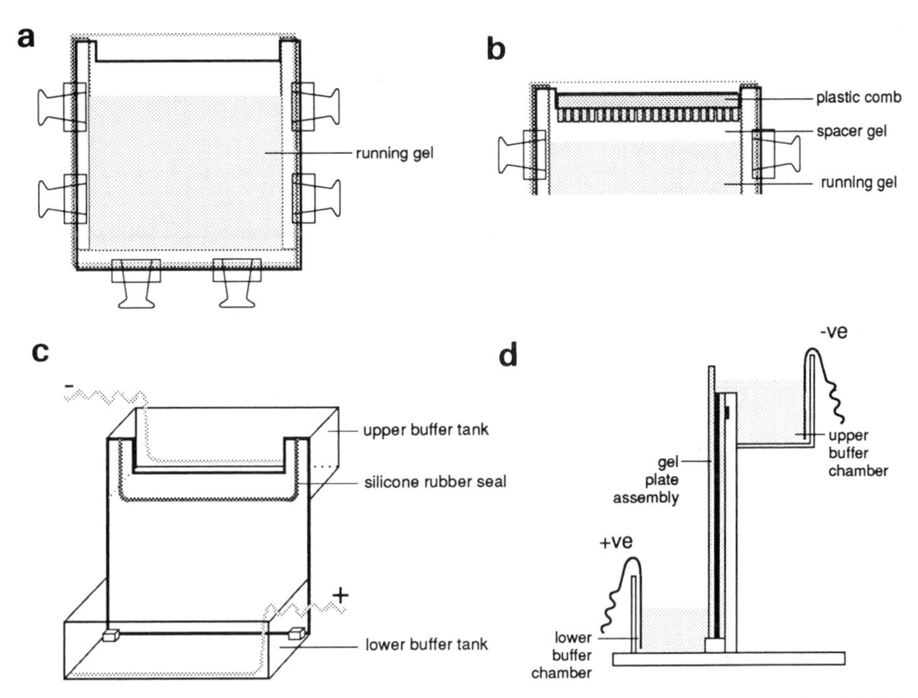

Figure 3. Apparatus and gel plate assembly for vertical slab gel electrophoresis of proteins. (**a**) Gel plate assembly. The two glass plates (20 cm square and 4 mm thick) are spaced as shown by strips of Perspex approximately 1 cm wide and 0.9 mm thick. The front plate has a section 16 cm × 2 cm cut out. (**b**) Gel plate assembly with sample well comb in place. The teeth of the Perspex sample well comb are of the same thickness as the spacers and form wells of 0.5 cm width, 1.5 cm depth and 2.5 mm apart. (**c**) Gel apparatus (front view). (**d**) Gel apparatus (side view). After removal of the lower spacer and the sample comb, the gel plate assembly is placed in the apparatus, resting on two small Perspex blocks in the lower buffer tank. The cut-out plate faces towards the upper buffer tank and a water-proof seal is formed between the glass plate and the silicon rubber seal.

reservoir. Switch on the current. A 20 cm^2 gel should be run at 40 mA until the samples (as indicated by the bromphenol blue) have entered the running gel and then either for 2–4 h at 30 mA, constant current, or for approximately 16 h (overnight) at 3–4 mA constant current. The voltage will rise as the run proceeds and the gel warms up. Gels can be run at 4°C in a cold room to minimize heating effects, or the gel apparatus may have a built-in cooling system. The bromphenol blue will migrate with the buffer front, so stop the gel when it reaches the bottom. Remove the gel assembly from the apparatus and lay it on a flat surface with the cut-out plate uppermost. Gently prise the two plates apart with a spatula, leaving the gel on the lower plate. Cut one lower corner off the gel to mark the orientation of samples. Blotting of the gel is described in the next section.

3.3 Gradient gels with Triton X-100

The stock solutions and apparatus for preparing gradient gels are listed in *Table 4*. The method below is described in detail only when it differs from that described

Table 4. Equipment and solutions for non-dissociating Triton X-100 gradient gel electrophoresis.

Stock solutions

A. 30% (w/v) acrylamide 0.8% (w/v) bisacrylamide.
B. 2 M Tris–glycine, pH 9.0 (131.4 g/litre glycine, 30.3 g/litre Tris).
C. 10% (w/v) Triton X-100.
D. 1.5% (w/v) ammonium persulphate.

Equipment

Power pack—capable of producing 8–10 mA, 500 V.
Gel apparatus, Glass plates, Spacers, Comb, Bulldog clips — usually purchased as a kit.
Gradient former
Magnetic stirrer
Peristaltic pump (optional)

Working solutions

1. Gel solution (for 35 ml total volume of gradient).

	Stock solution (ml)					
	A	B	C	D	Water	TEMED[a]
	(ml)	(ml)	(ml)	(ml)	(ml)	(μl)
Heavy gel (20%)	11.67	0.875	0.175	0.41	–[b]	6.0
or Heavy gel (15%)	8.75	0.875	0.175	0.41	–[b]	6.0
Light gel (3%)	1.75	0.875	0.175	0.41	14.3	6.0
Sample gel (3%)	1.5	0.75	0.15	0.35	12.25	5.0

Running buffer

100 mM Tris–glycine, pH 9.0, 0.1% Triton X-100
(50 ml of B + 10 ml of C diluted to 1 litre).

Sample buffer

20 mM Tris–glycine, pH 9.0, 2% (w/v) Triton X-100
(2 ml of C + 100 μl of B diluted to 10 ml: can add protease inhibitors if desired—see *Table 2*).

Preparation of samples

A crude membrane (or other) fraction is solubilized in sample buffer by homogenization (see *Table 3* for details) (~200 mg of wet weight of membranes in 1.0 ml of buffer gives an extract of ~10 mg protein/ml). Add glycerol to give ~10% (w/v) and apply to the gel.

[a] For gradient gels, the TEMED is added just before use. See protocol (Section 3.3).
[b] For the heavy gel solution, add 2.63 g of sucrose and make the volume to 17.5 ml.

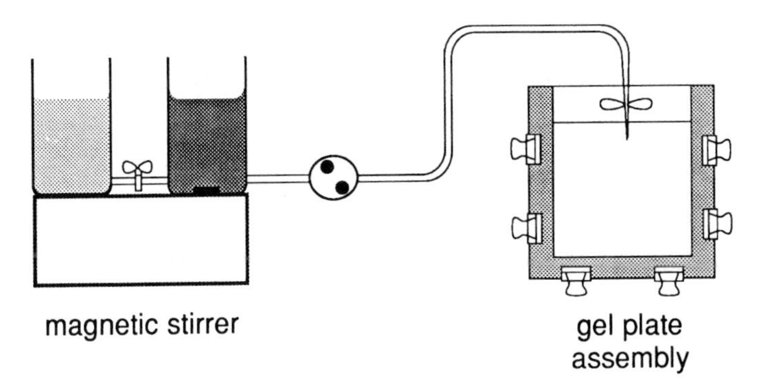

Figure 4. Preparation of a gradient gel. The gel solution is fed via a peristaltic pump from the gradient former into the gel plate assembly through a butterfly needle taped to the top of the back plate.

for SDS gels. This system can also be used with other non-ionic detergents, for example octylglucoside, or with 3-[(3-cholamidopropyl) dimethylammonio]-1-propane sulphonate (Chaps), a zwitterionic detergent.

(i) The gradient gel is poured using a gradient maker as shown in *Figure 4*. Put the light solution in the left-hand chamber and fill the tap between the chambers to prevent release of an air bubble later. Close the tap and put the heavy solution in the right-hand chamber. Check that the levels are the same. Add 6.0 μl of *N,N,N',N'*-tetramethylethylene diamine (TEMED) to each solution chamber and stir with glass rod. Switch on the stirrer gently (avoid forming a vortex) and open the tap between the chambers. Start the pump and fill the gel plate at approximately 3 ml/min through a butterfly needle taped to the top of the gel plate (see *Figure 4*). When the gel is poured, layer running buffer on the top.

(ii) Add 5.0 μl of TEMED to 15 ml of sample well gel solution and make the sample well gel.

(iii) When applying samples, load as small a volume as possible, because there is no stacking effect in this system (for example 100 μg of protein/10 μl of buffer in a sample well 0.5 cm × 1.0 mm gives reasonable resolution). Run the gel for 5–6 h at 4°C with 8–10 mA constant current. Bromphenol blue, which will run in a V-shape, will have partially run off the gel in this time, but most proteins will be trapped in the gradient gel.

4. TRANSFER OF PROTEINS FROM GEL TO MEMBRANE

The theoretical considerations for the process of transferring proteins from gels to a membrane have been discussed in detail in a review by Gershoni and Palade (13). We will describe the method in use in our laboratory, but three main variables should be considered.

(i) First is the choice of membrane. We have used nitrocellulose membranes obtained from Schleicher and Schüll, because they have given reproducible results and low background noise, but undoubtedly other membranes can be equally successful. Nylon membranes are more robust and can have a high capacity, but equally well will require more time for blocking the excess binding sites (see Section 5).

(ii) Second is the decision of whether or not to wash some of the detergent out of the gel before blotting. Since we can successfully blot the LDL receptor without this step, it is not included in the protocol, but some proteins may transfer better or retain their activity better if some of the detergent is removed first. This can be done by gently shaking the gel in transfer buffer for about 30 min. This may be particularly applicable when using gels containing Triton X-100, where the detergent can prevent proteins binding to the membrane.

(iii) Third is the choice of transfer buffer. An SDS gel is run in relatively high ionic strength buffer, but the electrophoretic transfer is carried out in low ionic strength buffer to permit the high voltages required for effective transfer. This change in ionic strength of the buffer would cause the gel to increase in size during blotting if methanol were not included in the transfer buffer to counteract this effect. The gel can be allowed to swell by pre-incubation in transfer buffer for about 2 h before blotting if you must omit methanol. All these variables are discussed in some detail in the review by Gershoni and Palade (13).

4.1 **Method for setting up the blot**

The solutions and apparatus required for blotting are shown in *Table 5*. Again, different commercially available blotting systems are available. These now include a 'dry-blotting' system in which transfer is more rapid than in the system described below and requires far less buffer solution. Although we have no direct experience of this blotting technique it should be equally applicable to ligand blotting. Disposable gloves should be worn for all the operations described below to prevent transferring proteins or grease from fingers.

(i) Cut membrane and pieces of Whatman 3MM paper to the size of the gel and

Table 5. Solutions and equipment for transfer.

Equipment

Power pack—capable of producing 200 mA (or more)
Blotting tank
Plastic holder
Two pieces of Scotchbrite
Whatman 3MM paper
(Blotting tank, plastic holder, Scotchbrite and 3MM paper: usually purchased as a kit (e.g. Bio-Rad Transblot Apparatus))
Nitrocellulose membrane (0.45 μm) (Schleicher & Schüll; Ref. No. 401196).

Transfer buffer

For SDS gels:	For Triton X-100 gradient gels:
25 mM Tris–HCl, 192 mM glycine containing 20% (w/v) methanol, pH 8.3	100 mM Tris–glycine, pH 9.0 (see *Table 4*)

pre-soak in transfer buffer. The membrane must be wetted by capillary action; that is, place it on the surface of the liquid and allow it to wet slowly from underneath, before it is immersed. Discard any nitrocellulose on which dry patches are visible after soaking.

(ii) After removing the gel from the apparatus, cut off the lane(s) required for protein staining with a razor blade and place in staining solution (see *Table 6*). Cut off one corner of the gel to mark the orientation of the samples.

(iii) Assemble the blotting sandwich, as shown in *Figure 5a*. With the gel lying on the bottom plate, place a sheet of soaked Whatman 3MM paper of the same size as the gel on top. Squeeze out any air trapped between the two by rolling with a clean pipette or glass rod. Place the wetted Scotchbrite on top, followed by the plastic holder. Carefully invert so that the glass plate is now on top and can be lifted off. It may be necessary to squirt some transfer buffer along the edge between the gel and the plate to reduce adhesion of the gel. Now, lay the sheet of pre-soaked membrane on the gel and roll as before to exclude air. Complete the sandwich with Whatman 3MM paper, Scotchbrite, and then close the plastic cover and secure with clips provided or with elastic bands. Tight contact between gel and membrane is essential. Some people

Table 6. Methods of staining gels and blots.

Protein stain for acrylamide gels

Stock solutions

Stain:	De-stain:
1 g of Coomassie blue R250	400 ml of methanol
62.5 ml of glacial acetic acid	100 ml of glacial acetic acid
93.75 ml of methanol	600 ml of water

Method

1. Dissolve the stain in methanol/acetic acid and then make to 500 ml with water.
2. Filter through Whatman No. 1 filter paper.
3. Stain the gel for 2 h or overnight in volume of stain sufficient to cover.
4. De-stain until the background is clear (12–48 h).

Protein stain for nitrocellulose membranes

Stock solutions

Stain:	De-stain:
0.1 g of Amido black 10B (Sigma N3005)	10% (w/v) glacial acetic acid
45 ml of methanol	
10 ml of glacial acetic acid	

Method

1. Dissolve the stain in methanol/acetic acid and then make up to 100 ml with water.
2. Filter through Whatman No. 1 filter paper.
3. Stain the nitrocellulose for 10 min, then de-stain until the background is clear (10–15 min).

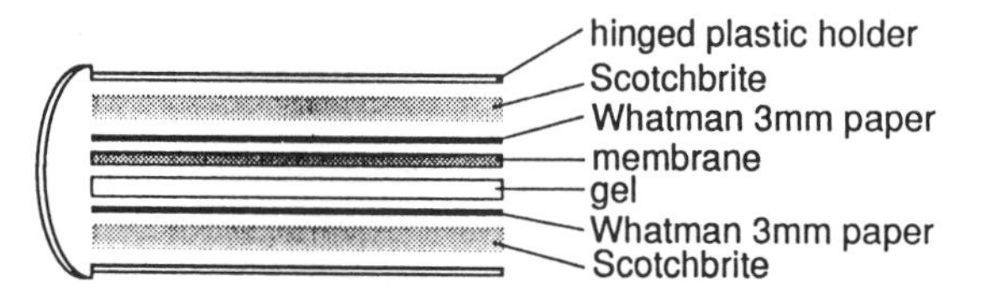

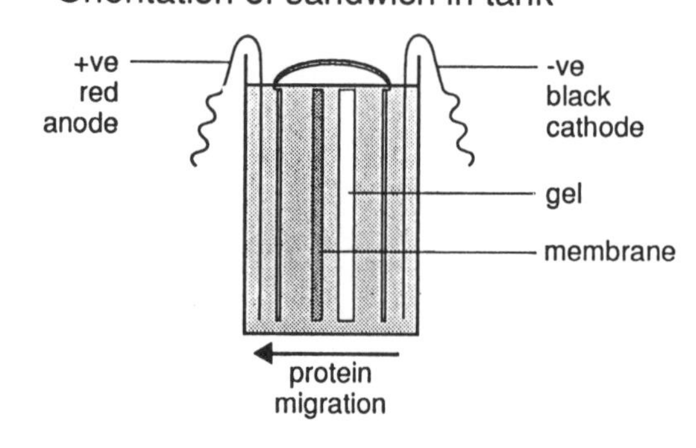

Figure 5. Electroblotting apparatus. (**a**) Assembly of the blotting 'sandwich'. (**b**) Orientation of the 'sandwich' in the blotting tank. For blotting of acidic proteins (e.g. those complexed with SDS) at pH 8.0, the gel must be adjacent to the black (negative) terminal and the membrane to the red (positive) terminal.

recommend assembling the sandwich with it fully immersed in buffer, but we have not found this necessary.

(iv) Place the sandwich in the apparatus (*Figure 5b*) and apply a constant current of 200 mA for 4–16 h at 4°C. We routinely blot for 16 h, but transfer may be complete in a shorter time. Other authors have used much higher currents, for example 600–1000 mA (13), when transfer can be complete in a few hours. The extent of transfer can be checked by either running pre-stained standards on the gel and noting the extent of transfer, or by staining the blot and the gel after transfer (*Table 6*). Large proteins transfer less well than small proteins.

5. BINDING AND DETECTION OF THE LIGAND

5.1 Blocking the membrane

After the proteins have been transferred from the gel to the membrane, excess protein binding sites on the membrane must be blocked to prevent non-specific binding of the ligand and thus avoid a high background signal. This is particularly important if the ligand is a protein or peptide. In general, bovine serum albumin (BSA) is used but some hydrophobic ligands might interact with this protein. In this event, some other inexpensive protein such as casein can be used (14). Many people have used commercial non-fat milk powder in which casein is the main blocking agent. Different membranes have different capacities to bind protein and the time required to block a particular membrane sufficiently to obtain negligible background signal should be determined.

5.2 Choice of ligand and detection system

In general, availability of pure ligand or monospecific antibodies to the ligand will determine what detection system to use. There are two main points to consider when choosing between the different possibilities, which are outlined in *Table 7*.

(i) First is the number of layers in the detection system. A ligand can be detected directly, or in a two-layer process in which unlabelled ligand is detected with a labelled monospecific antibody, or in a three-layer process with a second labelled antibody to the anti-ligand antibody. We have also detected biotin-modified ligands with labelled streptavidin in a two-step process. In theory, each successive layer in a multi-layer process should amplify the signal, but this may well be at the expense of the background. Multi-layer detection systems are obviously more time-consuming than direct detection of a labelled ligand, and require more reagents.

(ii) The second point to consider is the final detection method. A radiolabelled ligand or antibody can be detected with great sensitivity by autoradiography, but facilities are required for labelling proteins, usually with ^{125}I, to high specific activity. Alternatively, an enzyme-conjugated antibody or streptavidin can be used, in which a coloured band of enzyme product appears on the blot when it is incubated in substrate solution. This is probably less sensitive than detection of radioactivity, but the wide range of commercially-available antibodies and enzyme-conjugated second antibodies provide ample scope for detecting different ligands. Finally, colloidal gold-labelled protein ligands or antibodies are visible as a coloured band when bound to a protein on the blot. Although we have no practical experience of this method, it is claimed that it is as sensitive as autoradiographic methods when the colloidal gold is enhanced with silver.

5.3 Labelling of ligands and antibodies

It is beyond the scope of this chapter to describe all the possible methods of labelling ligands with radioactivity. For labelling antibodies with ^{125}I we use the Iodogen method which gives good reproducible labelling to high specific activity under mild conditions, and is probably applicable to most proteins and peptides (*Table 8*). However, for labelling lipoprotein ligands we use the iodine monochloride method originally described by McFarlane (15) to minimize the incorporation of label into the lipids in the particle. The best method for any particular ligand should be determined empirically, remembering to check that the labelling procedure does not adversely affect the ability of the ligand to bind.

Table 7. Methods for the detection of bound ligands.

Methods	*Advantages*	*Disadvantages*
One-step process		
1. Radiolabelled ligand → Autoradiography	Sensitive; simple one-step procedure	Need pure ligand, labelled to high specific activity; handling radioactivity; time for autoradiography
2. Colloidal gold ligand → Develop with silver	Sensitive; no radioactivity	Expense; need colloidal gold-labelled protein ligand
Two-step process		
Biotin ligand → Streptavidin conjugate → Develop in enzyme substrate	Fairly sensitive; rapid result with enzyme-linked streptavidin; no radioactivity required	Need pure biotin-modified ligand with binding activity
Three-step process		
Ligand → Specific antibody → Second labelled antibody → Autoradiography ([^{125}I] antibody) / Enzyme substrate (enzyme-linked antibody) / Develop colour (colloidal gold-labelled antibody)	Pure ligand not essential; can increase sensitivity with successive layers; conjugated second antibody commercially available. (Can be a two-step process if the specific antibody is labelled)	Need monospecific antibody to ligand; time-consuming

Table 8. Labelling of proteins with Iodogen.

Equipment and solutions

Iodogen-coated tubes

Sarstedt tubes (nominal volume 1.5 ml)
Iodogen (Pierce Chemical Co., Cat. No. 28600) 0.04 mg/ml in dichloromethane. Add 50 μl of Iodogen solution to each tube and evaporate the solvent under a gentle stream of nitrogen, to coat the tube with a thin layer. Tubes can be stored dry in the dark at −20°C.

Sephadex G-25 column

Prepare a small disposable column of Sephadex G-25 (bed volume ~10 ml) equilibrated with 0.05 M phosphate pH 7.0. Saturate any protein binding sites by applying 1.0 ml of 10 mg BSA/ml in the same buffer. Wash the column with several bed volumes of buffer.

$Na^{125}I$, 100 mCi/ml

Amersham International IMS-30

Protein to be labelled

2–40 mg/ml in 0.05 M phosphate buffer, pH 7.0 (labelling is more efficient with high protein concentration).

Methods

1. Place 25 μl of protein solution in an Iodogen tube and add 1 mCi of $Na^{125}I$ (10 μl).
2. Close the cap and mix by flicking the tube. Leave for 10 min at room temperature.
3. Add 0.5 ml of 0.05 M phosphate buffer pH 7.0 and leave for 10 min at room temperature.
4. Count 1 μl of this solution if you want to check the incorporation.
5. Remove unbound label on a Sephadex G-25 column.
6. Pool the tubes containing labelled protein and add BSA (or similar protein) to give ~10 mg/ml to stabilize the labelled protein.
7. Count 1 μl of this solution.

Incorporation of label should be ~80%. The stability of labelled proteins should be determined. For antibodies, add sodium azide to 0.01% (w/v) and store at 4°C.

Proteins can be covalently modified with biotin by reaction of their lysine residues with biotin–succinimide (16) or through sugar residues on glycoproteins by reaction with biotin–hydrazide (7). Obviously any modification with biotin should be such that it does not affect the binding site, yet sufficient biotin must be conjugated to allow sensitive detection of the ligand. The method has been used with success for lipoprotein ligands, which present a large surface area for modification. Enzyme-conjugated and colloidal gold-labelled antisera and streptavidin of good quality are now widely available commercially, and it is probably not worth preparing them yourself.

5.4 **Detailed protocol for incubating membranes**

5.4.1 *Blocking the membrane*

(i) Remove the sandwich from the blotting apparatus with gloved hands and

Table 9. Binding and detection of ligands for the LDL receptor.

Equipment

Plastic bags and bag sealer
Rocker platform
37°C incubator or water bath
Washing trays
Facilities for autoradiography

Solutions

1. Stock buffer. (This buffer may not be optimal for all systems. The ideal salt concentration and pH should be determined.)

50 mM Tris–HCl pH 8	6.06 g of Tris
90 mM NaCl	5.26 g of NaCl
2 mM $CaCl_2$	0.294 g of $CaCl_2 \cdot 2H_2O$
	Water to 1.0 litre

2. Blocking buffer/incubation buffer.
   Stock buffer plus 50 mg/ml BSA fraction V (Sigma A4503).
3. Washing buffer.
   Stock buffer plus 5 mg/ml BSA.
4. Horseradish peroxidase substrate (NB sodium azide inhibits peroxidase)
   10 ml of 4 mg/ml 4-chloro-1-naphthol (Sigma C-8890) in methanol
   90 ml of stock buffer
   Add 100 μl of 30% H_2O_2 solution just before use.
   } make fresh

Most commercially available enzyme-conjugated antibodies are supplied with instructions for substrate solutions.

discard the paper. (The gel can be stained for protein to check that transfer is complete—see *Table 6*),

(ii) Cut any lanes from the membrane that are required for protein staining. Lane(s) containing pre-stained standards need not be removed.
(iii) Cut one corner off the membrane to mark the orientation of the membrane with respect to the gel.
(iv) Place the membrane between two plastic sheets and seal three sides to form a bag just larger than the membrane. Leave the fourth side open and long enough for it to be sealed twice (~10 cm spare).
(v) Add blocking buffer (*Table 9*) to the bag, allowing approximately 20 ml per 100 cm^2 of membrane.
(vi) Squeeze out any air bubbles and seal the bag near the far edge.
(vii) Leave in a 37°C incubator or water bath for 1 h.

5.4.2 *Incubation with the ligand*

(i) Cut a corner off the fourth side of the bag and pour off the blocking buffer.
(ii) Add the buffer containing ligand, again allowing approximately 20 ml per 100 cm^2, and squeeze out the bubbles. Be careful if a radiolabelled ligand is used!
(iii) Re-seal the bag and leave at room temperature on a rocking platform for 1 h.

Note that the composition of the buffer depends on the optimum conditions for the binding of the ligand to the protein in question while minimizing non-specific binding. Inclusion of the blocking agent throughout the procedure may not be essential, or BSA could be replaced with a non-ionic detergent, such as Triton X-100 at a concentration that does not interfere with specific binding [e.g. 0.05–0.1% (w/v)]. The optimum concentration of ligand must be determined empirically, balancing the intensity of the specific band against an acceptable background density at the final detection step.

5.4.3 *Washing*

Remove the membrane from the bag and rinse it briefly with washing buffer (*Table 9*); then wash the membrane three times, for 20 min each time in 100–200 ml of washing buffer in a plastic sandwich box on a rocking platform at room temperature.

5.4.4 *Subsequent incubations*

For any subsequent incubations, re-seal the membrane in a fresh bag. Incubate with buffer containing the antibody or other required reagent (e.g. a streptavidin conjugate for biotin-labelled ligands or antibodies) as described in Section 5.4.2 above. Wash as described in Section 5.4.3 above between each incubation.

The optimum conditions for detecting a bound ligand with an antibody depend on the particular proteins involved, and the ideal concentration of primary antibody to use should be determined empirically. Most commercially available antibodies and enzyme-conjugated streptavidin complexes are supplied with some information about recommended dilutions, but in general the lowest concentration possible should be used to obtain low backgrounds.

5.4.5 *Detection by autoradiography*

Remove the membrane from the bag and let it dry in air, placed on paper towels. Wrap the dry membrane in Saranwrap (some plastic films may cause problems with static electricity, which can blacken film) and place it in direct contact with X-ray film (e.g. Kodak XAR-5). For maximum sensitivity with ^{125}I, autoradiography should be carried out at −70°C with the film and blot placed between intensifying screens (e.g. Cronex Lightning Plus Enhancing Screen, Dupont Ltd). If you want to scan the autoradiograph to quantify binding, the film must be pre-flashed so that the response is linear (17).

5.4.6 *Detection by enzyme reactions*

For enzyme conjugates, rinse the membrane briefly in buffer without albumin and then place it in enzyme substrate solution (see *Table 8*) until the bands are visible. Stop the reaction if desired. (If nothing appears, check that the enzyme is functional by adding a small amount of the buffer containing the enzyme conjugate to a tube containing substrate solution). For a permanent record, the blots should be photographed as soon as possible before fading or discoloration occurs. Store membranes wet in plastic bags in the dark to reduce fading.

5.4.7 *Detection and development of colloidal gold complexes*

A red colour appears on the blot when a colloidal gold-labelled protein is bound. The signal can be greatly enhanced by development with silver (18). Commercially available kits are now available for this procedure such as Janssen Life Sciences SE Kit.

6. QUANTIFICATION BY LIGAND BLOTTING

Ligand blotting can be used to estimate the relative amounts of a particular protein in an extract by scanning blots or autoradiographs. Care must be taken to ensure that the response is linear for any extract by applying different amounts of protein to the gel (*Figure 6*). Although precise quantification of protein is not possible (even if the pure binding protein is available, it may not behave on the blot identically to the same protein in crude extracts), the method can be a useful means of comparing distribution in different tissues or subcellular fractions, and for looking at physiological changes in the level of expression of a protein

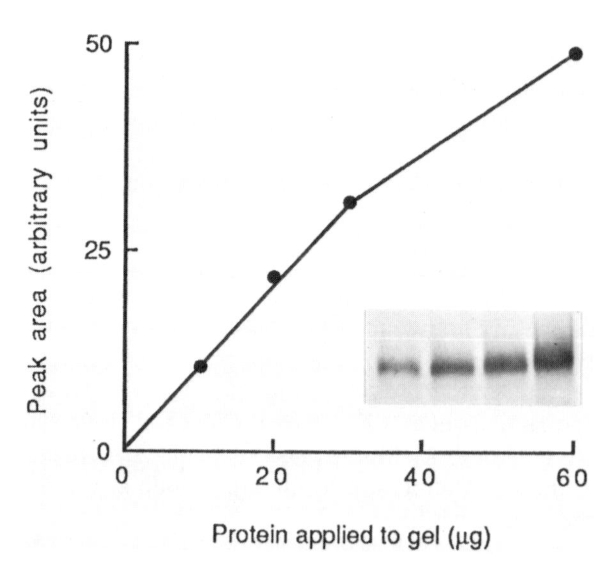

Figure 6. Quantification of LDL receptor protein by ligand blotting. Different volumes of an extract of bovine adrenal cortex membranes (1.0 mg protein/ml) were fractionated on a 10% polyacrylamide gel containing SDS. The separated proteins were transferred to nitrocellulose, and the LDL receptor was detected by incubation of the nitrocellulose membrane first with biotin-modified LDL (10 µg protein/ml) and subsequently with streptavidin-biotinylated horseradish peroxidase complex (Amersham International, 1:400 dilution) followed by development in peroxidase substrate solution (4-chloronaphthol). Bound ligand was quantified by densitometric scanning of a black and white photograph of the blot with a Joyce-Loebl Ultrascan in reflection mode. The details of the methods are given in the text.

(*Figure 7*). Ligand blotting can also be a useful assay for following a protein during purification procedures.

Autoradiographs (after exposure with pre-flashed film) can be scanned with a conventional gel scanner in 'transmission' mode, while blots must be scanned in the 'reflection' mode. In practice we have obtained better results by scanning black and white photographs of blots, rather than the blots themselves, which can be difficult to handle.

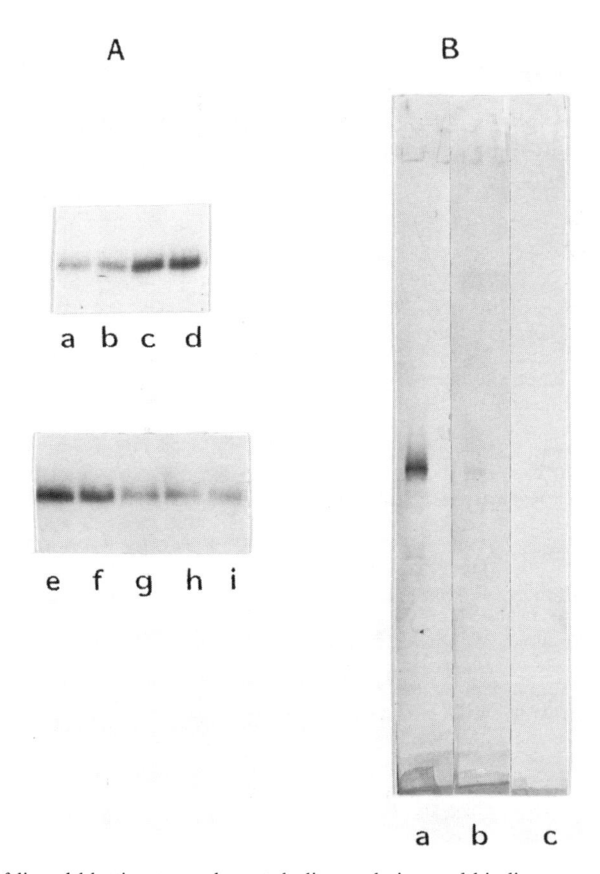

Figure 7. Use of ligand blotting to study metabolic regulation and binding properties of the LDL receptor in cultured human cells. (**A**) Stimulation of LDL receptor activity in cultured HepG2 cells by insulin, and suppression by LDL. Solubilized extracts (75 μg protein/lane) of whole cells incubated for 48 h in medium alone (**lanes a, b, e**) or with 100 mU insulin per ml (**lanes c, d**) or with medium containing LDL [LDL–cholesterol concentration of 10 μg/ml (**f**), 50 μg/ml (**g**), 250 μg/ml (**h**) and 500 μg/ml (**i**)] were fractionated by electrophoresis on a 10% polyacrylamide gel containing SDS. The LDL receptor was detected by ligand blotting with [^{125}I] βVLDL as described in the legend to *Figure 1*. (**B**) Saturability and calcium dependence of the binding of LDL to the LDL receptor. Solubilized bovine adrenal cortex membranes (40 μg protein/lane) were fractionated on a 10% polyacrylamide gel containing SDS and ligand blotted with biotin-modified LDL alone (**a**), or, in addition, with excess unmodified LDL (1 mg protein/ml) (**b**) or 10 mM EDTA (**c**).

7. OTHER APPLICATIONS

Some of the binding characteristics of a protein can be investigated by ligand blotting, for example the specificity for and competition between different ligands, as well as the effect of different inhibitors of binding (*Figure 7*). However, for both practical and theoretical reasons it is unlikely that reliable data on the affinity of the protein for its ligand can be obtained with this technique. For example, it has been shown that the affinity of the soluble acetylcholine receptor for α-bungarotoxin differs from that of the receptor immobilized on a nitrocellulose membrane (3).

8. ACKNOWLEDGEMENTS

We are grateful to Mrs E. Manson for typing the manuscript, to Mr Doig Simmonds and Louise Perks (Department of Medical Illustration, Royal Postgraduate Medical School) for preparing *Figures 3–5* and to Dr B. L. Knight for critical comments.

9. REFERENCES

1. Fernandez-Pol, J. A. (1982) *FEBS Lett.,* **143**, 86.
2. Islam, M. N., Briones-Urbina, R., Bako, G. and Farid, N. R. (1983) *Endocrinology,* **113**, 436.
3. Gershoni, J. M., Hawrot, E. and Lentz, T. L. (1983) *Proc. Natl. Acad. Sci. USA,* **80**, 4973.
4. Hawkes, J. (1982) *Anal. Biochem.,* **123**, 143.
5. Rendon, A., Filliol, D. and Jancsik, V. (1987) *Biochem. Biophys. Res. Commun.,* **149**, 776.
6. Daniel, T. O., Schneider, W. J., Goldstein, J. L. and Brown, M. S. (1983) *J. Biol. Chem.,* **258**, 4606.
7. Wade, D. P., Knight, B. L. and Soutar, A. K. (1985) *Biochem. J.,* **229**, 785.
8. Graham, D. L. and Oram, J. F. (1987) *J. Biol. Chem.,* **262**, 7439.
9. Miskimins, W. K., Roberts, M. P., McClelland, A. and Ruddle, F. H. (1985) *Proc. Natl. Acad. Sci. USA,* **82**, 6741.
10. Wright, J. M., Wiersma, P. A. and Dixon, G. H. (1987) *Eur. J. Biochem.,* **168**, 281.
11. Laemmli, U. K. (1970) *Nature,* **277**, 680.
12. Kuonen, D. R., Roberts, P. J. and Cottingham, I. R. (1986) *Anal. Biochem.,* **152**, 221.
13. Gershoni, J. K. and Palade, G. E. (1983) *Anal. Biochem.,* **131**, 1.
14. Dresel, H. A., Friedrich, E., Via, D. P., Schettler, G. and Sinn, H. (1985) *EMBO J.,* **4**, 1157.
15. Macfarlane, A. S. (1968) *Nature,* **182**, 53.
16. Guesdon, J.-L., Ternynck, T. and Avrameas, S. (1979) *J. Histochem. Cytochem.,* **27**, 1131.
17. Laskey, R. A. and Mills, A. D. (1975) *Eur. J. Biochem.,* **56**, 335.
18. Moeremans, M., Daneels, G., van Dijck, A., Langanger, G. and De Mey, J. (1984) *J. Immunol. Methods,* **74**, 353.

CHAPTER 4

Affinity labelling

ROBERTA F. COLMAN

1. INTRODUCTION

An important approach to identifying amino acid residues within the binding sites of enzymes is to exploit the specificity of the enzyme for a natural substrate or regulatory ligand to limit chemical modification to the immediate vicinity of that site. In this approach, termed affinity labelling, a reagent is designed that is structurally similar to the normal ligand of the enzyme, but which also features a functional group capable of reacting covalently with many different amino acids. The expectation is that the affinity label will initially form a reversible enzyme–reagent complex analogous to the enzyme–ligand complex and, once bound at the specific site, will react irreversibly with an amino acid residue within that site. The application of this technique can result in tagging of a particular binding site of a purified protein or in the labelling of one protein within a complex mixture of proteins. The general principles of affinity labelling have been summarized in books and review articles (e.g. 1, 2).

In applying the technique of affinity labelling to any given protein several steps are required.

(i) An appropriate reagent must be designed that has affinity for the natural ligand site, and which has a broadly reactive group whose extra bulk can be accommodated within the same binding site by the enzyme.
(ii) A practical synthesis of the affinity label must be devised.
(iii) Conditions for incubation of enzyme with affinity label must be selected that maximize stability of the enzyme, minimize decomposition of the reagent and yield measurable reaction rates.
(iv) An enzyme assay must be selected that allows monitoring of the functional changes in the enzyme during the period of incubation with the reagent.
(v) The rate constant for the reaction, k, may be monitored by the time dependence of the change in the functional enzyme assay and/or of the reagent incorporation. The value of k should be measured over a range of reagent concentrations to ascertain whether a limiting maximum rate constant is observed, indicating reversible enzyme–reagent complex formation prior to irreversible reaction.
(vi) Natural ligand(s) should decrease markedly the rate constant for modification by the affinity label, thereby indicating that the reaction target is at or near that ligand binding site.

(vii) A method for measuring the extent of reagent incorporation into the enzyme must be developed, and the reagent should be shown to react at a limited number of sites.

(viii) Finally, the labelled peptide(s) should be purified and characterized, and the modified amino acid(s) identified.

This chapter considers each of these steps, along with illustrative examples.

2. DESIGN OF AFFINITY LABEL

In selecting or designing a new affinity label for an enzyme, it is important to consider the structural requirements for binding to the particular ligand binding site. This may be accomplished by reviewing the available data on the range of alternate substrates that can be accepted by the enzyme, or on the various compounds that can serve as competitive inhibitors of a catalytic reaction or of binding to a regulatory site. Most enzymes have stringent requirements for part of the natural ligand (and these must be preserved in any effective affinity label), but are more tolerant of variation in other parts (and these are the regions where a reactive functional group may be inserted). *Figure 1* illustrates a series of fluorosulphonylbenzoyl derivatives that have been designed as affinity labels of purine nucleotide sites in enzymes: 5′-*p*-fluorosulphonylbenzoyl adenosine (5′-FSBA), 5′-*p*-fluorosulphonylbenzoyl guanosine (5′-FSBG) and 5′-*p*-fluorosulphonylbenzoyl-1,N^6-ethenoadenosine (5′-FSBεA). 5′-FSBA and 5′FSBG preserve the ribose and either the adenine or guanine of a natural adenosine or guanosine ligand; while in 5′-FSBεA the fluorescent 1,N^6-ethenoadenosine group provides a relatively conservative substitution. These compounds all have carbonyl groups adjacent to the 5′ position that are structurally similar to the first phosphoryl group of nucleotides. If these molecules are arranged in their extended conformations, the sulphonyl fluoride moiety is located in a position analogous to the terminal phosphate of ATP or to the ribose adjacent to the nicotinamide of NAD. Thus, although these compounds do not possess the negative charges of the phosphoryl groups of the nucleotides, they do preserve their over-all size and shape and might be expected to bind specifically to ATP or NAD sites (5′-FSBA), to GTP sites (5′-FSBG) and to either ATP or GTP sites (5′-FSBεA), depending upon the determinants of the enzyme's specificity. The sulphonyl fluoride can act

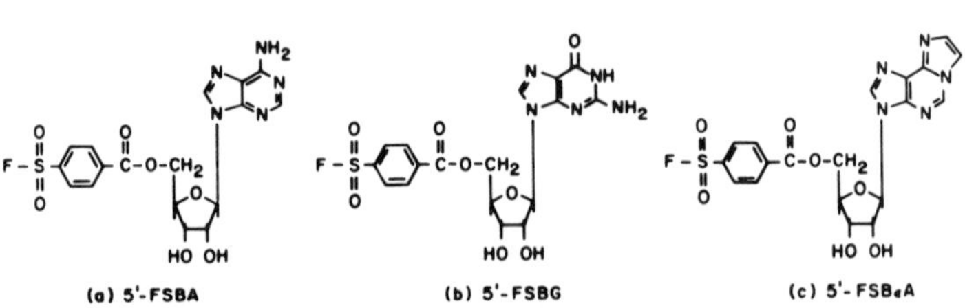

Figure 1. Fluorosulphonylbenzoyl nucleotide affinity labels. (**a**) 5′-*p*-fluorosulphonylbenzoyl adenosine; (**b**) 5′-*p*-fluorosulphonylbenzoyl guanosine; (**c**) 5′-*p*-fluorosulphonylbenzoyl-1,N^6-ethenoadenosine.

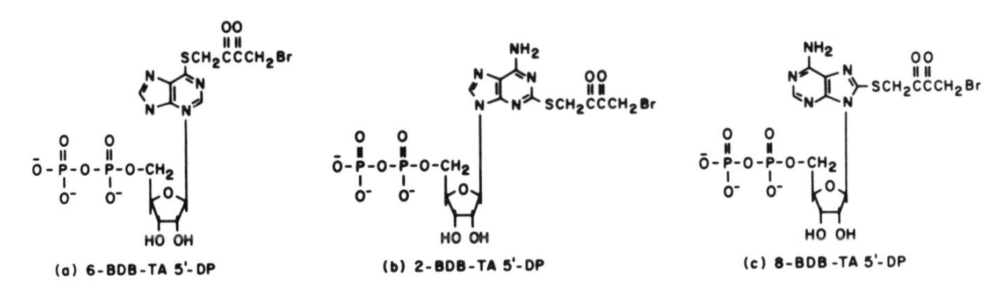

Figure 2. Bromo-2,3-dioxobutyl nucleotide affinity labels. (**a**) 6-[4-bromo-2,3-dioxobutylthio]-adenosine 5′-diphosphate; (**b**) 2-[4-bromo-2,3-dioxobutylthio]adenosine 5′-diphosphate; and (**c**) 8-[4-bromo-2,3-dioxobutylthio]adenosine 5′-diphosphate.

as an electrophilic agent in covalent reactions with the side-chains of several amino acids, such as tyrosine, lysine, cysteine, histidine and serine (2).

Figure 2 illustrates a different type of affinity label, the 4-bromo-2,3-dioxobutyl nucleotides, all of which preserve the ribose and 5′-diphosphates of ADP and are thus water-soluble and negatively charged at neutral pH. The three compounds have the same reactive functional group located in each at a distinct position of the purine ring: 6-[4-bromo-2,3-dioxobutylthio]-adenosine 5′-diphosphate (6-BDB-TA 5′-DP) (3), 2-[4-bromo-2,3-dioxobutylthio]-adenosine 5′-diphosphate (2-BDB-TA 5′-DP) (4) and 8-[4-bromo-2,3-dioxobutylthio]-adenosine 5′-diphosphate (8-BDB-TA 5′-DP) (5). These three can be used to probe systematically different sub-regions of a purine binding site in an enzyme. The bromoketo group is potentially reactive with several nucleophiles found in proteins including cysteine, glutamate, aspartate, histidine, lysine and tyrosine side-chains, and the diketo moiety adds the possibility of reaction with arginine residues.

The compounds shown in *Figure 3* represent a series of affinity labels in which the reactive group is maintained at the same relative position of the purine ring, while other structural features are changed. The positional distribution of the phosphates is varied between 2-[4-bromo-2,3-dioxobutylthio)-1,N^6-etheno-adenosine 2′,5′-diphosphate (2-BDB-TεA 2′,5′-DP) (6) and 2-[4-bromo-2,3-dioxobutylthio]-1,N^6-ethenoadenosine 5′-diphosphate (2-BDB-TεA 5′-DP) (7). With the 2′-phosphate characteristic of NADP binding sites in proteins, the fluorescent compound 2-BDB-TεA 2′,5′-DP might be expected to be specific for NADP-dependent dehydrogenases, while the fluorescent compound 2-BDB-TεA, 5′-DP with a 5′-diphosphate would more likely be directed to NAD or ADP binding sites. Compounds 2-BDB-TεA 2′,5′-DP and 2-[4-bromo-2,3-dioxobutylthio]-adenosine 2′,5′-diphosphate (2-BDB-TA 2′,5′-DP) (8) have the same distribution of phosphates, but in the latter the free 6-amino group of the adenine is retained, while in the former this group is incorporated into the extra etheno ring. Thus, a comparison of the abilities of 2-BDB-TεA 2′,5′-DP and 2-BDB-TA 2′,5′-DP to act as affinity labels of a particular NADP-dependent dehydrogenase indicates the importance of the 6-amino group in the binding of the nucleotide to that enzyme. 2-BDB-TεA 5′-DP and 2-[3-bromo-2-oxopropylthio]adenosine

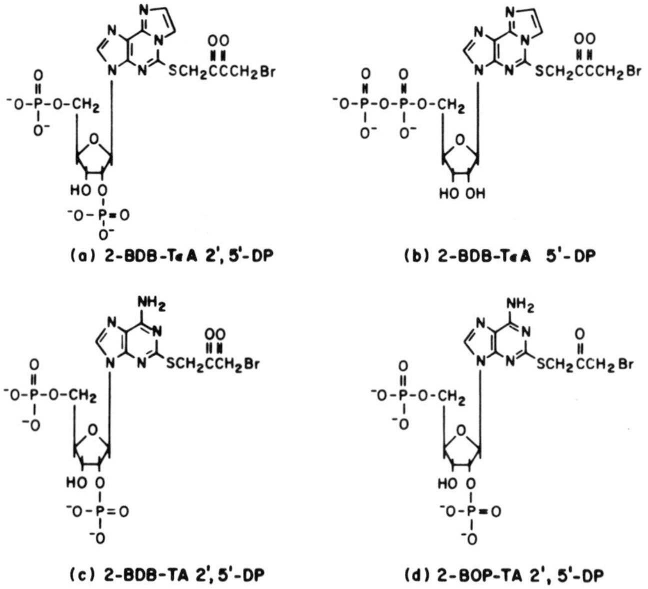

Figure 3. Additional bromo-2,3-dioxobutyl nucleotide affinity labels. (**a**) 2-[4-bromo-2,3-dioxobutylthio]-1,N^6-ethenoadenosine 2′,5′-diphosphate; (**b**) 2-[4-bromo-2,3-dioxobutylthio]-1,N^6-ethenoadenosine 5′-diphosphate; (**c**) 2-[4-bromo-2,3-dioxobutylthio]adenosine 2′,5′-diphosphate; and (**d**) 2-[3-bromo-oxopropylthio]adenosine 2′,5′-diphosphate.

2′,5′-diphosphate (8) are structurally identical except that the side-chains differ by one carbonyl group. The latter has a shorter reactive group, and a comparison betweeen the effectiveness of affinity labelling of a given enzyme by the two compounds may yield an estimate of the distance between the target amino acid and the enzyme-bound nucleotide.

Each of the nucleotide derivatives shown in *Figures 1–3* features a reactive functional group that is relatively indiscriminate in reactions with amino acids. In affinity labelling studies, the types of amino acid participants in the ligand binding sites are frequently unknown. Using a non-specific functional group improves the likelihood of covalent reaction after the binding specificity is determined by the rest of the affinity label.

3. SYNTHESIS OF AFFINITY LABEL

Once an appropriate affinity label has been designed for the enzyme of interest, it is necessary to plan a feasible synthesis, that is the target molecule should be obtainable in a reasonable amount of time in the quantity needed for the enzyme experiments, starting with materials that are not unduly expensive. In addition, in selecting a synthetic scheme, consideration should be given to possible methods for determining the amount of reagent incorporated into the enzyme. If a radiochemical method is selected, how will the radioactive tag be introduced into the reagent molecule from available starting materials? Will the radioactive tag be preserved during the synthesis? Three examples, one selected from the structures shown in each of *Figures 1–3*, will illustrate some of the approaches.

3.1 5′-*p*-Fluorosulphonylbenzoyl-1,N^6-ethenoadenosine (5′-FSBεA)

This compound [*Figure 1*, structure (c)] is synthesized from ethenoadenosine hydrochloride prepared by reaction of chloroacetaldehyde with adenosine (9).

(i) To 100 mg (0.31 mmol) of ethenoadenosine, dissolved in 0.7 ml of hexamethylphosphoramide with warming to 50°C, add 95 mg (0.43 mmol) of *p*-fluorosulphonylbenzoyl chloride in small amounts over a 5-min period with continual stirring.
(ii) After reacting for 30 min at room temperature, slowly add an additional 50 mg (0.23 mmol) of *p*-(fluorosulphonyl)benzoyl chloride.
(iii) Monitor the reaction by TLC on 200-μm silica gel plates (EM Reagents; Merck) with methyl ethyl ketone : acetone : water (65 : 20 : 15 by vol.) as the solvent.
(iv) After 2.5 h total the reaction should be complete, as indicated by the disappearance of the dark blue fluorescent spot due to ethenoadenosine.
(v) Extract the reaction mixture twice with 2 ml of petroleum ether.
(vi) To the resulting oil, slowly add 3 ml of a 1:1 (v/v) mixture of diethyl ether and ethyl acetate, which results in the formation of a fine white precipitate.
(vii) Collect the product by filtration and dry it under vacuum.
(viii) A typical yield is 140–150 mg of material (10).

Analysis of the material by TLC as in (iii) indicates two yellowish fluorescent compounds. The majority of the material migrates with an average R_f equal to 0.56 and is 5′-FSBεA, while the minor component has an average R_f of 0.76. The product is readily purified by preparative TLC on 500–1500 μm silica gel plates (Analtech Inc.). After development, the 5′-FSBεA is extracted from the silica gel with 95% ethanol. Alternatively, the product can be purified by high-performance liquid chromatography (HPLC) using a C_{18} column with 50% methanol–50% water as eluant.

Radioactive [2-^{3}H]5′-FSBεA has been synthesized by the addition of 5 mCi of [2-^{3}H]adenosine to a solution of 2.4 mmol of non-radioactive adenosine in 25 ml of aqueous chloracetaldehyde to generate etheno[2-^{3}H]adenosine (9), followed by reaction with *p*-fluorosulphonylbenzoyl chloride, as above. The tritium is completely retained during the synthesis. Radioactive [8-^{3}H]adenosine cannot be used for this preparation because tritium exchanges from the 8-position of the purine ring under the basic conditions used for this synthesis.

The concentrations of solutions of 5′-FSBεA can be determined spectrophotometrically at 275 nm using $\varepsilon_{275} = 7632\ M^{-1}cm^{-1}$ in ethanol. In using this compound as an affinity label, incorporation into enzyme can be measured either from the radioactivity or from the fluorescence of the compound, as described in Section 8.

3.2 8-[4-Bromo-2,3-dioxobutylthio]adenosine 5′-diphosphate (8-BDB-TA 5′-DP)

This compound [*Figure 2*, structure (c)] is synthesized from ADP by bromination to yield 8-bromoADP, which reacts with aqueous LiHS to give 8-thioADP. Condensation of 8-thioADP with 1,4-dibromobutanedione yields the final product, 8-BDB-TADP (5).

(i) Dissolve 1 g of ADP in 25 ml of water, and add 0.33 ml of bromine dropwise with vigorous stirring.
(ii) Stir the mixture at room temperature in the dark for 2 h, and then add 0.175 g of sodium metabisulphate.
(iii) Dry the solution by rotary evaporation and co-evaporate three times with ethanol.
(iv) Dissolve the solid material in 100 ml of water and apply to a 1.5 × 25 cm column of DEAE–Sephadex equilibrated with 0.01 M NH_4HCO_3 at 4°C.
(v) Elute the column with a linear gradient formed from 1.5 litres of 0.01 M NH_4HCO_3 and 1.5 litres of 0.5 M NH_4HCO_3.
(vi) Collect fractions of 10 ml and measure the absorbance at 260 nm.
(vii) Pool and evaporate the peak fractions (52–73).
(viii) The yield of 8-bromoADP as a white powder should be 80%.
(ix) Dissolve the 8-bromoADP (100 mg) in 2.5 ml of 1 M LiHS (prepared by bubbling H_2S into 1 M LiOH until the pH is lowered to 9.0).
(x) Leave the reaction solution at room temperature overnight.
(xi) Dilute the mixture with approximately 10 volumes of water and purify it on a DEAE–Sephadex column as in (iv).
(xii) Collect fractions of 10 ml and measure the absorbance at 295 nm.
(xiii) Pool the peak fractions (119–143) and dry them.
(xiv) Re-dissolve the solid material in water and lyophilize it repeatedly to remove the salt.
(xv) The 8-thioADP is recovered as a white powder in 60% yield.
(xvi) Convert the 8-thioADP (50 mg) to its free acid form by dissolving it in 2 ml of water.
(xvii) Apply the solution to a 1.8 × 50 cm column of AG50W-X4 (H^+), and elute with distilled water at 4°C.
(xviii) Collect fractions of 10 ml and measure the absorbance at 308 nm.
(xix) Pool and dry the nucleotide-containing fractions (4–10), and co-evaporate with methanol three times on a rotary evaporator.
(xx) For a single reaction, dissolve 8-thioADP in 0.5 ml of methanol by gradual addition of triethylamine, to form the soluble triethylammonium salt of 8-thioADP at 26 mM.
(xxi) Add sufficient triethylamine to adjust the pH to approximately pH 5.3, as estimated by pH paper.
(xxii) Dissolve 1,4-dibromobutanedione in 0.5 ml of methanol to give 820 mM.
(xxiii) Add this solution (with rapid mixing) to 0.5 ml of the 8-thioADP solution (26 mM).
(xxiv) The reaction occurs immediately and can be assessed spectrophoto-

metrically by the decrease in absorbance at 308 nm and by the increase at 278 nm.

(xxv) Cool the product to 0°C and precipitate it by the addition of 10 ml of diethyl ether.

(xxvi) Collect the precipitate by centrifugation, redissolve it in 0.5 ml of methanol, and precipitate it again with diethyl ether to give a white powder in 70% yield.

The $\varepsilon_{278\,\mathrm{nm}}$ is 1.9×10^4 $\mathrm{M^{-1}\ cm^{-1}}$ for 8-BDB-TADP in 0.05 M Mes buffer, pH 6.0.

3.3 **2-[4-Bromo-2,3-dioxobutylthio]adenosine 2′,5′-diphosphate (2-BDB-TA 2′,5′-DP)**

This compound [*Figure 3*, structure (c)] is synthesized from the relatively inexpensive starting material $NADP^+$ (8). The nucleotide 2′-phosphoadenosine 5′-diphosphoribose (PADPR) is generated enzymatically and is converted to PADPR-1-oxide by reaction with *m*-chloroperoxybenzoic acid. Treatment with NaOH followed by reaction with carbon disulphide yields 2-thioadenosine 2′,5′-diphosphate (2-TA 2′,5′-DP). Condensation of TA 2′,5′-DP with 1,4-dibromo-butanedione gives the final product, 2-BDB-TA-2′,5′-DP.

3.3.1 *Preparation of PADPR*

(i) Dissolve $NADP^+$ (800 mg) in 6 ml of 100 mM potassium phosphate buffer, pH 7.6, and adjust the pH to 7.5 with 1 M KOH.

(ii) Dissolve three units of crude *Neurospora* NAD^+ hydrolase (NADase, Sigma Chem. Co.) in 1.5 ml of 100 mM potassium phosphate buffer, pH 7.6, and add to the $NADP^+$ solution; incubate the reaction mixture for 20 h at 35°C.

(iii) Follow the progress of the reaction by analytical TLC on cellulose aluminium-backed sheets (EM Reagents: 0.1 mm thickness) using isobutyric acid:concentrated NH_4OH:water (66:1:33 by vol.) as the solvent system. Unreacted $NADP^+$ has an R_f of 0.29, compared with R_f values of 0.15 and 0.75 for PADPR and nicotinamide, respectively.

(iv) Apply the reaction mixture to a 2 × 31 cm column of DEAE–cellulose (DE-52, Whatman) equilibrated with 0.01 M ammonium bicarbonate at room temperature.

(v) Elute the column with a linear gradient (800 ml of 0.01 M ammonium bicarbonate and 800 ml of 0.4 M ammonium bicarbonate).

(vi) Collect fractions of 2.2 ml and monitor the absorbance at 260 nm. Nicotinamide elutes between fractions 50 and 70, while unreacted $NADP^+$ and PADPR elute between fractions 260 and 275 and fractions 290 and 360, respectively.

(vii) Pool and lyophilize the fractions containing PADPR. Re-dissolve the sample in water and lyophilize repeatedly until all the ammonium bicarbonate is removed. The yield should be 85–90% (6).

3.3.2 *Preparation of PADPR-1-oxide*

(i) Stir a biphasic mixture containing 720 mg of PADPR, 8 ml of 1 M sodium acetate, 8 ml of 1 M acetic acid, 15 ml of ethyl acetate and 2 g of *m*-chloroperoxybenzoic acid for 48 h at room temperature.
(ii) Separate the aqueous (lower) layer, and add to it chloroform (5 ml) and 1 M HCl (8 ml); stir the mixture at room temperature for 1 h.
(iii) Recover the aqueous (upper) layer, add isopropyl alcohol (500 ml), and keep the cloudy white solution overnight at 4°C.
(iv) Concentrate material under vacuum and dry it to a white powder.
(v) TLC analysis as in Section 3.3.1 (iii) should exhibit a single UV-absorbing spot for this PADPR-1-oxide, which should be obtained in 90% yield (8).

3.3.3 *Preparation of 2-TA 2′,5′-DP*

(i) Dissolve the PADPR-1-oxide in 6 ml of 2 M NaOH.
(ii) Add the resultant yellow solution to 6 ml of refluxing 2 M NaOH.
(iii) Ten minutes later, cool the reaction flask on ice and adjust the pH to 11 by adding the cation-exchange resin AG50W-X4 (H^+ form, 100–200 mesh, Bio-Rad).
(iv) Filter the resin and wash with water until the absorbance at 260 nm of the washing is less than 0.01.
(v) Apply the filtrate and washings to a 1.5 × 45 cm column of the anion-exchange resin AG1-X2 (formate form, 100–200 mesh, Bio-Rad), and wash with distilled water until the absorbance at 260 nm is less than 0.02.
(vi) Elute the column with a linear gradient (1 litre of water and 1 litre of 2 M formic acid), and monitor the absorbance at 260 nm. The product is the second of three peaks to be eluted and is located between fractions 186 and 251 (4.5 ml/fraction).
(vii) Pool the fractions and concentrate them under vacuum to give a clear glass.
(viii) Dry this material by co-evaporating twice with 10 ml of methanol plus 10 ml of isopropyl alcohol.
(ix) The product (5-amino-1-β-D-ribofuranosylimidazole-4-carboxamide oxime 2′,5′-diphosphate) should be obtained as a white powder in 48% yield.
(x) UV absorption spectra should show a λ_{max} at 285 nm at pH 1 and a λ_{max} at 260 nm at pH 11.
(xi) TLC analysis as in Section 3.3.1 (iii) should exhibit a single UV-absorbing spot with $R_f = 0.12$, whereas the corresponding 5′-diphosphate and 5′-monophosphate analogues have R_f values of 0.13 and 0.24, respectively.
(xii) Add 30 ml of methanol to the product from step (ix).
(xiii) Add sufficient triethylamine to solubilize the material in the form of the triethylammonium salt.
(xiv) Add pyridine (20 ml) and carbon disulphide (15 ml), and reflux the mixture for 6 h.

(xv) Allow the yellow mixture to cool to room temperature and remove the solvents under vacuum.

(xvi) Dissolve the crude product in water and centrifuge to remove any insoluble material.

(xvii) Apply the supernatant to a 2 × 35 cm column of DEAE–Sephadex (A-25, Pharmacia) equilibrated with 0.01 M NH_4HCO_3 at 4°C.

(xviii) Elute the column with a linear gradient from 0.1 to 0.5 M NH_4HCO_3 (4 litres of each). The product (TA 2′,5′-DP) is the second of two peaks to elute and should be located in fractions 241–310 (16 ml/fraction).

(xix) Pool this TA 2′,5′-DP peak, and evaporate it to dryness under vacuum.

(xx) De-salt by repeatedly re-dissolving it in water and evaporating to dryness under vacuum. The product should be obtained as a yellowish oil in 85% yield.

(xxi) The UV absorption spectrum should exhibit λ_{max} at 292 and 256 nm in 50 mM Mes buffer, pH 6.0.

(xxii) TLC analysis as in Section 3.3.1 (iii) should show a single UV-absorbing spot at R_f = 0.14 (8).

3.3.4 *Preparation of 2-BDB-TA 2′,5′-DP*

(i) Dissolve TA 2′,5′-DP in 2 ml of water and apply to a 1 × 26 cm column of AG50W-X4 resin (H^+ form, 100–200 mesh) at 4°C.

(ii) Elute with distilled water and monitor the absorbance at 240 nm.

(iii) Collect the product (free acid) and evaporate to dryness under reduced pressure.

(iv) Synthesize 2-BDB-TA 2′,5′-DP by reaction of the TA 2′,5′-DP with 1,4-dibromobutanedione as described in Section 3.2 (xx)–(xxvi) for the reaction of 8-thioadenosine 5′-diphosphate.

The yield for steps (i)–(iv) should be 80%, and the over-all yield of 2-[4-bromo-2,3-dioxobutylthio]adenosine 2′,5′-diphosphate from NADP is usually 26% (8). The product should exhibit a single UV-absorbing spot of R_f 0.37 in the TLC system described in Section 3.3.1 (iii). Its extinction coefficient at 270 nm is $11.5 \times 10^3\ M^{-1}cm^{-1}$.

In the cases of the bromodioxobutyl derivatives of nucleotides, such as those illustrated in Sections 3.2 and 3.3, it is possible to introduce a radioactive tag *after* the reaction of enzyme with the affinity label. Once the modified enzyme is separated from excess reagent, $[^3H]NaBH_4$ is used to reduce the dioxo groups of the covalently bound reagent. The approach is unsatisfactory for use as a direct quantitative measure of the amount of reagent incorporated into an enzyme because of the likelihood of isotopic selection in the reduction process; however, it is a useful strategy for providing a radioactive tracer of modified peptides during their purification (11). The incorporation of 8-BDB-TADP or 2-BDB-TA 2′,5′-DP into an enzyme can be determined from an assay of the phosphorus content of most isolated, modified enzymes, as discussed in Section 8.

4. CONDITIONS FOR REACTION OF ENZYME AND AFFINITY LABEL

The selection of reaction conditions for studying the affinity labelling of an enzyme is important and may determine the specificity of the reaction, the rate constant and/or the ease of interpretation of the kinetic data. Many affinity labels, such as those shown in *Figures 1–3*, are attacked by the nucleophilic side-chains of amino acid residues, and the reaction rates are generally higher with the basic form than with the protonated species of the target amino acids. Thus, reaction rates tend to increase with increasing pH, in the region of the p*K* of the target amino acid. However, the stability of the reagent may also be influenced by pH and frequently decreases with increasing pH. The fluorosulphonylbenzoyl derivatives, such as those in *Figure 1*, hydrolyse with release of fluoride ion when dissolved in aqueous solution, and the rate of hydrolysis depends upon pH. For example, the half-life for decomposition of 5′-FSBG is 90 min and 223 min at pH 7.93 and pH 7.65, respectively, while below pH 7.6 these compounds are markedly more stable (2). Corrections can be made for the reagent decomposition in analysing the kinetics of chemical modification reactions (2). The ester linkage between the ribose and the benzoyl groups exhibits satisfactory stability under the usual conditions for reaction between enzymes and reagents, but ester hydrolysis rates increase below pH 6 and above pH 9; this characteristic may become an important consideration in subsequent treatment of the enzyme–reagent complex. The bromodioxobutyl derivatives are both more reactive than the fluorosulphonyl analogues and less stable in aqueous solution. Thus, the half-time for loss of bromide from 6-BDB-TA 5′-DP (*Figure 2*) is only 61 min in aqueous solution at pH 7.1 and 25°C (12), but is increased to 9.9 h at pH 6.1 (13); the bromo-derivatives are quite stable at pH 5.0, and solutions of this pH can be stored at −80°C (8). Because of the rapid decomposition rates of these compounds at higher pH values, studies of reactions of the bromodioxobutyl derivatives with enzymes have been restricted to pH 7.1 or below.

Once a pH has been chosen for the affinity labelling, a particular buffer must be selected to maintain the pH. Any buffer with the potential to react chemically with the affinity label should be avoided, including those with primary amino groups, such as Tris or ammonium bicarbonate. Sulphydryl compounds (such as mercaptoethanol or dithiothreitol) are frequently added to buffers to stabilize enzymes, but they react readily with the functional groups of many affinity labels and so must be removed completely prior to the initiation of any enzyme modification reaction. Phosphate buffers might be appropriate buffers in some cases; however, for nucleotide binding enzymes, phosphate may compete weakly for binding to the nucleotide binding sites and therefore may influence the course of an affinity labelling reaction. Tertiary amines, such as triethanolamine, *N*-2-hydroxyethyl piperazine-*N*′-2-ethanesulphonic acid (Hepes) and piperazine-*N*,*N*′-bis-(2 ethanesulphonic acid) (Pipes), are relatively unreactive and have proved to be satisfactory buffers for affinity labelling experiments for a wide range of enzymes.

Another choice to be made is that of the temperature of the reaction to be measured. The rates of affinity labelling reactions generally increase with increasing temperature, but are limited by the stability of the enzyme and of the reagent. Temperatures of 20–37°C are most often used.

Many affinity labels have a significant hydrophobic moiety added to the natural structure, as exemplified by the fluorosulphonyl nucleosides. Such compounds have limited water solubility. They are usually dissolved in organic solvents such as ethanol, dimethylformamide or dimethylsulphoxide and are added to the reaction mixture to yield final concentrations of organic solvent of 1–15%. It is important to evaluate the effect of the organic solvent on the enzyme's stability and structure in order to evaluate fully the results of affinity labelling conducted in a mixed aqueous–organic solvent.

5. SELECTION OF FUNCTIONAL PROPERTY TO MONITOR AFFINITY LABELLING REACTION

The goal of an affinity labelling study is usually to identify amino acids in an enzyme as participants in a particular binding site. Therefore select a functional property of the enzyme that can be assayed during the period of incubation with the affinity label. Such a time-dependent functional change allows measurement of the rate constant for reaction of the enzyme with the affinity label. The change in the functional property can also be evaluated for possible correlation with reagent incorporation.

In the case of reaction at the active site, an assay of the maximum velocity is appropriate. *Figure 4* illustrates the time-dependent inactivation of rabbit muscle pyruvate kinase upon incubation with 2-[4-bromo-2,3-dioxobutylthio]adenosine

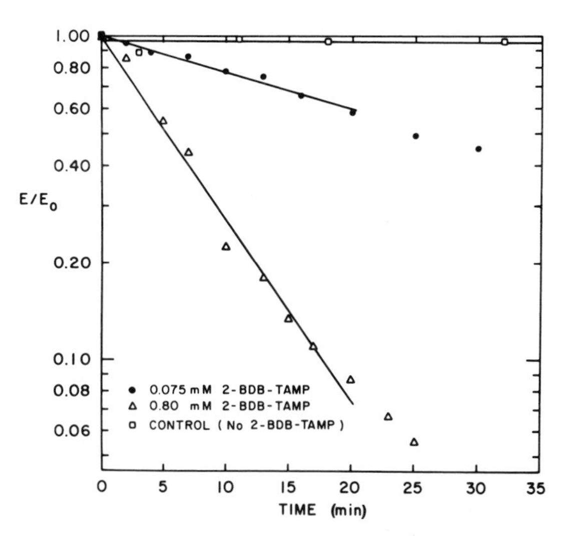

Figure 4. Inactivation of pyruvate kinase by 2-BDB-TAMP. Rabbit muscle pyruvate kinase (0.33 mg/ml) was incubated with the indicated concentrations of 2-BDB-TAMP at 25°C in 50 mM Hepes buffer, pH 7.0. E and E_0 represent the maximum enzymatic activities at the given time and at zero time, respectively. From (14), with permission.

5′-monophosphate (2-BDB-TAMP) (14). The enzymatic activity was assayed at saturating concentrations of phosphoenolpyruvate, ADP, Mg^{2+} and K^+. Linear semi-logarithmic plots of the fraction of enzyme activity remaining versus time can be used to calculate the pseudo-first-order rate constant for the reaction.

For affinity labelling of an allosteric site, the maximum velocity of the catalytic reaction may be unaffected by the chemical modification reaction; however, loss of allosteric regulation may be monitored as the time-dependent functional change. Bovine liver glutamate dehydrogenase is activated by ADP. Incubation of the enzyme with 2-BDB-TAMP causes a marked decrease in the activation by ADP and elimination of one of the two ADP sites of the native enzyme (15). Curve A of *Figure 5* shows no change during reaction with 2-BDB-TAMP in the maximum velocity when assays are conducted in the absence of regulatory compounds, indicating that 2-BDB-TAMP does not react at the active site. In

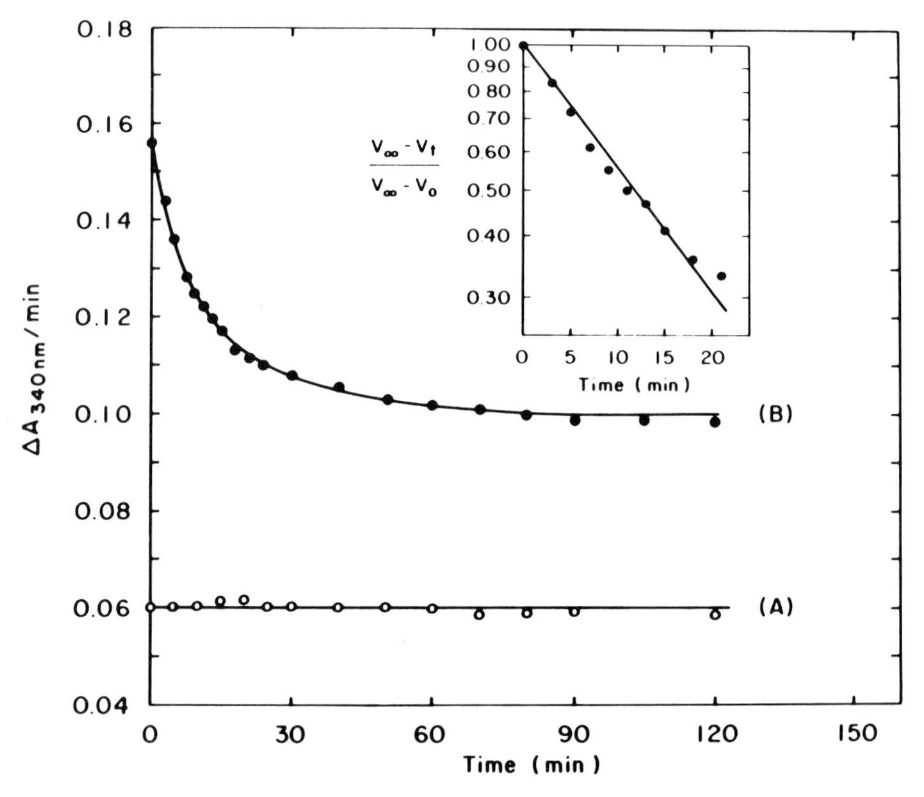

Figure 5. Reaction of 1.2 mM 2-BDB-TAMP with bovine liver glutamate dehydrogenase (1 mg/ml) at 25°C in 50 mM potassium phosphate buffer, pH 7.1. Aliquots were withdrawn at various times and assayed with saturating substrate concentrations in the absence (**A**) and in the presence (**B**) of the allosteric activator ADP (100 μM). Inset, determination of the pseudo-first-order rate constant from the decrease in enzyme activity assayed in the presence of ADP. Plot of ln $(V_\infty - V_t)/(V_\infty - V_0)$ versus time where V_t and V_0 are the enzymatic velocities (in the presence of ADP) at a given time and at zero time, respectively, and V_∞ is the constant maximal velocity reached at the end of the reaction. From (15), with permission.

contrast, Curve B of *Figure 5* illustrates the time-dependent decrease in the activity as assayed in the presence of a constant concentration of ADP. A pseudo-first-order rate constant of 0.060 min^{-1} can be calculated from the semi-logarithmic plot in the inset.

Glutamate dehydrogenase is allosterically inhibited by GTP, which binds to two sites per enzyme subunit. The compound 5′-FSBεA (*Figure 1*) reacts covalently at one of the two GTP sites causing a striking decrease (but not an elimination) of the GTP inhibition (16). Curve A of *Figure 6* indicates that the enzyme is not inactivated by covalent reaction with 5′-FSBεA; that is no change is seen in the maximum velocity in assays measured in the absence of allosteric ligands. However, As shown in Curve B of *Figure 6*, a time-dependent increase is observed in the activity of the enzyme as assayed in the presence of a constant concentration of the inhibitor GTP. The reaction rate can be determined from the time-dependent desensitization to GTP inhibition caused by 5′-FSBεA (see inset to *Figure 6*).

Glutamate dehydrogenase is also inhibited by high concentrations of NADH (600 μM) by binding at a regulatory site distinct from the catalytic site. *Figure 7* illustrates the time-dependent increase in activity, as assayed in the presence of 600 μM NADH, during the covalent reaction of the enzyme with 6-BDB-TA 5′-DP (12). This compound reacts covalently at the NADH inhibitory site and relieves the inhibition by high concentrations of added NADH. The time-dependent desensitization to NADH inhibition is the basis for a convenient method for measuring the rate of reaction of glutamate dehydrogenase with

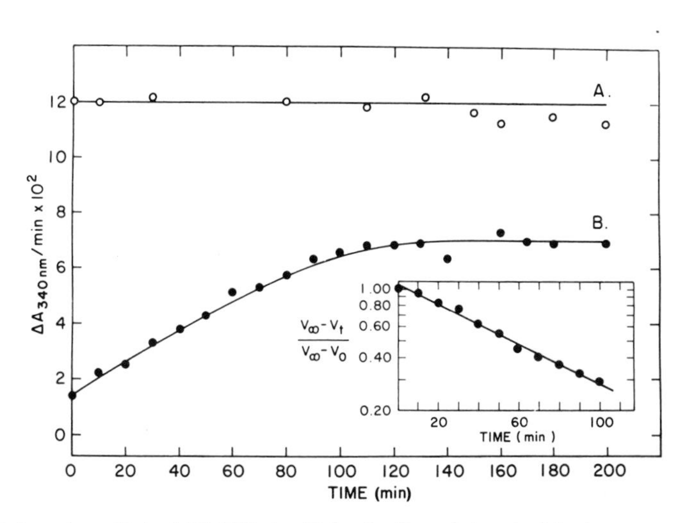

Figure 6. Reaction of 1.4 mM 5′-FSBεA with bovine liver glutamate dehydrogenase at 30°C in 0.01 M sodium barbital buffer, pH 8, containing 0.2 M KCl and 10% dimethylformamide. At various times aliquots were withdrawn and assayed with saturating substrate concentrations in the absence (**A**), and in the presence (**B**) of 1 μM GTP. Inset, determination of the pseudo-first-order rate constant from the increase in enzyme activity assayed in the presence of GTP. The increase in activity as a function of time is expressed as ln $(V_\infty - V_t)/(V_\infty - V_0)$ where V_t and V_0 are the velocities at a given time and at time zero, and V_∞ is the constant velocity reached at complete reaction. From (16), with permission.

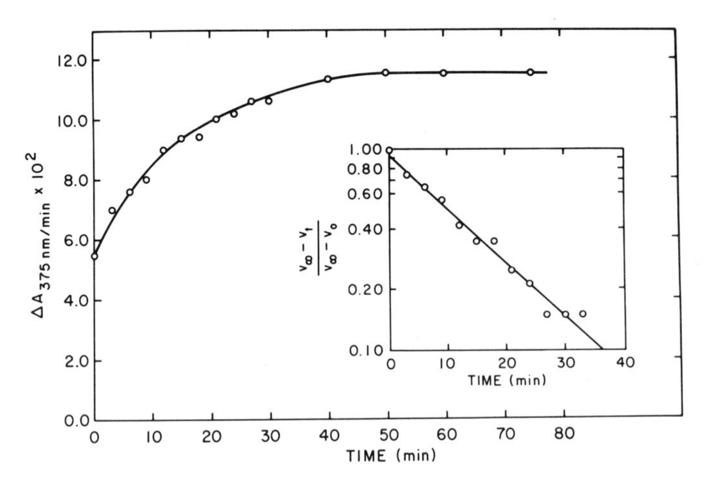

Figure 7. Reaction of 0.25 mM 6-BDB-TA 5′-DP with glutamate dehydrogenase at 25°C in 0.05 M potassium phosphate buffer, pH 7.1. Aliquots were withdrawn at various times, diluted and assayed in the presence of 600 μM NADH. From (12), with permission.

6-BDB-TA 5′-DP (12). *Figures 4–7* thus provide representative examples of properties which can be exploited to measure the reaction rate constant for an affinity label at a specific site of an enzyme.

6. DEPENDENCE OF RATE CONSTANT ON REAGENT CONCENTRATION

Regardless of the method chosen to measure the rate of the affinity labelling reaction, it is desirable to examine the dependence of the rate on reagent concentration. An affinity label characteristically forms a reversible enzyme–reagent complex prior to the irreversible modification. This behaviour can be expressed as:

$$E + R \underset{k_{-1}}{\overset{k_1}{\rightleftharpoons}} ER \xrightarrow{k_{max}} ER' \qquad (1)$$

where E represents the free enzyme, R is the affinity label, ER the reversible enzyme–reagent complex, and ER′ the covalently modified enzyme. The existence of a reversible enzyme–reagent complex is indicated by a 'rate saturation effect' in which the rate of modification increases with increasing reagent concentration until the enzyme site is saturated with reagent; further increases in reagent concentration do not enhance the rate of modification. The observed rate constant (k_{obs}) at a particular reagent concentration is described by the equation:

$$k_{obs} = \frac{k_{max}}{1 + (K_I/[R])} \qquad (2)$$

where

$$K_I = \frac{k_{-1} + k_{max}}{k_1}$$

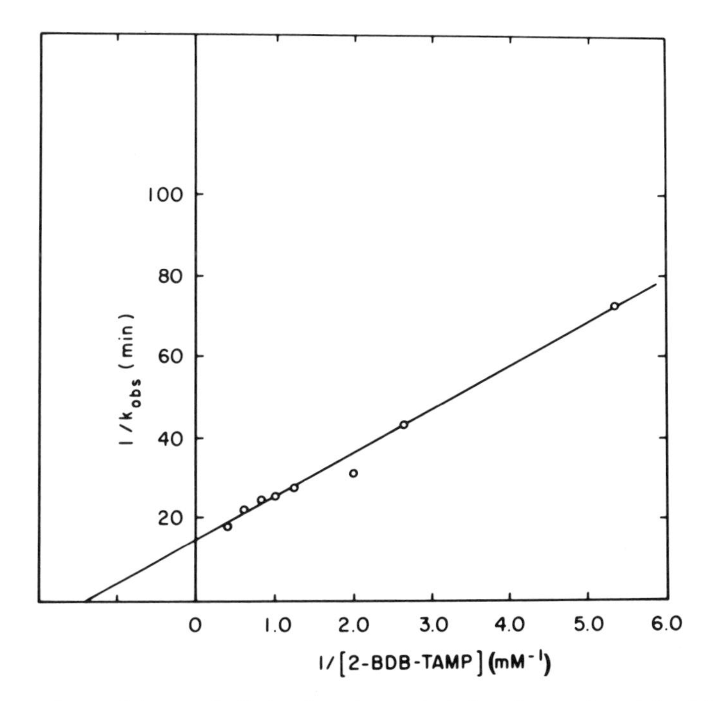

Figure 8. Dependence of the pseudo-first-order rate constant, k_{obs}, for reaction at the ADP site of glutamate dehydrogenase on the concentration of 2-BDB-TAMP. Glutamate dehydrogenase was incubated with varying concentrations of 2-BDB-TAMP under the conditions described in *Figure 5*, except that the enzyme concentration was 0.2 mg/ml. The value of k_{obs} was calculated as illustrated in the inset of *Figure 5*. From (15), with permission.

and is the apparent dissociation constant of the enzyme–reagent complex, and k_{max} is the maximum rate of modification at saturating concentrations of reagent. The reciprocal form of Equation 2 is:

$$\frac{1}{k_{obs}} = \frac{1}{k_{max}} + \frac{K_I}{k_{max}} \frac{1}{[R]} \tag{3}$$

A double reciprocal plot of $1/k_{obs}$ versus $1/[R]$ provides values for K_I and k_{max}. *Figure 8* shows such a plot for the rate of modification of the ADP site of glutamate dehydrogenase by 2-BDB-TAMP over the range 0.19–2.45 mM reagent. From this data, $K_I = 0.74$ mM with $k_{max} = 0.069\ min^{-1}$.

It is often observed that affinity labels react more rapidly with the enzyme than does a corresponding reactive compound that lacks the structural similarity to the natural substrate. For example, the NADP-dependent isocitrate dehydrogenase is rapidly inactivated by 2-BOP-TA 2′,5′-DP (*Figure 3*) with rate constants for inactivation that exhibit a non-linear dependence on reagent concentration (8). A double reciprocal plot reveals $K_I = 190\ \mu M$ with $k_{max} = 1.05\ min^{-1}$. Bromoacetone might be considered a reasonable model for the reactive group of 2-BOP-TA 2′,5′-DP. Bromoacetone inactivates isocitrate dehydrogenase with a linear dependence of the rate constant on reagent concentration over the concentration range 100–500 μM. At 200 μM, the value of k_{obs} for 2-BOP-TA 2′,5′-DP is 32 times greater than for bromoacetone, indicating the great rate enhancement achieved by placing the reactive bromoacetone group on a nucleotide specific for the co-enzyme binding site (8). However, it is not always easy to demonstrate kinetically that an affinity label reacts more rapidly than does the corresponding simple compound; sometimes the affinity label and the model compound inactivate an enzyme at similar rates, but the inactivation by the model compound is the result of modification of multiple enzymatic amino acids, while inactivation by the affinity label is attributable to attack at a single site. The essential feature of an affinity label is that the extent of modification of amino acid residues of an enzyme is more limited than that produced by a structurally unrelated reagent with the same functional group.

7. EFFECT OF LIGANDS ON RATE OF MODIFICATION BY AFFINITY LABEL

An evaluation of the effects of adding substrates or regulatory compounds to the incubation mixture of enzyme and affinity label can yield evidence for the specificity of the reaction and provide insight into the binding site that is the target. The rate constant for the modification of the enzyme is first measured at a given concentration of the affinity label under an established set of conditions in the absence of any ligands. Individual ligands are then added to identical incubation mixtures at the same concentration as the affinity label, and the rate constant for the chemical modification is then measured again. If the natural ligand binds at the same site as is under attack by the affinity label, the two will compete for binding at the site, and the rate constant for modification will be decreased to an extent dependent on the ratio of the natural ligand concentration to its dissociation constant, and the ratio of the affinity label concentration to K_I. In other words, in screening to ascertain whether an affinity label is reacting at one of several sites on an enzyme, it is advisable initially to choose concentrations of ligands that are high relative to known dissociation constants for the enzyme's natural ligands under the reaction conditions used.

This strategy is illustrated in *Table 1*, which shows the effect of adding ligands to the incubation mixture of 2-BDB-TAMP and glutamate dehydrogenase. Relatively little decrease in the rate constant is caused by the substrate α-keto-glutarate or by the co-enzyme NADH when present at a concentration (200 μM), sufficient to bind to the catalytic site. The inhibitory nucleotide GTP does not affect the rate constant when present alone, and the addition of GTP together with 200 μM NADH does not reduce the rate constant more than does NADH alone; it is known that GTP binds to both of its inhibitory sites under these conditions. A concentration of NADH sufficiently high to occupy the inhibitory site (5.0 mM) causes only a 3-fold decrease in the rate of modification by 2-BDB-TAMP. If a ligand causes a relatively small decrease in the rate constant, this could indicate that either (i) it binds weakly to the site of attack and competes poorly with the affinity label, or (ii) it saturates a different binding site and, by changing the conformation of the enzyme, indirectly decreases the rate of attack at the first site.

Table 1. Effect of substrates and allosteric ligands on reaction of 2-BDB-TAMP with glutamate dehydrogenase.[a]

Additions to reaction mixture	$k_{obs} \times 10^3$ (min^{-1})
None	60.5
α-Ketoglutarate	57.8
NADH (200 μM)	36.4
GTP (50 μM)	63.0
NADH (200 μM) + GTP (10 μM)	36.7
NADH (5.0 mM)	20.1
ADP (5.0 mM)	No reaction

[a] Glutamate dehydrogenase (1 mg/ml) was incubated with 1.2 mM 2-BDB-TAMP at 25°C in 50 mM potassium phosphate buffer, pH 7.1. Ligands were added as indicated. The pseudo-first-order rate constants were determined as illustrated in *Figure 5*. From (15), with permission.

These two possibilities can be distinguished by varying the concentration of the ligand causing a small decrease in k: if increasing the concentration of the ligand further decreases k, then (i) is the correct interpretation (i.e. that the concentration of ligand chosen initially was not sufficient to saturate its binding site); whereas, if changing the ligand concentration does not further change k, then (ii) is the appropriate interpretation (i.e. the ligand under evaluation produces a limited decrease in the rate constant as a result of saturating a distinct binding site). In this case of glutamate dehydrogenase, concentrations of NADH from 1.0 to 8.0 mM produced the same rate constant of 0.020 min^{-1}, indicating that the inhibitory NADH site is not the target of 2-BDB-TAMP. Only the allosteric activator ADP protects completely against the loss of ADP activation, reducing the rate constant below the level detectable by the experimental methods used (*Table 1*, last line). This result suggests that 2-BDB-TAMP reacts covalently at an ADP activation site.

Having used saturating concentrations of various natural ligands to indicate which ligand provides effective protection against modification by the affinity label, it is often helpful to vary the concentration of the protecting ligand. Non-saturating concentrations of the ligand cause a smaller decrease in the rate constant for modification. Measurement of the decreasing k for modification as a function of ligand concentration allows the calculation of a dissociation constant for the enzyme–ligand complex at the site being attacked by the reagent. An example of this type of experiment appears in a study of the inactivation of the pig heart NAD-dependent isocitrate dehydrogenase by 3-bromo-2-ketoglutarate (17). This reagent is structurally related to the enzyme's substrate, and isocitrate provides a high degree of protection against inactivation. The dependence of the rate constant for modification on the total DL-isocitrate concentration allows the calculation of 280 μM as the dissociation constant of the enzyme–isocitrate complex from the site at which isocitrate protects. This value is in good agreement with an independently determined dissociation constant of 320 μM for isocitrate from the active site, which strengthens the interpretation that 3-bromo-2-keto-glutarate is acting as an affinity label of the substrate binding site of isocitrate dehydrogenase (17).

8. MEASUREMENT OF REAGENT INCORPORATED

The primary reason for designing an affinity label is to limit the extent of chemical modification of the enzyme; thus, it is essential to develop a method for determining the amount of reagent incorporated covalently into the enzyme in order to evaluate the effectiveness of the affinity label. For most reactions, the modified enzyme must first be separated from the excess unreacted compound. This separation can be accomplished by various techniques such as dialysis, gel filtration through a column of Sephadex G-25 or column centrifugation through Sephadex G-50 (18, 16, 4). The first two methods are slower and less thorough in removing all of the non-covalently bound reagent, but they are more convenient for handling large amounts of modified enzyme. The column centrifugation method is rapid, thorough and allows the convenient preparation of multiple small samples of enzyme incubated with reagent for various time periods. It is important to test whether all of the non-covalently bound reagent is removed by the procedure used. One method that we have found useful is to decompose the reactive functional group of the affinity label (e.g. by the addition of dithiothreitol to a bromodioxobutyl-nucleotide) immediately prior to adding it to the enzyme; since the reagent is now unable to react covalently with the enzyme, it must be removed completely from the enzyme for the separation method used to be deemed acceptable. Furthermore, in order to relate the functional assay at a particular incubation time to the amount of reagent incorporated, the recovery of protein by the separation method must be high and representative of all of the modified enzyme; for example be careful to avoid using a gel filtration technique if inactivated enzyme tends to precipitate selectively in the incubation mixture. Once the free reagent has been removed from the sample, a method for measuring the protein content is required. Methods dependent on the UV absorption spectrum of the enzyme should be avoided since many affinity labels either contribute directly to the UV absorption spectrum to an extent dependent on incorporation, or change the conformation of the enzyme and indirectly alter the enzyme's UV absorption spectrum. Dye-binding methods such as the Bio-Rad Protein Assay [based on the method of Bradford (19)] have proved to be satisfactory when the unmodified enzyme being investigated is used to establish the standard protein curve.

A variety of specific methods can be used to measure the amount of reagent incorporated; these will obviously depend upon the characteristics of the affinity label. The most general approach is to use a radioactive reagent synthesized from commercially available radioactive starting material for the particular purpose of measuring the amount of reagent incorporated. It is essential that the radioactive label (particularly tritium) not be located in a position from which it is exchanged

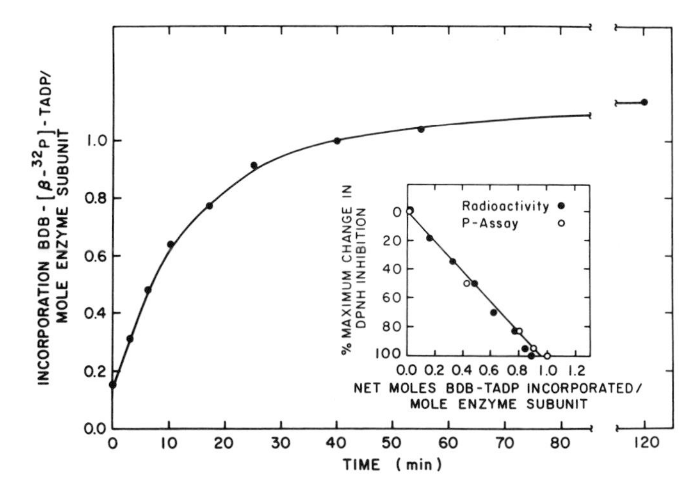

Figure 9. Incorporation of 6-[4-bromo-2,3-dioxobutylthio]adenosine 5′-diphosphate into glutamate dehydrogenase as a function of time. Glutamate dehydrogenase (0.25 mg/ml) was incubated with 0.25 mM [β-^{32}P]6-BDB-TADP under the conditions described in *Figure 7* and at various times the enzyme was isolated by column centrifugation. The incorporation was determined either from the covalently bound radioactivity after reaction with ^{32}P-labelled reagent or from an assay of the phosphorus content of the modified enzyme (P-assay). Inset, relationship between the % maximum change in NADH inhibition (calculated from the inset of *Figure 7*) and the reagent incorporated. From (12), with permission.

during the synthesis. This possibility can be evaluated by comparing the specific activity of the starting material with the specific activity of the product affinity label: the two should be the same. The fluorosulphonylbenzoyl nucleotides have all been synthesized with tritium in the purine moiety, and a few have also been prepared with ^{14}C in the benzoyl group by starting from [carboxy-^{14}C]*p*-aminobenzoic acid (2). In some cases, it has been possible to conduct most of the synthetic steps using non-radioactive material, and to add the radioactive tag toward the end of the synthetic scheme. The [β-^{32}P]6-BDB-TADP provides an example of this strategy: prior to coupling with 1,4-dibromobutanedione, the [β-^{32}P]6-mercaptopurine ribonucleoside 5′-diphosphate was synthesized from 6-mercaptopurine ribonucleoside 5′-monophosphate by phosphorylation using [^{32}P]phosphoric acid (12). *Figure 9* illustrates the time-dependent incorporation of radioactive 6-BDB-TADP into glutamate dehydrogenase. The use of a radioactive reagent often provides the most sensitive method for measuring reagent incorporation; however, other methods offer the advantage of not requiring an additional synthesis. In the case of a non-phosphorus containing enzyme, incorporation of a nucleotide affinity label can be measured by quantitation of the organic phosphorus content of the derivatized enzyme (4). *Figure 9* shows that data for incorporation of 6-BDB-TADP into glutamate dehydrogenase are comparable when obtained by phosphorus determination and by specific radioactivity of the modified enzyme.

The fluorescence of an affinity label is another property that can be used to measure the covalently bound reagent. The ethenoadenosine derivatives exhibit a fluorescence emission maximum around 412 nm when excited at 308 nm. This property has been used to measure the incorporation of 5′-FSBεA into pyruvate kinase (10) and glutamate dehydrogenase (20) and of 2-BDB-TεA 2′,5′-DP into NADP-dependent isocitrate dehydrogenase (6). The amount of reagent incorporated was determined by comparison of the fluorescence of the modified enzymes with that of the fluorescent reagent. In these cases the modified protein was denatured (e.g. in 5 M guanidine hydrochloride) before making the fluorescence measurements in order to minimize any protein environmental effects on the fluorescence of covalently linked reagent; the assumption was made that the fluorescence of covalently bound reagent on the denatured enzyme is the same as that of the free reagent under the same conditions. It is important to evaluate the validity of this assumption by some other method, and that has been done using the phosphorus method for incorporation of 2-BDB-TεA 2′,5′-DP into isocitrate dehydrogenase (21) and the radioactivity method for incorporation of 5′-FSBεA into glutamate dehydrogenase (20).

Regardless of the procedure used for measuring the amount of reagent that becomes covalently bound to the enzyme, the expectation for an affinity label is that it will be incorporated into a limited number of sites on the enzyme, that the incorporation will increase with incubation time and that it will be directly related to the functional change produced by the affinity label, as discussed in Section 5. All of these features are illustrated by the data shown in *Figure 9*. The inset relates the functional change and the incorporation: the percentage maximum change in NADH inhibition is linearly proportional to the incorporation of 6-BDB-TADP and extrapolates to 0.96 mol reagent bound per subunit when the enzyme becomes unresponsive to NADH inhibition.

In many cases, ligands that protect the enzyme against functional changes produced by the affinity label will also decrease the amount of reagent incorporated, and the difference between the amount of reagent incorporated in the absence and presence of the most effective protectants will be a measure of the specific site(s) labelled. In some cases, however, ligands that protect against labelling of a specific site may also change the conformation of the enzyme and increase the reaction rate at a different site; in that situation, the measured reagent incorporation may not decrease despite the fact that the enzyme modified in the presence of protecting ligands retains full activity. The most definitive determination of the number and identification of specifically modified sites is by isolating and comparing the peptides that react with the affinity label in the absence and presence of protecting ligands.

9. ISOLATION OF SPECIFICALLY LABELLED PEPTIDES(S)

Once an enzyme has been modified specifically with an affinity label, it is usually treated with iodoacetate to convert unreacted cysteine residues to their corresponding carboxymethyl derivatives and digested under mild conditions with an appropriate proteolytic enzyme, such as trypsin, chymotrypsin or thermolysin. In preparing to isolate peptides that are chemically modified, thought must be given to the method by which those labelled will be detected among the large number of unreacted peptides. If a radioactive reagent has been synthesized, the radioactivity can be used to locate the peptides of interest. It is important to consider whether the structural portion of the affinity label that bears the radioactive tag will remain covalently linked to the peptide under all the conditions used during the purification procedures. For example, peptides have been isolated from several proteins labelled with 5′-*p*-fluorosulphonylbenzoyl adenosine in which the adenosine contains the label. As long as the pH is maintained between 6 and 8, the radioactivity remains associated with the peptide; but at lower or higher pH values, the ester linkage between the benzoyl and adenosine moieties is hydrolysed, with consequent loss of the radioactive tag. A similar limitation may exist in using as a radioactive probe ^{32}P in the β or γ phosphate of a nucleotide affinity label. For affinity labels that have keto groups or that form Schiff bases in reaction with proteins, non-radioactive reagents can be used for the modification reaction; after the excess reagent is removed from the modified enzyme, $[^3H]NaBH_4$ can be used to reduce the keto group or Schiff base and thereby introduce a radioactive tracer at the location of the altered amino acid residue. This strategy has been used successfuly in labelling and then isolating a peptide of isocitrate dehydrogenase reacted with 3-bromo-2-ketoglutarate (22) and peptides of glutamate dehydrogenase (11) and isocitrate dehydrogenase (21) reacted with bromo-2,3-dioxobutyl nucleotides. In addition to radioactive labels, fluorescent tags, such as those introduced using 5′-FSBεA or 2-BDB-TεA 2′,5′-DP can be used to follow peptides of interest through chromatographic procedures. In some cases, the fluorescence may be dependent on pH (as in the case of ethenoadenosine derivatives); if a pH below the p*K* of the fluorophore is used for chromatography, it may be necessary to adjust the pH of chromatographic fractions in order to detect the fluorescence.

Numerous general methods can be used to separate peptides of proteins modified by affinity labels. These include ion-exchange chromatography and gel filtration. Most effective among these general purification methods has been high-performance liquid chromatography (HPLC), particularly protocols employing reverse-phase columns (e.g. 11, 22). In addition, special procedures can be used, dependent on the properties of the affinity label, to select for peptides modified by these compounds. One such example is the use of dihydroxyboryl-substituted cellulose or polyacrylamide columns to purify peptides labelled with nucleotide analogues (11, 23). The modified peptides are preferentially bound to such columns in phosphate or ammonium bicarbonate buffers at pH 8 by reversible interaction of the *cis*-diol moiety of the reagent's ribose with the column-bound boronate, while most of the unmodified peptides are removed from the column; elution of the nucleotidyl-peptide is accomplished with de-ionized water. Another approach for a nucleotidyl-peptide is to subject the proteolytic digest to chromatography on DEAE–cellulose, followed by treatment with a phosphatase (to decrease the negative charge by removing phosphate groups from covalently bound reagent) and re-chromatography on the same DEAE–cellulose column. This strategy was used in the isolation of a peptide of isocitrate dehydrogenase labelled with 2-BDB-TεA 2′,5′-DP (21). Other purification schemes could readily be devised featuring affinity chromatography aimed at interaction of the column support with the covalently bound affinity label.

Once a single peptide has been isolated, it must be characterized in terms of amino acid composition and amino acid sequence. The derivatized amino acid can be identified either by comparison of the properties of an altered amino acid with those of model amino acid derivatives synthesized by reaction of simple amino acids with the affinity label or by comparison of the modified peptide's sequence with the order of amino acids in unmodified enzyme. For example, to aid in the characterization of peptides labelled by fluorosulphonylbenzoyl nucleotides, N^{ε}-(4-carboxybenzenesulphonyl)lysine (CBS–Lys) and *O*-(4-carboxybenzenesulphonyl)tyrosine (CBS–Tyr) were synthesized by reaction of *p*-fluorosulphonylbenzoic acid with either N^{α}-acetyl lysine or N^{α}-acetyl tyrosine (24). These are the stable products of acid hydrolysis of fluorosulphonylbenzoyl nucleotide derivatives of lysine and tyrosine, respectively. The CBS–Lys and CBS–Tyr standards have been used to identify these modified residues in several enzymes including glutamate dehydrogenase (23, 24, 25) and pyruvate kinase (26). For the bromodioxobutyl nucleotides, a relatively stable cysteine derivative was prepared by reaction with 6-BDB-TA 5′-DP, and the characteristics of that standard, in addition to the known amino acid sequence of the bovine liver glutamate dehydrogenase, were used to identify the particular cysteine residue modified in the enzyme by the same reagent (11). Esters, formed by reaction of aspartate or glutamate residues with affinity labels, may regenerate the original amino acid during the acid hydrolysis preceding amino acid analysis; this occurred in the case of the 13-membered peptide of NADP-dependent isocitrate dehydrogenase containing a glutamate modified by 2-BDB-TεA 2′,5′-DP (21). However, in that case gas-phase sequencing of the radioactively labelled peptide revealed, at cycle 5, no detectable phenylthiohydantoin (PTH)-derivative but rather the highest amount of radioactivity. Since PTH–Glu was the only amino acid missing from the sequence determination but observed upon amino acid analysis after acid hydrolysis, glutamate at position five of the isolated peptide was designated as the target of 2-BDB-TεA 2′,5′-DP in this enzyme (21). These approaches can readily be extended to the identification of other modified amino acids.

10. CONCLUSIONS

The technique of affinity labelling has very general applicability to the identification and probing of specific catalytic and regulatory sites in purified enzymes, and can also be used to identify a single protein with a particular ligand binding site within a complex mixture of proteins. The approaches and examples cited in this chapter are representative only. By irreversibly occupying the active or allosteric site, the affinity label can yield information on the effect of occupying that site on the conformation of the enzyme, on the reactivity of other sites, or on subunit–subunit interactions. Fluorescent affinity labels can be used to introduce fluorescent probes into specific sites on enzymes with the goal of monitoring conformational changes or making estimates of distances between known functional sites on enzymes (6, 20, 27). Affinity labelling studies can be used to provide a rational basis for the design of site-directed mutagenesis experiments. By itself, affinity labelling can yield important insights into the identification of functional sites in enzymes, but it is most powerful in revealing new information on the role of identified amino acid residues when it is used in conjunction with other techniques described in this volume.

11. REFERENCES

1. Jakoby, W. B. and Wilchek, M. (eds) (1977) *Methods in Enzymology*. Vol. 46. Academic Press, Inc., New York.
2. Colman, R. F. (1983) *Annu. Rev. Biochem.*, **52**, 67.
3. Colman, R. F., Huang, Y-C., King, M. M. and Erb, M. (1984) *Biochemistry,* **23**, 3281.
4. Huang, Y-C., Bailey, J. M. and Colman, R. F. (1986) *J. Biol. Chem.*, **261**, 14100.
5. DeCamp, D. L., Lim, S. and Colman, R. F. (1988) *Biochemistry,* **27**, 7651.
6. Bailey, J. M. and Colman, R. F. (1985) *Biochemistry,* **24**, 5367.
7. DeCamp, D. L. and Colman, R. F. (1988) *Biochemistry,* **27**, 3097.
8. Bailey, J. M. and Colman, R. F. (1987) *Biochemistry,* **26**, 6858.
9. Secrist, J. A., Barrio, J. R., Leonard, N. J. and Weber, G. (1972) *Biochemistry,* **11**, 3499.
10. Likos, J. J. and Colman, R. F. (1981) *Biochemistry,* **20**, 491.
11. Batra, S. P. and Colman, R. F. (1986) *Biochemistry,* **25**, 3508.
12. Batra, S. P. and Colman, R. F. (1984) *Biochemistry,* **23**, 4940.
13. Huang, Y-C. and Colman, R. F. (1984) *J. Biol. Chem.*, **259**, 12481.
14. Kapetanovic, E., Bailey, J. M. and Colman, R. F. (1985) *Biochemistry,* **24**, 7586.
15. Batra, S. P. and Colman, R. F. (1986) *J. Biol. Chem.*, **261**, 15565.
16. Jacobson, M. A. and Colman, R. F. (1982) *Biochemistry,* **21**, 2177.
17. Bednar, R. A., Hartman, F. C. and Colman, R. F. (1982) *Biochemistry,* **21**, 3681.
18. Penefsky, H. S. (1979) In *Methods in Enzymology*. Vol. 56. Fleischer, S. and Packer, L. (eds). Academic Press, New York, p. 527.
19. Bradford, M. M. (1976) *Anal. Biochem.*, **72**, 248.
20. Jacobson, M. A. and Colman, R. F. (1983) *Biochemistry,* **22**, 4247.
21. Bailey, J. M. and Colman, R. F. (1987) *J. Biol. Chem.*, **262**, 12620.
22. Ehrlich, R. S. and Colman, R. F. (1987) *J. Biol. Chem.*, **262**, 12614.
23. Schmidt, J. A. and Colman, R. F. (1984) *J. Biol. Chem.*, **259**, 14515.
24. Saradambal, K. V., Bednar, R. A. and Colman, R. F. (1981) *J. Biol. Chem.*, **256**, 11866.
25. Jacobson, M. A. and Colman, R. F. (1984) *Biochemistry,* **23**, 6377.
26. DeCamp, D. L. and Colman, R. F. (1986) *J. Biol. Chem.*, **261**, 4449.
27. Bailey, J. M. and Colman, R. F. (1987) *Biochemistry,* **26**, 4893.

CHAPTER 5

Cross-linking of protein subunits and ligands by the introduction of disulphide bonds

ROBERT R. TRAUT, CARLOS CASIANO and NICK ZECHERLE

1. INTRODUCTION

Biological structures frequently contain assemblies or complexes of protein subunits. They may be relatively stable, such as multi-enzyme complexes and ribosomes, or dynamic or transitory, such as receptor–ligand complexes. Determination of the spatial arrangement of the polypeptides in a multi-component complex is a major step in understanding its structure and its relation to function. 'Cross-linking' represents the introduction of covalent bridges between neighbouring subunits in the native complex, and it provides a simple and rapid experimental approach for obtaining extensive information on quaternary structure. Cross-linking generally involves the use of protein-modifying reagents to form a collection of *inter*-molecular covalent dimers (cross-links), each of which contains a pair of proximal subunits. In addition, both inter-molecular cross-links and *intra*-molecular cross-links within a single polypeptide can be useful in providing information on tertiary structure and possible conformational changes. This chapter deals with inter-molecular cross-linking, the goal of which is to identify the different *dimers* that represent the proximity of one subunit with another. The yield of cross-link formation is kept low: cross-links consisting of more than two components are difficult to analyse and give less information on the spatial arrangement of the subunits. The cross-linking reactions should take place under conditions that maintain the native structure of the protein complex. Since each individual polypeptide subunit could form a cross-link with more than one of its neighbours, analysing the ensemble of pair-wise interactions (dimers) permits reconstruction of a network, or map, specifying the relative locations of the components in the assembly.

The analysis of cross-linking experiments on structures with multiple non-identical subunits is facilitated if the cross-linking bridge has within its structure a cleavable bond that will permit regeneration of the original monomeric protein subunits from separated cross-linked complexes. Sodium dodecyl sulphate–polyacrylamide gel electrophoresis (SDS–PAGE) is the technique most frequently used to analyse for cross-link formation, on the basis of molecular weight. The existence of cleavable bonds in the cross-bridge makes possible the use of a two-dimensional gel electrophoresis technique in which the dimers are separated in the first dimension, following which the bridges are cleaved and the monomeric subunits separated in the second dimension. The technique is called 'diagonal' SDS–PAGE (1, 2) since those subunits that never formed cross-links fall on a diagonal line, while those that did, fall below it. Disulphide bonds, readily cleavable by mild reduction, are advantageous for cross-linking studies and have been used frequently.

Cystine disulphide linkages occur naturally within single polypeptide chains and between two chains. Inter-molecular disulphide bonds occur between subunits in many biological structures such as peptide hormones, receptors and toxins. The cleavage of the disulphide bond in cystine to give two cysteine residues is readily achieved by mild reduction. These cystine cross-links show the proximity of the polypeptides thus linked; moreover, the location of the cysteine residues within each primary structure indicates the contact region in each chain. Recombinant DNA techniques can be employed to introduce new cysteine sulphydryl groups at pre-determined locations of interest in a protein. The new cysteine residues can be used as sites for disulphide bond formation in studies of protein conformation or topography. The relative rarity of cystine cross-links in proteins makes them of relatively little use in providing a general method for determining the spatial arrangement of protein subunits in a structure. Cross-linking sites that are present in greater abundance than cysteine are needed in order to obtain a general, inclusive spatial map relating the locations of most of the multiple polypeptides in a complex supramolecular assembly. It is necessary to use chemical modification reagents of either broad reactivity, or specificity for amino acid residues that are relatively abundant and exposed. Lysine-specific reagents are ideal because of the relative abundance of exposed lysine residues in most proteins. These considerations led to the development of cross-linking methods using lysine-specific bifunctional reagents with a disulphide bond already in their structure or, alternatively, for converting lysine residues to sulphydryl derivatives for subsequent oxidation to disulphide cross-links. This chapter will present a brief general survey of disulphide cross-linking approaches and then describe detailed procedures for using the reagent 2-iminothiolane (2-IT) to produce inter-molecular disulphide cross-links and for the analysis of such cross-linked complexes by diagonal SDS–PAGE.

1.1 **Cross-linking reagents**

Cross-linking procedures generally employ bifunctional reagents that modify amino acid side-chains of proteins. Bifunctional reagents are classified on the basis of the following:

(i) their chemical specificity;
(ii) the length of the cross-bridge formed;
(iii) whether both reactive cross-linking groups are the same or not; that is whether they are homo-bifunctional or hetero-bifunctional;
(iv) whether the groups react chemically or photochemically; and
(v) whether the reagents contain a cleavable bond.

In addition, certain cross-linking reagents can be radiolabelled. A large number of reagents of diverse structures have been used and excellent and comprehensive reviews describing them are available (3–7). Many useful reagents are commercially available (e.g. from Pierce Chemical Company). The structures of some useful reagents are presented in *Table 1*.

In general, the chemical reactions of bifunctional reagents involve modification of protein amino groups or sulphydryl groups, that is, lysine or cysteine residues. The two groups may be the same or different: homo-bifunctional reagents, such as reagent 1, or hetero-functional reagents, such as reagents 3–10 (*Table 1*). Reagent 2, 2-IT, although monofunctional, behaves like a hetero-bifunctional reagent in that it generates a new functional group, a sulphydryl group, when it reacts with protein amino groups (see Section 2.2). A very useful type of hetero-bifunctional reagent has one functional group that first is allowed to react *chemically* with lysine or cysteine residues on a peptide ligand or on a separated polypeptide chain of a complex. Subsequently the modified component is bound to the receptor or complex and the second functional group is irradiated to allow it to react *photochemically* with another polypeptide chain in the receptor or binding site in the complex. The photochemical reactions of the second functional group of such hetero-bifunctional reagents generally involve the light-dependent generation from aryl azides of highly reactive nitrenes that react with little specificity with either the amino acid side-chains or polypeptide backbones of proteins. The advantage of hetero-bifunctional reagents is that the two reactions can be carried out separately.

1.2 Analysis of cross-linking by gel electrophoresis

1.2.1 *Non-cleavable cross-linking reagents*

A major advantage of the cross-linking method is the simplicity of experimental analysis by molecular weight using SDS–PAGE. The cross-linked protein complexes (cross-links) appear on stained one-dimensional SDS–PAGE gels as new, more slowly migrating bands due to their elevated molecular weights. For oligomeric complexes consisting of a repeating subunit, the number of new bands reveals the number of subunits in the multi-subunit structure. For asymmetric structures with non-identical subunits, the molecular weights of cross-links may still suffice to identify their component polypeptide chains, if the number of different subunits is not large and if they have distinct molecular weights. However, for complicated structures such as ribosomes, cross-linking with non-cleavable bifunctional reagents is of limited value because it is difficult to determine the composition of the cross-links from their behaviour in SDS–PAGE.

1.2.2 *Cleavable cross-linking reagents*

For asymmetric multi-polypeptide structures in which the number of different subunits is large, such as the ribosome, it is advantageous to use procedures in which the cross-link contains a cleavable bond. This facilitates the identification of the components of a cross-link because it can be cleaved to form the original monomeric components. Cleavable bonds used in different cross-linking reagents

Table 1. Selected cleavable cross-linking reagents.

A. LYSINE (NH_2)-SPECIFIC REAGENTS

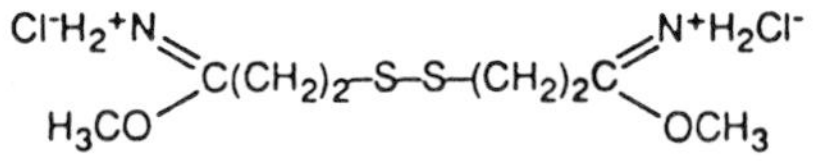

1. DTBP
Dimethyl-3,3′-dithiobispropionimidate (8)

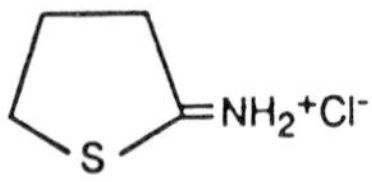

2. 2-IT
2-Iminothiolane (9)

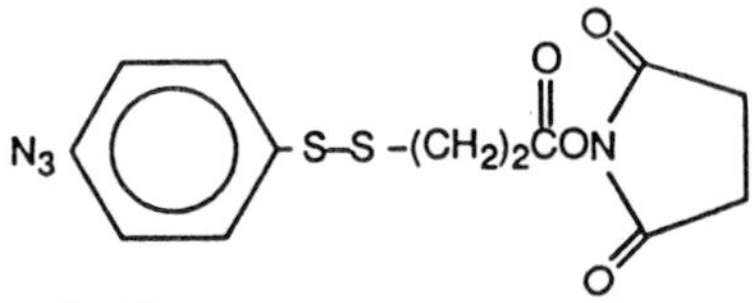

3. SADP
N-Succinimidyl-(4-azidophenyl)-1,3-dithiopropionate (10)

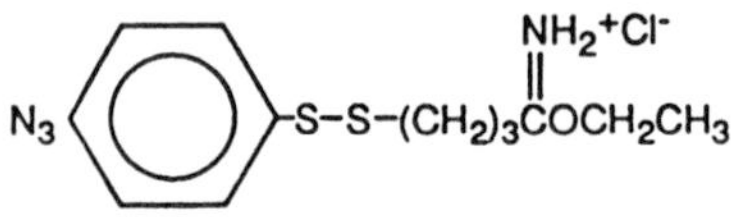

4. EADB
Ethyl-4-azidophenyl-1,4-dithiobutyrimidate (11)

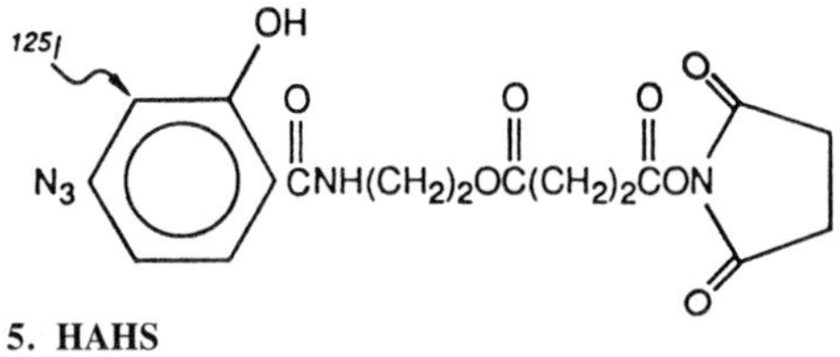

5. HAHS
1-[*N*-(2-hydroxy-5-azidobenzoyl)-2-aminoethyl]-4-(*N*-hydroxysuccinimidyl)-succinate (12)

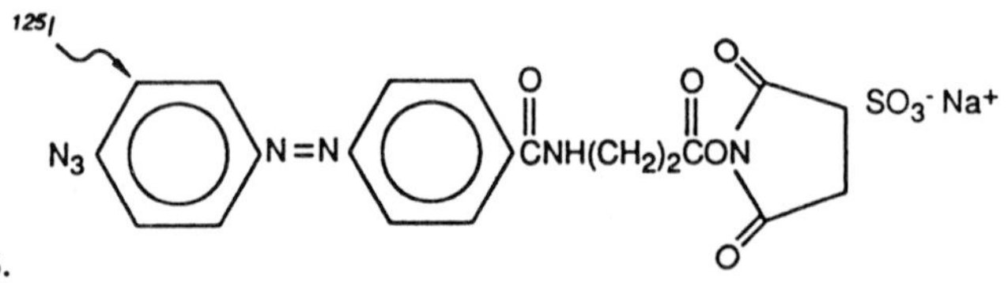

6.
N-[-4-(*p*-azidophenylazo)benzoyl]-3-aminopropyl-*N*′-oxysulphosuccinimide ester (13)

B. CYSTEINE (SH) SPECIFIC REAGENTS

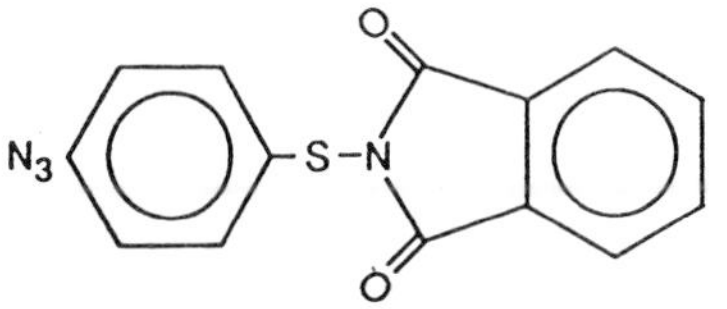

7. APTP
N-(4-azidophenyl)phthalimide (14)

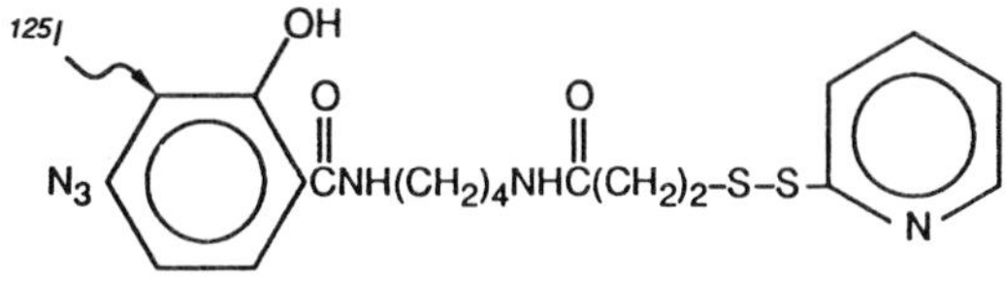

8. APDP
N-[4-(*p*-azidosalicylamido)butyl]-3′-(2′-pyridyldithio)propionamide
(E. Fujimoto, Pierce Chemical Co.)

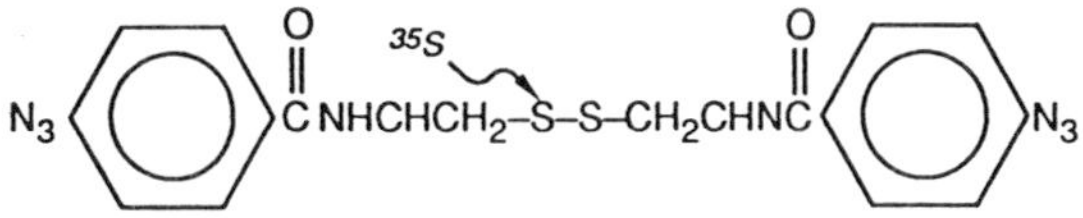

9. $(ABC)_2$
N,N′-bis(4-azidobenzoyl)cystine (15)

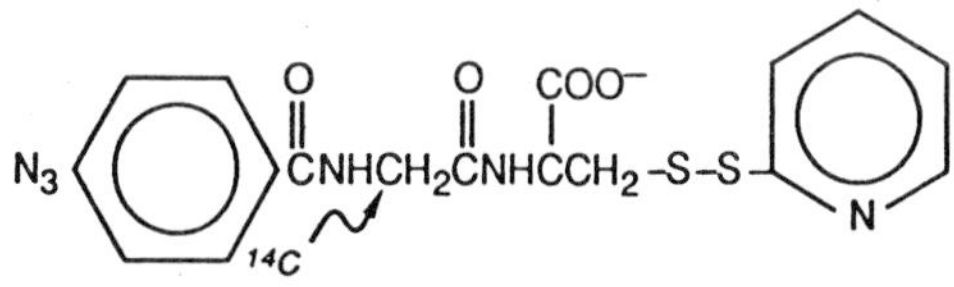

10. AGTC
N-(4-azidobenzoylglycyl)-*S*-(2-thiopyridyl)cysteine (16)

---

include glycol bridges, cleavable by periodate; azo bridges, cleavable by dithionite; ester linkages, cleavable by base; sulphones, cleavable by reduction; and disulphide bonds, also cleavable by reduction (5). Cross-linked species can be excised from gels, cleaved and analysed as monomeric components by SDS–PAGE or other suitable analytical systems; or, as described below in Section 3, an intact one-dimensional SDS gel with resolved cross-linked and non-cross-linked components can be subjected in its entirety to the appropriate cleavage condition and then used as the sample for a second SDS–PAGE step to reveal the monomeric products of the cleaved cross-links, a method called 'diagonal SDS–PAGE', or simply diagonal gel analysis.

Table 1 presents a selected list of cleavable cross-linking reagents with key references for their availability and usefulness. Part A lists reagents specific for lysine, with either an imidoester or succinimidyl moiety as one of the functional groups. The first reagent listed, dithiobispropionimidate (DTBP), is a symmetrical homo-bifunctional reagent with two imidoester functions. The next reagent, 2-IT, is unique among those listed in that it has only one reactive centre. The reagent converts lysine residues to sulphydryl derivatives and is thus functionally equivalent to a hetero-bifunctional reagent (see Section 2.2). The other reagents are hetero-bifunctional with an aryl azide as the second photo-activated moiety. Two have a disulphide as the cleavable bond, and the other has a base-labile ester bond.

Part B of *Table 1* lists hetero-bifunctional reagents of which one functional group reacts with protein sulphydryl groups. A protein sulphydryl displaces a leaving group in each reagent to form a new mixed-disulphide as the cross-linking bridge. The second reactive group of each is an aryl azide that reacts photochemically upon irradiation.

1.2.3 *Radiolabel transfer to the target polypeptide in the binding site by cleavable cross-linking reagents*

In situations where one purified component of a multi-protein complex can be reconstituted to occupy its original location in the complex, or in which a ligand or factor can be bound to a receptor complex, suitably designed hetero-bifunctional, cleavable cross-linking reagents with a radiolabelled moiety in their structure may be used to attach covalently the radiolabel to the target binding site component. *Figure 1A* depicts the generalized structure of such a cross-linking reagent and *Figure 1B* its application. One functional group of the labelled hetero-bifunctional reagent reacts first, usually chemically, with the purified protein subunit or ligand to form an affinity-labelled derivative, which is then allowed to bind specifically to the target receptor complex. The second functional group is then photo-activated to form a cross-link to the polypeptide(s) of the reconstituted complex that interact closely with the previously modified, affinity-labelled ligand. One of the reagents listed in *Table 1A* (reagent 6) and one in *Table 1B* (reagent 8, APDP) can be readily labelled with ^{125}I in a single step to modify the final compound immediately before use. Two other reagents, 9 [$(ABC)_2$], and 10 (AGTC), can be labelled with ^{35}S and ^{14}C, respectively, during the course of their synthesis. For all of the reagents the radiolabel is located on the target side of the cleavable bond in the bifunctional reagent; therefore, cleavage results in the transfer of the radiolabelled moiety from the cross-linking reagent to the monomeric binding site component(s).

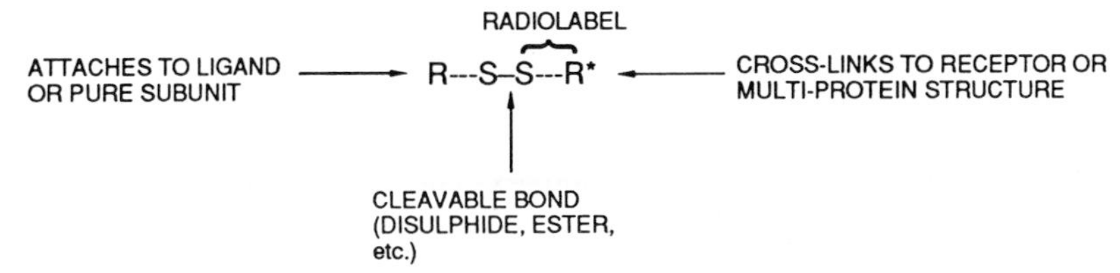

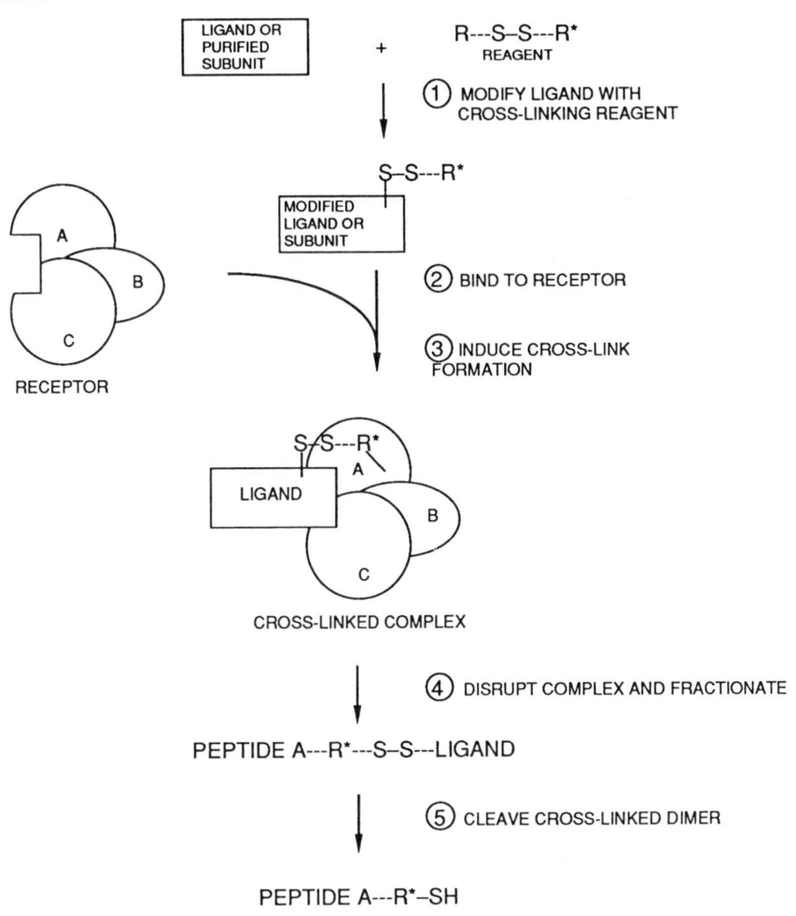

Figure 1. Radiolabelled, cleavable photoaffinity cross-linking reagents for the identification of binding site proteins.

2. REVERSIBLE DISULPHIDE CROSS-LINKING STRATEGIES

2.1 **Ribosomes as objects for cross-linking studies**

Ribosomes are large, highly asymmetric organelles that contain both RNA and many different protein components. The small and large ribosomal subunits from bacteria contain about 20 and 30 different proteins, respectively; eukaryotic ribosomes are significantly more complex. The functional properties of ribosomes in protein synthesis are the result of cooperative interactions among RNA and protein components. A number of different approaches have been taken to investigate ribosome structure and function. These include X-ray crystallography of intact ribosomes and individual components, neutron diffraction, NMR, fluorescence energy transfer, affinity labelling, reconstitution with single components omitted, immune electron microscopy with protein-specific antibodies, the study of mutants of both RNA and proteins, and both protein–protein and protein–RNA cross-linking. A method for reversible protein–protein cross-linking through the formation of disulphide bonds between proximal ribosomal proteins was developed in this laboratory (17) and used to determine the overall relative spatial arrangement of most of the protein components of both the large and small ribosomal subunits from *Escherichia coli* (18) and to investigate the protein infrastructure of ribosomal functional domains (19). This method has also been used to identify proteins at the interface between the two ribosomal subunits and to identify ribosomal proteins at the binding sites for initiation and elongation factors. The method to be described in detail here involves the modification of lysine amino groups in ribosomes with 2-IT (reagent 2 of *Table 1A*) to form sulphydryl derivatives and the formation of disulphide cross-links between neighbouring proteins by mild oxidation.

These cross-linking methods are of general applicability and are appropriate for the investigation of other biological structures that contain multiple protein components.

2.2 **Strategies for disulphide cross-linking**

Three strategies for introducing disulphide cross-links between neighbouring polypeptide chains are shown schematically in *Figure 2*, which depicts a multi-protein complex of six different subunits, none of which have endogenous sulphydryl groups. The simplest approach, step 1, employs a homo-bifunctional, lysine-specific reagent that contains a pre-formed disulphide bond. Incubation of the multi-protein complex with the reagent at a pH between 7.5 and 8.5 leads to modification of a lysine residue by one of the functional groups, and, if prior to hydrolysis the other end of the reagent reacts with another lysine on an adjacent protein subunit, to the formation of an inter-molecular disulphide cross-link. One of the first reagents of this type to be used is the bisimidoester DTBP, the structure of which is shown in *Table 1A*. Other cleavable reagents with *N*-succinimidyl moieties as the lysine-specific modifying function are also available. DTBP has been used to study the spatial arrangement of proteins in the erythrocyte

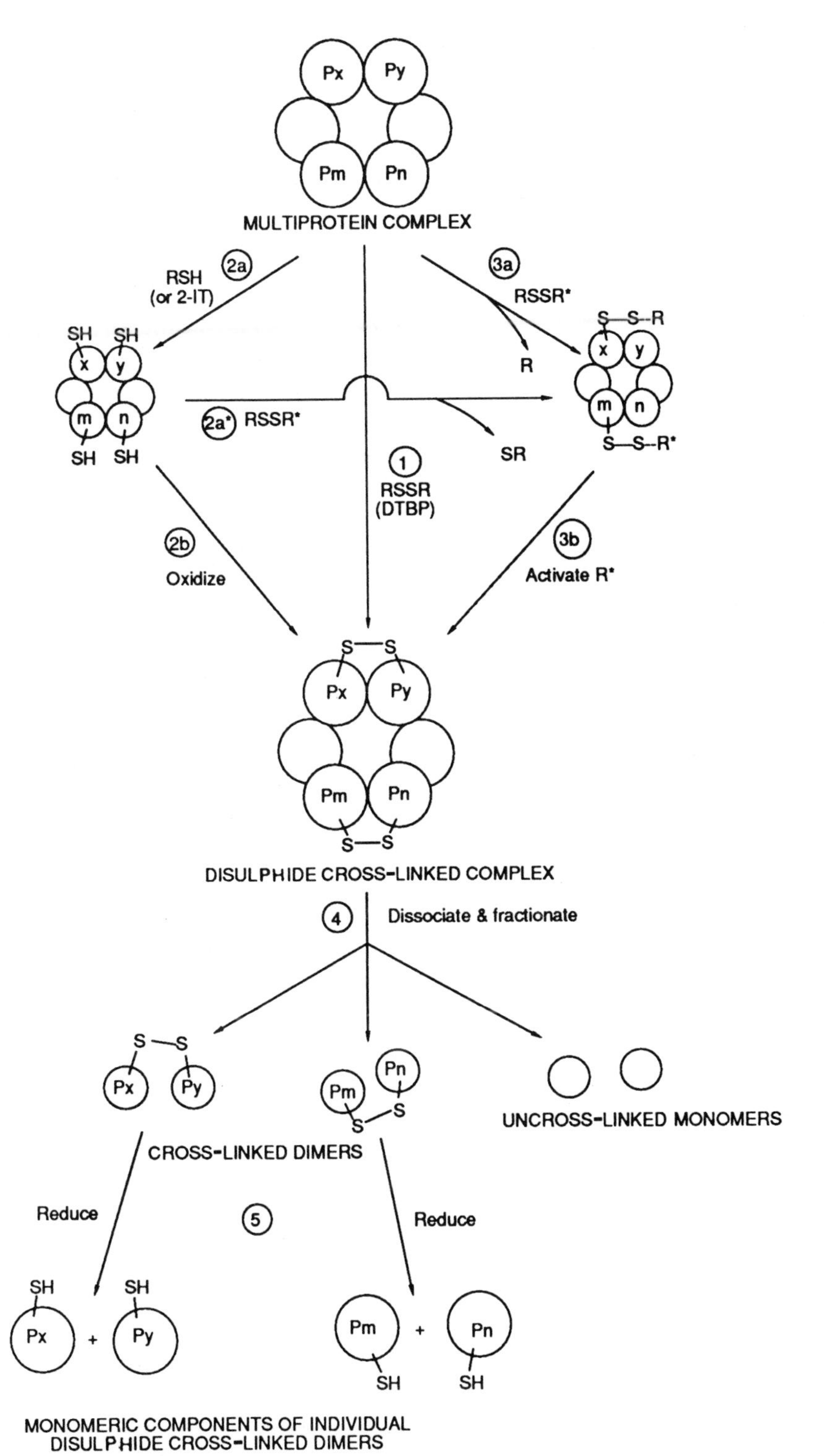

Figure 2. Outline of reversible disulphide cross-linking.

membrane (8), in nucleosomes (20, 21) and in ribosomes (22). Since disulphide bonds are readily cleaved by reducing agents, it is obvious that reducing agents must be absent from buffers in which the cross-linking reactions are carried out and in which the cross-links are fractionated or analysed.

The method that has been emphasized in this laboratory is numbered as steps 2a–2b in the *Figure 2* and consists of the addition of SH groups to the complex *prior* to inducing disulphide bond formation. Several modifying reagents lead to the formation of sulphydryl derivatives of lysine: acetyl mercaptosuccinic anhydride (23); *N*-acetyl homocysteine thiolactone (24); *N*-succinimidyl-3-(2-pyridyldithio) propionate (25); and aliphatic mercaptoimidates such as 3-mercaptopropionimidate (26). The reagent used in the procedures described here is 2-IT (18). Its reaction with protein amino groups is shown in *Figure 3*. As with DTBP, lysine residues are modified to form amidine derivatives. 2-IT is formed by the cyclization of 4-mercaptobutyrimidate, as which it was at one time mistakenly identified. The carbon adjacent to the imino group is attacked by a lysine amino group, with concomitant cleavage of the S–C bond and generation of a new sulphydryl group, which is the second cross-linking site. Advantages over the non-cyclic mercaptoimidate (27) include: first, that it is a less reactive reagent than the open chain imidate, permitting better control of the extent of modification, and second, that it is more convenient to use because it is more soluble in water and more stable. Both the imidate esters and 2-IT form amidinated lysyl residues that retain the cationic charges of the original molecule, a property that may be of importance for the maintenance of the native conformation of the structure. This property contrasts with alternate methods for the introduction of sulphydryl groups.

In the first step of the two-step procedure (step 2a of *Figure 2*), exogenous

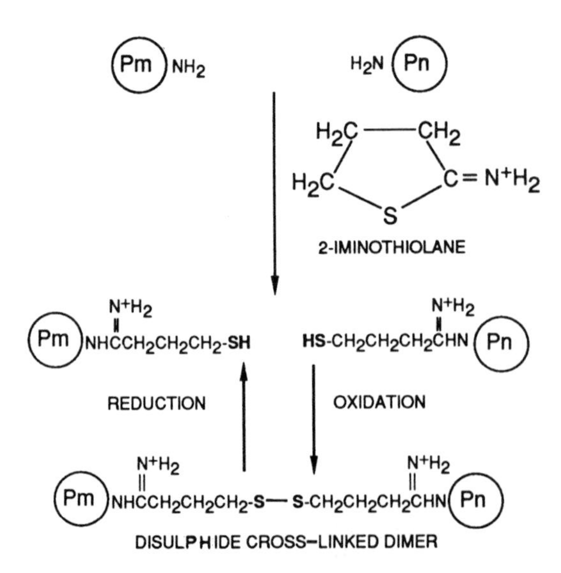

Figure 3. Conversion of protein amino groups to sulphydryl derivatives with 2-iminothiolane.

sulphydryl groups are introduced by random, or undirected, chemical modification of lysine residues in proteins of the structure under investigation. The intermediate product of reaction 2a is an SH-charged derivative of the multi-protein complex. The location and number of the added exogenous SH-groups depends upon the reactivity of individual lysine residues and the extent of the modification reaction. The latter is controlled by the concentration of 2-IT and the time of reaction. The pH of 8 was chosen as a compromise between maintaining the integrity of the ribosomes and enhancing the nucleophilic character of the lysine amino groups. The extent of the modification reaction should be such that each component of the complex has at least one potential cross-linking site. For the ribosome, conditions were chosen to give on average the addition of 2–3 SH groups per protein subunit, determined by the reaction of each with [^{14}C]iodoacetamide (28). The actual cross-linking (step 2b) takes place through the oxidation of proximal sulphydryl groups to disulphide cross-links, promoted by the addition of an oxidizing agent such as hydrogen peroxide. The formation of disulphide cross-links in a second controlled step has a distinct advantage over a bifunctional disulphide reagent like DTBP. The disadvantage of DTBP is that the intermediate mono-adduct formed by reaction of one of its imidoester groups with a protein might not form a cross-link, because the second imidoester of the mono-adduct may hydrolyse instead. With 2-IT, each modification generates a sulphydryl that is a potential cross-linking site. Studies on ribosomes showed higher yields and more cross-links with 2-IT than with DTBP.

An alternative route for forming the disulphide cross-linked complex involves the use of hetero-bifunctional reagents, the second functional group of which is photo-activable, frequently an aryl azide. Examples of photo-activable, hetero-bifunctional reagents specific for sulphydryl groups are shown in *Table 1B*. In step 2a* of *Figure 2*, a sulphydryl displaces SR from the bifunctional reagent, forming a new disulphide bond and attaching the photo-activable group to the protein. If the protein subunits of the complex have endogenous cysteine residues (see Section 5.2), these can also be similarly used as cross-linking sites. Hetero-bifunctional, photo-activable reagents that contain a disulphide linkage and are specific for lysine residues are shown as numbers 3, 4 and 6 in *Table 1A*. In reaction 3a of *Figure 2*, lysine amino groups in the unmodified complex react chemically with the lysine-specific functional group of a hetero-bifunctional reagent. It contains a disulphide bond as part of its structure, plus a second, photo-activable group. The lysine amino group displaces R, forming a disulphide derivative with a photo-activable functional group attached. The intermediates formed chemically either by reactions 3a or 2a* are then photolysed to form nitrenes from the protein-linked aryl azides, which can react with other proximal proteins to form disulphide cross-links.

The next step is the dissociation of the cross-linked proteins from the complex. The total protein will most likely contain monomers, dimers and some higher oligomeric complexes. For simple complexes with only a few polypeptide subunits, SDS–PAGE in one dimension may be sufficient to identify the composition of various cross-links. For more complicated structures it may be necessary to fractionate the protein mixture by high-pressure liquid chromatography (HPLC), ion-exchange chromatography or electrophoresis on the basis of charge, prior to SDS–PAGE. The final step in the determination of the composition of the cross-links is cleavage and regeneration of the monomeric components (steps 4 and 5 in *Figure 2*). The necessity of avoiding reducing agents that will prematurely cleave either the disulphide cross-linking reagent or the protein cross-link has already been mentioned. In addition, precautions must be taken to avoid possible disulphide interchange between the components of disulphide cross-links. For example, a small number of unoxidized sulphydryls remain on the ribosome following the oxidation step and could initiate disulphide interchange, with formation of cross-links between distant or unrelated proteins. This possibility is eliminated by alkylating any sulphydryls in the sample with iodoacetamide before and during extraction of cross-linked proteins from the native ribosome structure. Alternatively, extraction of the proteins under acidic conditions prevents disulphide interchange. The subsequent steps of the procedure involve the separation of cross-linked species from each other and from non-cross-linked chains and the identification of the members of the individual cross-linked species. This analysis is facilitated by taking advantage of the cleavability of the cross-link and is the subject of Section 5.

3. DIAGONAL SDS–GEL ELECTROPHORESIS

3.1 **Overview**

The disulphide bonds in the protein–protein cross-links formed by any of the strategies outlined above are readily cleavable by reduction. Diagonal SDS–PAGE uses this property of the cross-links to separate *inter*-molecular cross-links from non-cross-linked monomeric proteins, even though the latter may have *intra*-molecular disulphides. Diagonal gel electrophoresis is also applicable to the analysis of cross-links formed by any cleavable reagent. The method employs two electrophoretic steps; the cleavage of the labile bond in the cross-link takes place *after* the first step and *before* the second. First, as shown schematically in *Figure 4A*, the sample of total protein extracted from the cross-linked multi-component complex under non-reducing conditions is separated by SDS–PAGE, again under non-reducing conditions. SDS–PAGE separates proteins on the basis of molecular weight: polypeptides of low molecular weight migrate more rapidly than those of high molecular weight. Three disulphide protein dimers (cross-links) are depicted in *Figure 4* in boxes. The other vertical bands in the gel indicate non-cross-linked protein subunits. Although in general the cross-links migrate more slowly than the non-cross-linked polypeptides, certain monomers among the ribosomal proteins have molecular weights higher than many of the dimers. After electrophoresis, the proteins are reduced *in situ* to cleave the disulphide bonds and convert cross-links into monomers. The first gel becomes the origin of the second gel slab. Protein subunits that had been part of a cross-link in the first electrophoresis migrate more rapidly as monomers in the second electrophoresis.

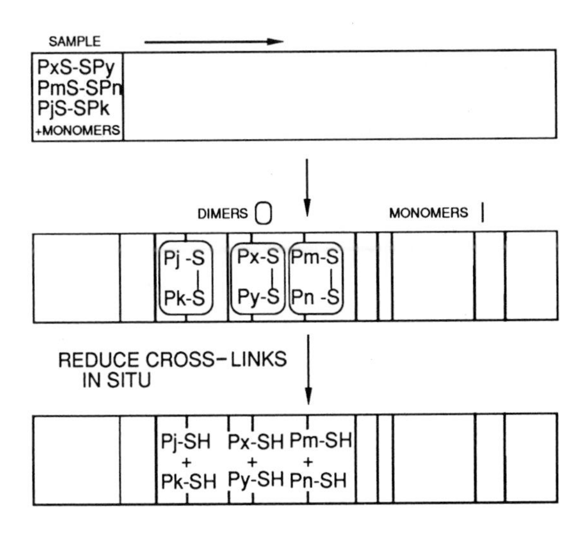

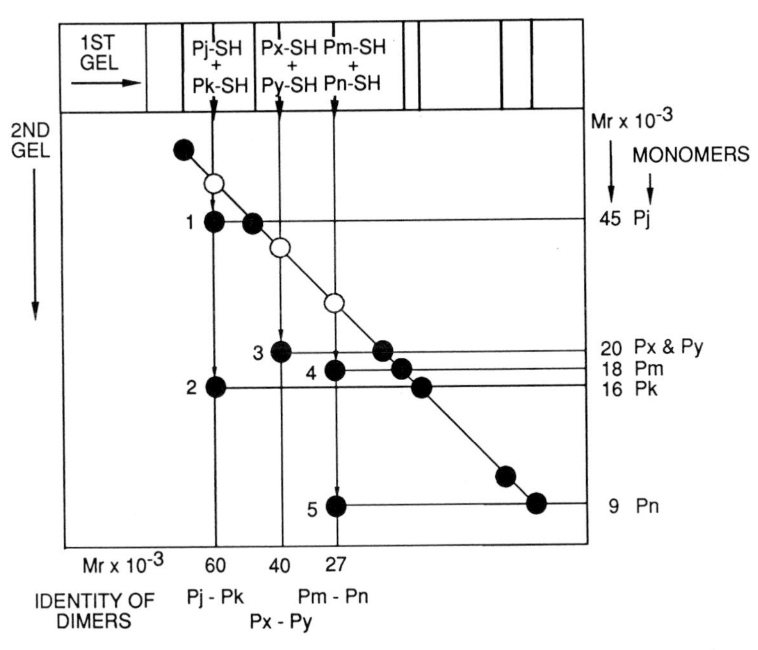

Figure 4. Analysis of disulphide cross-linked protein mixture by diagonal SDS–gel electrophoresis. (**A**) First dimension electrophoresis of cross-links under non-reducing conditions. The boxes indicate cross-linked dimers and the lines indicate non-cross-linked proteins. (**B**) Second dimension electrophoresis following reduction of first dimension gel. The shaded circles on the diagonal represent proteins that have not been cross-linked. The open circles on the diagonal represent the position of cross-links prior to reduction. The black circles represent reduced monomeric components from disulphide cross-links.

Non-cross-linked protein subunits retain the same relative mobility in both electrophoretic steps. The pattern of proteins that results from the two-dimensional analysis is shown schematically in *Figure 4B*. The proteins that had not been members of a cross-link fall on a diagonal line. Cross-linked protein subunits fall below this diagonal as a result of their increased mobility following cleavage of the cross-link.

3.2 **Identification of members of cross-links**

The SDS–PAGE system gives a linear relationship between log [apparent molecular weight] and relative mobility for both cross-links and non-cross-linked proteins in each of the two steps. Both electrophoretic steps can be calibrated using commercially available molecular weight markers or individual ribosomal proteins as standards. Protein subunits that had been members of the same unique cross-linked complex necessarily have the same mobility in the first electrophoresis and therefore appear after the second electrophoresis below the diagonal on the same vertical line with a common intercept on the diagonal. Within the range of molecular weights for which the relationship between mobility and molecular weight holds (depending on the acrylamide composition of the gel), the sum of the apparent molecular weights of the monomeric protein subunits appearing below the diagonal arising from a putative cross-link has been found empirically to fall within 7% of the apparent molecular weight of the parent cross-link estimated from its mobility in the first dimension. Thus in *Figure 4* proteins Pj and Pk (spots 1 and 2) have apparent molecular weights of 45 000 and 16 000, while the cross-link Pj–Pk has an apparent molecular weight of 60 000; and proteins Pm and Pn, with molecular weights of 18 000 and 9000 arise from a cross-link of molecular weight 27 000. The cross-link Px–Py, 40 000, gives rise to two monomers with the same molecular weight of 20 000. This situation arises frequently in the analysis of ribosomal proteins. In the absence of additional information on the stoichiometry of components in the original native complex the result could also be consistent with the presence of dimers Px–Px, or Py–Py. For ribosomal proteins the analysis is even more difficult because often three or more proteins have nearly the same molecular weight.

4. EXPERIMENTAL PROCEDURES FOR CROSS-LINKING WITH 2-IMINOTHIOLANE

These methods will be given in detail as applied to ribosomal subunits from *E.coli*; however, they are general in their applicability. Certain steps pertaining specifically to ribosomes because of the presence of RNA and the relative insolubility of ribosomal proteins will be pointed out, and they should be unnecessary for other structures consisting only of protein.

The preparation of ribosomal subunits from *E.coli* can be found elsewhere (2). Ribosomal subunits isolated from *E.coli* strain MRE600 are assayed for purity and integrity by analytical sucrose density gradient centrifugation, prior to cross-linking. This is repeated following cross-linking to confirm that the native structure is retained. The ribosome concentration is determined from the absorbance at 260 nm (A_{260}); at a ribosome concentration of 1 mg/ml the A_{260} is 15.

4.1 **Cross-linking subunits with 2-IT**

The compositions of the solutions for the cross-linking procedure are given in *Table 2*.

(i) Incubate the ribosomal subunits, re-suspended in solution 1 (*Table 2*) at a concentration of 10–20 mg/ml, for 15 min at 37°C. This step is necessary to ensure that the ribosomes are in their native conformation, with all cysteine residues reduced, prior to the cross-linking procedure.

(ii) Adjust the ribosomal subunit concentration to 3 mg/ml with solution 2. Since the RNA:protein ratio in the ribosome is 2:1, the final concentration of ribosomal proteins after dilution is 1 mg/ml. The concentrations of ribosomal subunits are 1.9 μM for the 50S and 3.5 μM for the 30S subunit. The use of more concentrated solutions of ribosomal subunits during the cross-linking procedure may lead to a significant formation of inter-subunit cross-links, resulting in the identification of protein–protein cross-links that do not necessarily reflect proximity within the ribosomal subunit.

(iii) Add 24 μl of solution 4 per ml of ribosome solution. The final concentration of 2-IT is 12 mM. Incubate the mixture for 2.5 h at 0°C. The extent of modification can be controlled by adjusting the final concentration of 2-IT and/or changing the incubation time.

(iv) Add 4.5 μl of 30% hydrogen peroxide (H_2O_2) per ml of reaction mixture. The final concentration of hydrogen peroxide is 40 mM. Incubate for 30 min at 0°C to promote cross-linking between adjacent sulphydryl groups by disulphide bond formation.

(v) Remove unreacted H_2O_2 by adding 15 μg of catalase per ml of reaction mixture. Incubate for 15 min at 0°C.

(vi) Add iodoacetamide to a final concentration of 40 mM (7.4 mg per ml of reaction mixture) to alkylate free sulphydryl groups that were not oxidized. Incubate the mixture for 30 min at room temperature.

(vii) Concentrate the ribosomal solution by centrifugation of the reaction mixture at 50 000 r.p.m. for 5 h at 4°C in a Beckman Type 60 Ti rotor, followed by re-suspension of the pelleted ribosomes in solution 3 to a concentration of approximately 20 mg/ml. Store the re-suspended subunits at −70°C or use them immediately to extract the ribosomal proteins. This step may not be

Table 2. Solutions for the cross-linking procedure.

1. 100 mM NH_4Cl; 10 mM Tris–HCl, pH 7.2; 10 mM $MgCl_2$; 14 mM 2-mercaptoethanol.
2. 50 mM KCl; 50 mM Tris–HCl, pH 8.0; 1 mM $MgCl_2$; 5 mM DTT.
3. 100 mM NH_4Cl; 10 mM Tris–HCl, pH 7.2; 10 mM $MgCl_2$; 40 mM iodoacetamide.
4. 500 mM 2-iminothiolane–HCl (2-IT, from Pierce Chemical Co.); 500 mM Tris–HCl, pH 8.0.

required for material other than ribosomes, which require separation of RNA from proteins.

4.2 Extraction of proteins from cross-linked subunits

Several methods for the extraction of proteins from cross-linked ribosomes can be employed to separate the proteins from RNA. These include:

(i) treatment of ribosomal subunits with 66% (v/v) acetic acid, centrifugation of the precipitated RNA, dialysis of the supernatant fraction containing the ribosomal proteins against 6% (v/v) acetic acid and lyophilization;
(ii) treatment with 8 M urea/6 M LiCl/40 mM iodoacetamide, centrifugation of the precipitated RNA, extensive dialysis of supernatant fraction to remove LiCl and lyophilization;
(iii) treatment with 8 M urea/6 M LiCl/40 mM iodoacetamide, precipitation of proteins with 10% trichloroacetic acid (TCA) and recovery of the precipitated proteins by centrifugation; and
(iv) combinations of the methods described above.

Comparison of diagonal gels of proteins extracted by all these methods show no differences.

In general, any method to prepare a relatively salt-free, acid-free and RNA-free protein sample can be employed for the preparation of protein samples for diagonal SDS–PAGE. The most recent procedure used in this laboratory to prepare the ribosomal protein samples is described here. In this method the ribosomes are first treated with 8 M urea to disrupt the RNA–protein and protein–protein interactions. This step is followed by the addition of acetic acid and $MgCl_2$ to precipitate the rRNA while still maintaining the ribosomal proteins in solution. Disulphide interchange is avoided during the entire procedure by alkylating free sulphydryl groups with iodoacetamide, or alternatively *N*-ethylmaleimide, at pH 8.0, and by extraction of the proteins under acidic conditions.

(i) Mix the concentrated cross-linked ribosomal subunits from Section 4.1 (vii) with an equal volume of a solution containing 8 M urea and 40 mM iodoacetamide (the solution should be made fresh just before use). Incubate for 2 h at 0°C.
(ii) Increase the $MgCl_2$ concentration to 0.1 M and then add two volumes of glacial acetic acid. Incubate for 4 h at 0°C.
(iii) Remove the precipitated RNA by centrifugation at 10 000 r.p.m. for 30 min in a Sorvall SA-600 rotor and dialyse the supernatant containing the ribosomal proteins against four changes of 6% acetic acid (500 vols per change) during 24 h at 4°C.
(iv) Determine the protein concentration in the dialysed supernatant fraction, distribute the solution in aliquots and lyophilize them.

4.3 **Diagonal SDS–gel electrophoresis**

Virtually any kind of laboratory gel electrophoresis equipment can be used for this technique. The first dimension may be a tube gel or a lane excised from a slab gel. Procedures are given in detail here for tube gels (length = 11 cm, diameter = 0.4 cm) for the first dimension, and slabs gels (length = 23.5 cm, width = 13 cm, spacers = 1.5 or 3.0 mm) for the second dimension. A variety of electrophoresis chambers for tube and slab gel electrophoresis are commercially available; for example, those from Bio-Rad and Idea Scientific have been used successfully. A home-made slab gel apparatus is described in reference 29.

4.3.1 *First dimension SDS–gel electrophoresis*

Stock solutions and recipes for the preparation of polyacrylamide–SDS gels are described in *Tables 3* and *4*.

The first dimension of the diagonal SDS–PAGE is run under non-reducing conditions to separate the cross-links from non-cross-linked monomeric proteins. Polyacrylamide concentrations ranging from 13.5% to 17.5% give a good separation of complex mixtures of proteins in the molecular weight range between 10 000 and 60 000. The amount of cross-linked protein sample to be loaded in the first dimension depends on the yield of cross-linking, the complexity of the cross-linked sample and the size of the second dimension gel, especially its thickness. In general, the lower the yield of cross-links in the sample to be analysed, the greater the amount of protein that has to be loaded in the first dimension. The same situation applies when very complex mixtures of cross-linked proteins are analysed. In the procedures described here, the appropriate amounts of protein sample to be loaded in the first dimension for the analysis of *E.coli* cross-linked ribosomal proteins are 300 μg and 400 μg of cross-linked 30S ribosomal proteins (TP30) and 50S ribosomal proteins (TP50), respectively, when 1.5 mm spacers are used in the second dimension; and 600 μg and 800 μg of TP30 and TP50, respectively, when 3 mm spacers are used. Proportionally less protein is needed if a mini-gel apparatus is used. The procedure for electrophoresis in the first dimension is as follows.

(i) Boil several glass tubes (length = 14 cm, i.d. = 0.4 cm) in 0.1% SDS, or other strong cleaning agent, for 30 min. Rinse them thoroughly with

Table 3. Stock solutions for the preparation of SDS–polyacrylamide gels.

A. 30% (w/v) acrylamide, 0.8% (w/v) bisacrylamide
B. 1.5 M Tris–HCl, pH 8.8
C. 0.5 M Tris–HCl, pH 6.8
D. 10% (w/v) SDS
E. 10% (w/v) ammonium persulphate
F. TEMED
G. 2% (w/v) bromphenol blue, sodium salt, in water

Table 4. Recipes for SDS–polyacrylamide gels.

Stock solution (Table 3)	*Sample buffer*	*Electrophoresis buffer*	*1st dimension stacking gel*	*1st dimension separation gel*	*2nd dimension separation gel*	*2nd dimension cementing solution*
A	—	—	2.5 ml	14.5 ml	35 ml	—
B	—	16.7 ml	—	7.5 ml	15 ml	1.66 ml
C	1.0 ml	—	3.4 ml	—	—	—
D	2.0 ml	10 ml	0.15 ml	0.3 ml	0.6 ml	1.0 ml
E	—	—	0.45 ml	0.3 ml	0.6 ml	—
F	—	—	5 μl	0.01 ml	0.02 ml	—
G	0.1 ml	—	—	—	—	2.0 ml
Distilled water	0.5 ml	973 ml	8.7 ml	7.5 ml	9.0 ml	95 ml
Iodoacetamide	37 mg	—	—	—	—	—
Ultra-pure urea	1.9 g	—	—	—	—	—
Glycine	—	14.3 g	—	—	—	—
Agarose	—	—	—	—	—	1.5 g

The electrophoresis buffer is the same in both dimensions.
The final concentrations in the solutions are:

Sample buffer: 100 mM Tris–HCl, pH 6.8, 4% (w/v) SDS, 6 M urea, 40 mM iodoacetamide, 0.04% (w/v) bromphenol blue.
Electrophoresis buffer: 190 mM glycine, 0.1% (w/v) SDS, 25 mM Tris–HCl, pH 8.8.
1st dimension stacking gel: 5.0% (w/v) acrylamide, 0.1% (w/v) SDS, 114 mM Tris–HCl, pH 6.8, 0.3% (w/v) ammonium persulphate, 0.03% (v/v) TEMED.
1st dimension separation gel: 14.5% (w/v) acrylamide, 0.1% (w/v) SDS, 375 mM Tris–HCl, pH 8.8, 0.1% (w/v) ammonium persulphate, 0.03% (v/v) TEMED.
2nd dimension separation gel: 17.5% (w/v) acrylamide, 0.1% (w/v) SDS, 375 mM Tris–HCl, pH 8.8, 0.1% (w/v) ammonium persulphate, 0.03% (v/v) TEMED.
2nd dimension cementing solution: 1.5% (w/v) agarose, 25 mM Tris–HCl, pH 8.8, 0.1% (w/v) SDS, 0.04% (w/v) bromphenol blue.

distilled water and ethanol. Let them air dry. Seal the bottom of each tube with Parafilm and place them vertically in a rack.

(ii) Dissolve the cross-linked protein sample in 30–50 μl of sample buffer and incubate for 15 min at 65°C.

(iii) Centrifuge the sample for a few minutes in a microcentrifuge to eliminate possible insoluble or aggregated proteins that could deposit on the top of the gel. Carefully remove the supernatant fraction. The presence of insoluble aggregated material is common in cross-linked samples. This, as well as residual RNA and salts, should be eliminated from samples for PAGE. All three can cause streaking and tailing of the cross-linked proteins in the first dimension with loss of resolution.

(iv) Filter and de-gas gel solutions prior to use. Initiate polymerization by adding ammonium persulphate and *N,N,N′,N′*-tetramethylethylene diamine (TEMED).

(v) Using a long Pasteur pipette, fill the glass tubes with separation gel solution to a height of 10 cm. Avoid formation of bubbles at the bottom of the gel. With a micropipette layer a few microlitres of distilled water on top of the separation gel. This step is necessary to ensure uniform polymerization at the surface of the separation gel. Polymerization is complete after approximately 15 min at room temperature.

(vi) Remove the water from the top of the separation gel and replace with 120 μl of stacking gel solution. Again with a micropipette, place a few microlitres of distilled water on top of the stacking gel. Allow to polymerize. The length of the stacking gel should be 1 cm.

(vii) Place the tube gel in the electrophoresis apparatus for first dimension electrophoresis. Fill the upper and lower reservoirs with electrophoresis buffer. Check for leaks and remove air bubbles from the gel surfaces. Apply the sample to the top of the stacking gel.

(viii) Run the gel at a constant voltage of 150 V for 4 h toward the anode at room temperature. The voltage gradient in the gel is 15 V/cm.

(ix) If marker proteins are desired for calibration of the second dimension, dissolve them in sample buffer and apply them to the origin of the first dimension approximately 20 min prior to the completion of electrophoresis. Stop electrophoresis after the marker proteins enter the separation gel.

(x) Remove the gel from the tube by carefully smashing the glass. On the other hand, gels with a polyacrylamide concentration below 10% can be easily removed from the tube by squirting a stream of water through a fine needle between the tube wall and the gel. The gel is now ready for the reduction and equilibration steps prior to the second dimension electrophoresis.

4.3.2 *Reduction and equilibration of first dimension gels*

After electrophoresis in the first dimension, the gel must be equilibrated in a buffer containing 2-mercaptoethanol to cleave the disulphide bonds of the cross-links, thereby converting them into monomers. After the reducing step, the gel is equilibrated in a buffer containing no reducing agents to remove the mercaptoethanol. The presence of residual reducing agent on the surface of the first dimension gel may inhibit its attachment to the second dimension gel; this leads to the appearance of horizontal streaks of proteins, instead of compact spots in the second dimension. Note that the pH of the washing buffer, in step (ii) below, is 6.8, not 8.8. This pH difference between the sample gel and the separation gel provides a stacking effect when the proteins start migrating into the second dimension slab gel, which does not have a stacking gel.

(i) Reduce the cross-links in the first dimension gel by placing it in a glass test tube containing 25 ml of a buffer consisting of 50 mM Tris–HCl, pH 8.8, 1% (w/v) SDS and 3% (v/v) 2-mercaptoethanol. Cover the tube with Parafilm, and heat it in a water bath at 65°C for 15 min.
(ii) Decant the reducing buffer and equilibrate the gel against three 50-ml changes of a buffer containing 50 mM Tris–HCl, pH 6.8, and 0.1% (w/v) SDS for a total of 45 min at room temperature with constant shaking. The gel is now ready to be embedded as the origin of the second polyacrylamide gel.

4.3.3 *Second dimension SDS–gel electrophoresis*

The apparatus for the second dimension gel is of the type described in Figure 3 of Chapter 3 (30) and consists of a sandwich made with two glass plates (length = 24 cm, width = 16 cm) separated by plastic spacers (1.5 or 3.0 mm), and clamped to a plexiglass electrophoresis chamber with upper and lower reservoirs for electrophoresis buffer. The upper edge of one of the plates may be bevelled to form a trough to facilitate the attachment of the first dimension tube gel to the second dimension slab gel. Recipes for the preparation of second dimension polyacrylamide–SDS gels are given in *Tables 3* and *4*.

The procedure for running the second dimension gel is as follows.

(i) Wash the glass plates with a mild detergent and rinse them with distilled water and ethanol prior to use. Dry them with a clean Kimwipe.
(ii) Assemble the glass plate sandwich.
(iii) Filter and de-gas the separation gel solution before use. Initiate polymerization by adding ammonium persulphate and TEMED.
(iv) Pour the separation gel between the glass plates. Immediately place the reduced gel from the first dimension on top of the separation slab gel. Before the second dimension slab gel solidifies, remove any air bubbles trapped under the sample gel by slightly lifting the gel with a thin spatula. Alternatively, the first dimension gel can be placed at the bottom of the upside-down glass plate sandwich, which is bounded by a horizontal spacer across the bottom, and the separation gel poured over it, thereby forming a bond between the two gels. Following polymerization, the slab is turned rightside-up and the spacer above the first dimension gel is carefully removed. This method is particularly effective if the first dimension gel is a lane from a slab gel instead of a tube gel. Polymerization is completed within approximately 30 min at room temperature.
(v) After polymerization, apply cementing solution (*Table 4*) to the top of the

gel to ensure proper bonding of the first dimension gel to the second dimension. Allow hardening of the agarose for 15 min.

(vi) Clamp the glass plate sandwich to the plexiglass electrophoresis unit. Fill the upper and lower reservoirs with electrophoresis buffer (*Table 4*).

(vii) Electrophoresis is toward the anode at a constant voltage of 120 V and room temperature for 16 h. The voltage gradient is approximately 4 V/cm. Electrophoresis is stopped when the bromphenol blue is about 1 inch from the bottom of the separation gel.

(viii) Place the gel slab in a tray containing 100 ml of a staining solution made of 0.04% (w/v) Coomassie blue R-250 in five volumes of methanol, five volumes of distilled water and one volume of acetic acid. Stain the gel for at least 8 h with gentle rocking. Soak the gel in 10% (w/v) TCA for 3 min to fix the protein spots. De-stain by gently rocking the gel in a tray containing 100 ml of a de-staining solution, made of 5% methanol and 7% acetic acid in distilled water, until the intensity of the spots is optimal relative to an almost clear background.

4.3.4 *Interpretation of the diagonal gel*

Figure 5 shows stained diagonal gels for ribosomal proteins from cross-linked 30S and 50S subunits. The patterns are very complex due to the large number of protein constituents in ribosomes, but many cross-links can be identified by the criteria mentioned previously and illustrated in *Figure 4B*: members of a cross-link fall on the same vertical line, since they arose from a single unique disulphide-linked complex in the first dimension; the sum of the apparent molecular weights of the two monomers is equal to that of the cross-link. For ribosomes many more cross-links are present than can be readily assigned from these diagonal gels of total cross-linked protein. This arises not only because of the number of components, but because many ribosomal proteins have the same or similar molecular weights and are not well resolved along the diagonal or on the vertical axis.

In most applications of the method to other more simple multi-protein structures, it should be possible to identify the monomeric components of cross-links directly from their position with respect to the diagonal line of total non-cross-linked protein or to proteins used to calibrate the second dimension SDS–gel electrophoresis.

Because of the complexity of the mixtures of cross-linked proteins from ribosomal subunits, procedures were developed to separate the mixtures into simpler fractions before diagonal gel analysis. Methods used in the analysis of *E.coli* ribosomes were successive extractions of the cross-linked ribosomal subunit with increasing concentrations of LiCl, and preparative electrophoresis of cross-linked protein mixtures in acid/urea polyacrylamide gels (2). Other methods for protein purification, such as ion-exchange chromatography, reverse-phase HPLC and isoelectric focusing, are also appropriate. The only prerequisites are that the procedures avoid the introduction of reducing agents and that the sample for diagonal gel analysis be relatively salt-free.

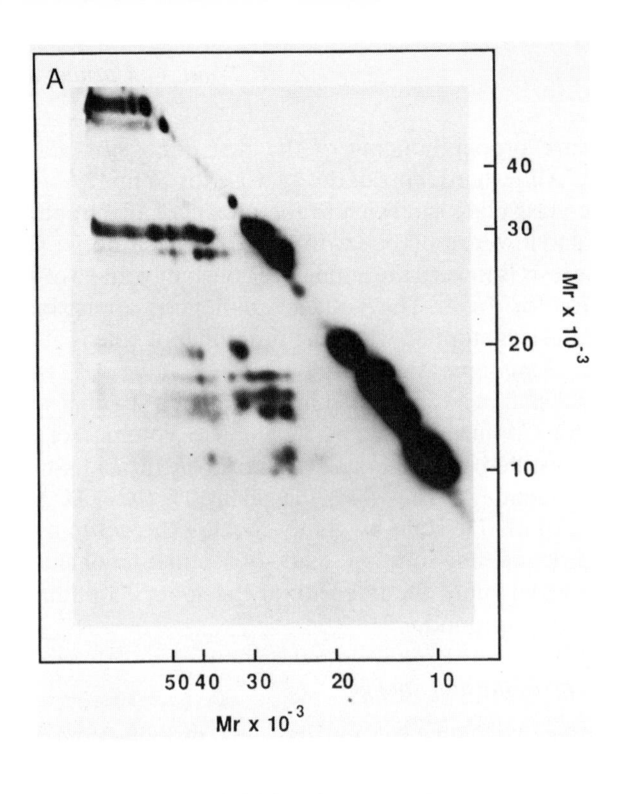

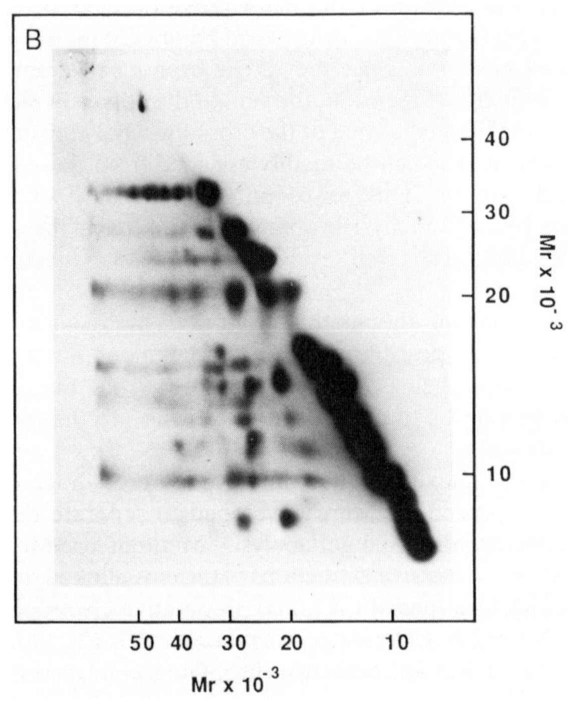

Figure 5. Diagonal gel patterns for ribosomal protein cross-linked with 2-iminothiolane. (**A**) Total protein from 30S ribosomal subunits. (**B**) Total protein from 50S ribosomal subunits.

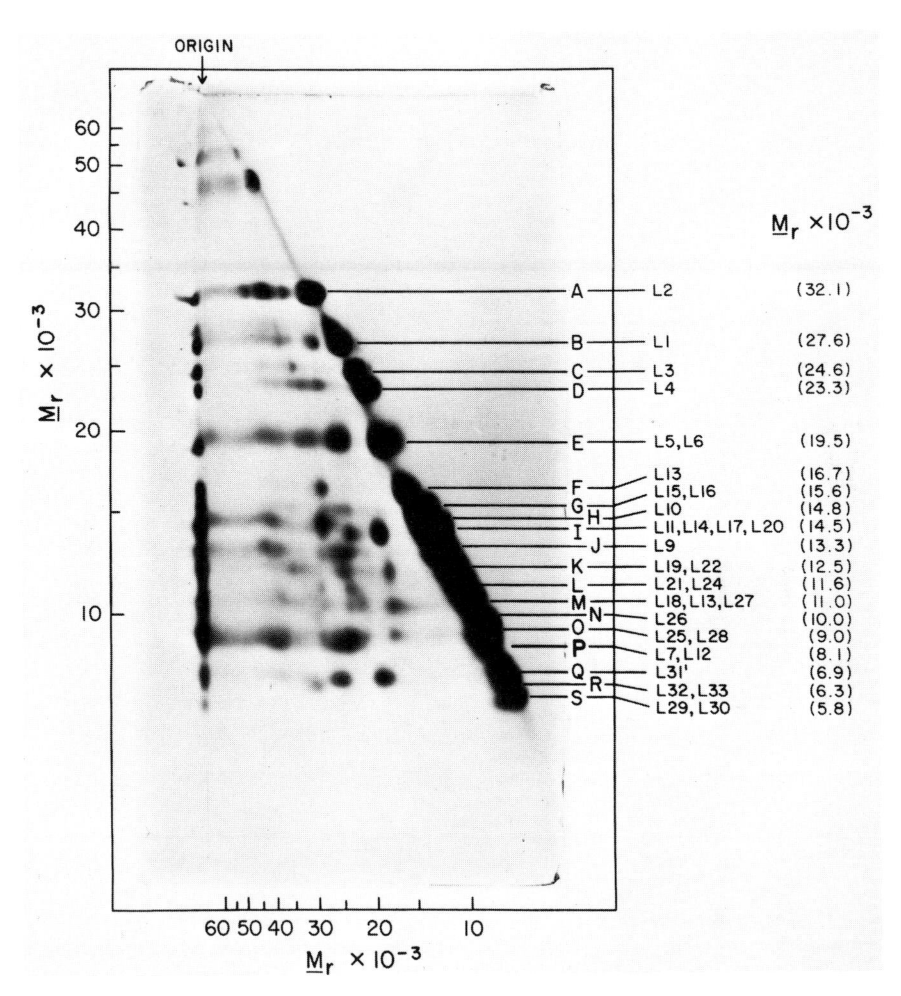

Figure 6. Diagonal gel pattern for total protein from cross-linked 50S ribosomal subunits showing the location and molecular weight of each component. 50S ribosomal proteins are numbered from L1 to L34.

Figure 6 shows a diagonal gel for total cross-linked 50S protein and indicates the location and molecular weights of each individual polypeptide component. *Figure 7* shows the results of diagonal gel analysis of two simplified fractions from the total cross-linked proteins. The assignments of the monomeric components marked by vertical arrows to the lettered zones along the diagonal are unambiguous and in many cases are sufficient to define specific cross-links. In other cases in which the lettered zone represents the mobility of two or more proteins, a further step must be taken to complete the identification. For the ribosomal studies, the stained spots in question were eluted from the diagonal gel, radioiodinated, mixed with total non-radiolabelled ribosomal protein and analysed by two-dimensional PAGE in systems capable of separating all the

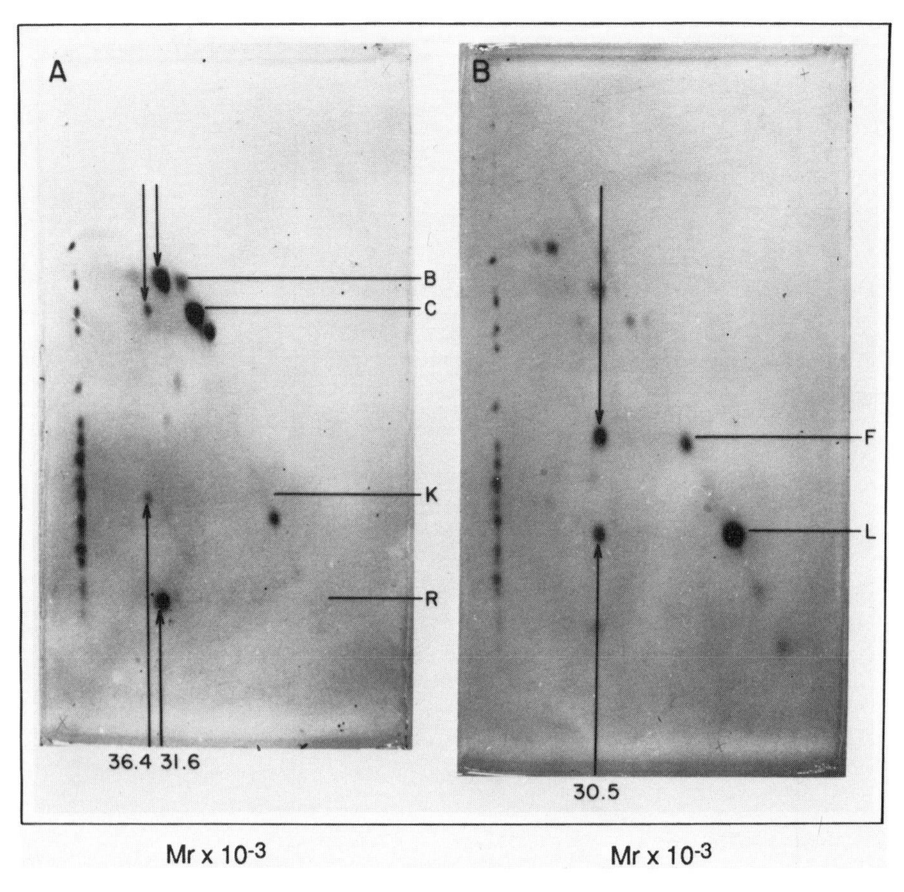

Figure 7. Diagonal gel patterns of two fractions purified from total cross-linked 50S proteins prior to electrophoresis. The vertical arrows indicate conspicuous protein pairs arising from a single cross-link and the letters correspond to the zones shown in *Figure 6*.

ribosomal proteins. These procedures (2) are pertinent specifically to ribosomes and consequently are beyond the scope of this discussion.

The diagonal gel patterns for the total 50S and 30S *E.coli* ribosomal proteins, although difficult to analyse in detail directly, are highly reproducible and represent characteristic 'fingerprints' of the overall protein topography of each ribosomal subunit. The protein patterns beneath the diagonal are independent of the method used to prepare the cross-linked samples, provided that the precautions suggested for avoiding disulphide interchange are taken. The majority of protein components have been identified in one or more cross-links, and this has led to the reconstruction of maps of the relative locations of most of the proteins in these complex organelles. These maps can be integrated with other biochemical, physical and electron microscopic results with which they correlate well, and which they significantly extend for types of ribosomal subunits not yet so extensively studied by other methods.

There are differences in the intensities of various spots below the diagonal that represent members of cross-linked complexes. The yield of no single cross-link exceeds 10%; the major fraction of any single polypeptide remains monomeric and lies on the diagonal. Explanations for the variable and limited yields of the different cross-links include:

(i) the use of cross-linking conditions optimized for dimer formation as opposed to higher aggregates;
(ii) differences in the abundance and reactivity toward 2-IT of lysine residues in different protein subunits;
(iii) the relative 'isolation' of certain modified lysines on one polypeptide from those on another;
(iv) possible compositional or conformational heterogeneity in the population of structures studied;
(v) competition between alternative cross-linking paths;
(vi) competition between inter-molecular cross-link formation and intra-molecular disulphide bond formation.

Nonetheless the two-step disulphide cross-linking method with 2-IT appears to produce a collection of cross-links that together represent the majority of protein proximity relationships in the ribosome structure.

5. ALTERNATE STRATEGIES FOR DISULPHIDE CROSS-LINKING COMBINING *IN VITRO* SITE-DIRECTED MUTAGENESIS AND PHOTO-ACTIVABLE CROSS-LINKING REAGENTS

5.1 Advantages and disadvantages of lysine-based cross-linking with homo-bifunctional reagents

Lysine residues are relatively abundant in proteins and relatively exposed. The use of reagents that react with lysine takes advantage of this and provides a fairly general method for cross-linking proximal protein subunits. On the other hand, it is still possible that certain proteins may lack lysines appropriately situated for cross-link formation.

Cross-linking may not only be used to determine which protein subunits are proximal in a multi-protein structure, but also to define at higher resolution the contact surface between adjacent subunits, by identifying the cross-linked amino acid residues. The abundance of lysine residues, the presence of some lysines that may have reacted mono-functionally but not formed a cross-link, and the presence of intra-molecular cross-bridges make the identification of cross-linked sites based on lysine derivatives technically difficult. A much simpler way to identify the precise site of cross-linking would be to utilize reagents with reactivity toward a less abundant amino acid or one whose location is readily established.

Cysteine residues in proteins are ideal for inter-molecular cross-linking when the goal is to establish within the primary structures of neighbouring protein subunits those regions that constitute the contact domains. Two proximal cysteines may form a disulphide cross-link by oxidation, or they may react with cysteine-specific bifunctional reagents. The relatively low abundance of cysteine residues would simplify identification of the site of cross-link formation. On the other hand this scarcity of cysteine keeps it from serving as the basis of a general approach for mapping the location of protein subunits. These difficulties can be overcome in two ways. First, site-directed mutagenesis can be used to place a cysteine residue at any location desired (Chapter 11). Second, the introduced cysteine may serve as a specific site for attaching a sulphydryl-specific hetero-bifunctional reagent. If the second functional group of such a reagent is photo-activable, it will react generally with any proximal protein subunit regardless of the amino acid composition of its contact surface.

5.2 **Cysteine site-directed mutagenesis**

Oligonucleotide-directed mutagenesis is described elsewhere in this volume (Chapter 11). The placement of cysteine residues at specific locations in proteins through the alteration of cloned DNA sequences, termed cysteine mutagenesis (30), and subsequent measurement of intra-peptide disulphide bond formation has been applied to studies of protein folding (31) and the thermal stability and proteolytic resistance of proteins (32). Inter-peptide disulphide bond formation between pairs of introduced cysteines can be used to study the topography of oligomeric proteins or of multi-protein structures, and possible conformational changes therein (30). The application of cysteine mutagenesis to the dimeric ribosomal protein L7/L12 will be described as an example of this approach. Whether or not a disulphide bond is formed between members of the dimer provides information on the structure of the dimer. Attachment of photo-activable, cleavable hetero-bifunctional cross-linking reagents to the introduced sulphydryl, followed by binding of the modified mutant protein to the ribosome (receptor) and photo-cross-linking, provides a method for determining the location of a specific region of the protein with respect to other ribosomal proteins with which it interacts.

5.2.1 *Introduction of a sulphydryl into ribosomal protein L7/L12 by cysteine site-directed mutagenesis*

E.coli ribosomal protein L7/L12 of the large ribosomal subunit is unique in its presence in four copies, constituting two dimers per ribosome. The protein is evolutionarily conserved, and homologous multi-copy, acidic proteins are found in ribosomes from other bacterial, archaebacterial and eukaryotic sources. The protein is involved in the translocation step of protein synthesis. The two dimers occupy different locations: one or both of them constitutes a uniquely mobile structural element in the ribosome, and L7/L12 undergoes protein–protein cross-linking to other proteins both within and beyond the immediate common site at which the two dimers are anchored by their N-terminal domains to the particle. Previous cross-linking studies involved the random modification of ribosomes with 2-IT and oxidative disulphide cross-linking. To define more precisely within the L7/L12 structure the site of cross-link formation, a serine to cysteine substitution was made by oligonucleotide-directed mutagenesis of the gene for L7/L12, *rpl* L. Since wild-type L7/L12 has no cysteine residues, the mutant protein contains a unique sulphydryl for disulphide cross-linking and for derivatization with sulphydryl-specific hetero-bifunctional reagents. Residue 89, in the C-terminal globular domain of L7/L12, was chosen for cysteine mutagenesis. When the mutant gene replaced the wild-type, there was no effect on the growth rate of the cells. Protein L7/L12 represents a useful model system for an investigation by cross-linking of binding site/receptor components because it can readily be dissociated from and re-associated with the ribosome.

5.2.2 *Oxidation to form disulphide cross-links at specific sites*

The two residue 89 positions in the members of a dimer appeared from the crystal structure to be sufficiently exposed so as to permit disulphide formation with other proximal proteins, but near enough that intra-dimer disulphides might also form. This was tested by oxidation of purified L7/L12Cys89 both in solution and bound to ribosomes. Purified L7/L12Cys89 forms covalently cross-linked homo-dimers, as seen by SDS–PAGE, upon oxidation with either H_2O_2 or with a complex of Cu(II) and 1,10-phenanthroline. The procedure for the oxidative cross-linking of the purified protein is detailed below.

(i) Resuspend lyophilized L7/L12Cys89 at 3 mg/ml in 50 mM sodium phosphate, pH 8.0. Add 10 μl of 2-mercaptoethanol (final concentration 0.14 M) to each ml of the protein solution and incubate at 30°C for 30 min. Remove excess reducing agent by gel filtration (de-salting) on Bio-gel P-2 equilibrated with 50 mM sodium phosphate, pH 8.0, without reducing agent. Follow the separation of the reduced protein from mercaptan spectrophotometrically at A_{595} using the Bradford assay for protein (33) and at A_{412} using the Ellman assay for mercaptan (35). Combine the fractions containing protein and place them on ice. Determine the protein concentration and, if necessary, adjust the concentration to 1 mg/ml with 50 mM sodium phosphate, pH 8.0.

(ii) Place 10 μg (10 μl) of protein into each of three microcentrifuge tubes. Tube 1 is an unoxidized control. Add 1.4 μl of 1% H_2O_2 (prepared by mixing 2 μl of 30% H_2O_2 with 60 μl of double-distilled water immediately before use) to Tube 2. The final H_2O_2 concentration is 0.12% (35 mM). Add 1.5 μl of a solution of 10 mM $CuSO_4 \cdot 5H_2O$ and 30 mM 1,10-phenanthroline (31) to Tube 3. The final concentration is 1.3 mM Cu(II)(1,10-phenanthroline)$_3$. Place the tubes on ice.

(iii) Remove 3 μg (3 μl) from each tube at 15, 30 and 60 min intervals and immediately add an equal volume of SDS–gel sample buffer, containing 40 mM iodoacetamide (*Table 4*), to each and incubate at 65°C for 15 min. Analyse the proteins by SDS–PAGE in gels containing 15% acrylamide. Lanes containing oxidized proteins are marked by the appearance of covalently linked L7/L12Cys89 dimers, which are not present in the unoxidized controls. Under the conditions used, the Cu(II) complex produced a slightly higher yield of cross-linked L7/L12Cys89 dimer than

H_2O_2, and the oxidation appeared to be nearly complete in both cases within 15 min.

In an attempt to generate disulphide cross-links between L7/L12Cys89 and the endogenous sulphydryls of other ribosomal proteins, a similar experiment was carried out with ribosomes containing four copies of L7/L12Cys89. Ribosomes containing either L7/L12w$^+$ or L7/L12Cys89 were oxidized with H_2O_2 as described in Section 4.1 for ribosomes modified with 2-IT. Proteins from the oxidized ribosomes were separated by one-dimensional SDS–gel electrophoresis, blotted onto nitrocellulose and probed (Chapter 3) with a monoclonal antibody to L7/L12. An L7/L12 dimer was found in high yield in oxidized ribosomes containing L7/L12Cys89, but was absent from ribosomes containing L7/L12w$^+$ (*Figure 8B*, lanes 2 and 3). The L7/L12Cys89 dimer was the only cross-linked species detected by this method, indicating that the L7/L12Cys89 sulphydryls do not lie in close proximity to the endogenous sulphydryls of other ribosomal proteins.

5.2.3 *Coupling of hetero-bifunctional photo-activable cross-linkers to the cysteine site-directed mutant protein*

The introduced cysteine sulphydryl group serves as a site for reaction with hetero-bifunctional reagents like those shown in *Table 1B*. As shown schematically in *Figure 2*, the cysteine-SH reacts by disulphide interchange to displace its leaving group. Photoreactive reagents, as their name implies, become highly reactive when excited by light of the appropriate wavelength. The wavelength of excitation of aryl azides depends on the nature of the ring substituents and the local chemical environment of the azide, and is generally between 250 and 400 nm. It is important to choose conditions for the use of photoreagents that do not destroy the photolabile function until the appropriate time. Complete darkness, illumination from red darkroom safety lights and indirect lighting have been suggested for the use of particular photoreagents. The compounds APTP and APDP (a generous gift of the Pierce Chemical Co.) (see *Table 1B*) have been found to be stable in acetonitrile for at least 5 days when exposed to normal laboratory fluorescent light (avoiding sunlight) in a sealed borosilicate glass tube. Similar stability has been reported for other compounds (35), and this feature of the reagents increases their practicality. The stability of new reagents should be experimentally determined before their use, as should the conditions for photolysis of the azide. The stability of photoreagents, as well as the conditions for their photolysis, can be determined experimentally by following the UV absorbance profile of the reagent under various conditions of storage and irradiation. For example, APTP and APDP have an absorbance maximum around 270 nm. Decomposition of the azide can be readily followed by monitoring decreases in absorbance of the reagent, or protein derivatized with the reagent, at this wavelength. Photolysis can be carried out with a variety of light sources. The Mineralight UVGL-58 hand-held light source provides either short-wave UV (254 nm) or long-wave UV (366 nm) light. The short-wave lamp is closer to the wavelength of excitation of most aryl azides and may give the most efficient

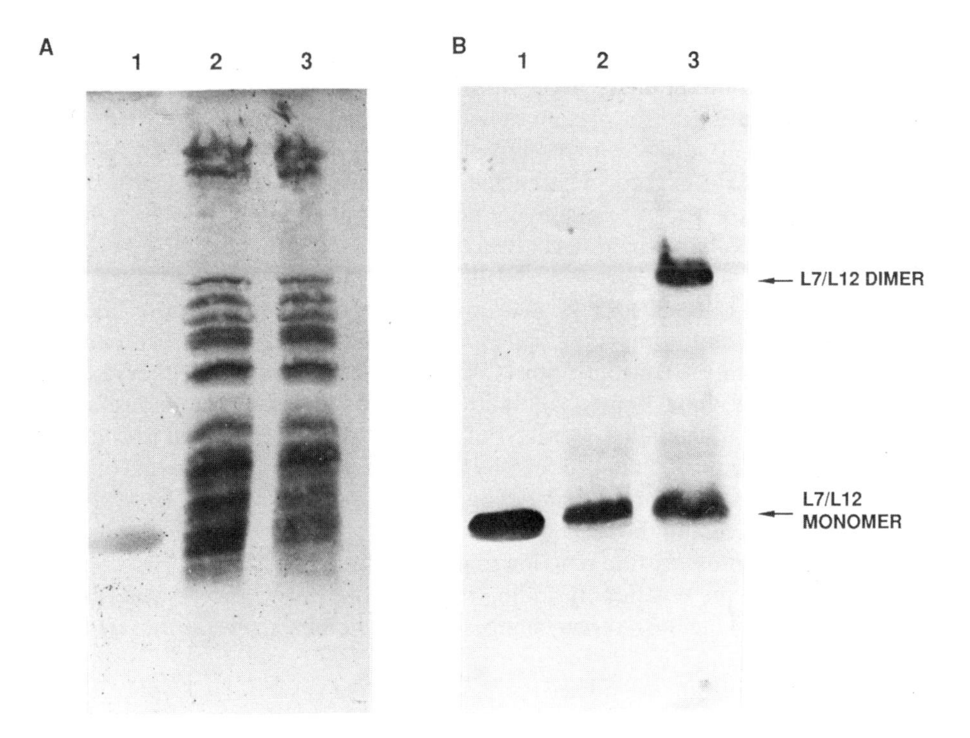

Figure 8. Disulphide cross-links introduced between cysteine residues. (**A**) amido black-stained blot of a 15% polyacrylamide–SDS gel. **Lane 1**, L7/L12Cys89. **Lane 2**, total proteins from 70S ribosomes containing four copies of L7/L12w$^+$, oxidized with H_2O_2.**Lane 3**, total proteins from 70S ribosomes containing four copies of L7/L12Cys89, oxidized with H_2O_2. (**B**) immunoblot of a gel identical to that described in **a**, probed with a mouse monoclonal antibody to L7/L12 and a second antibody (goat anti-mouse) coupled to horseradish peroxidase. Colour was developed with 4-chloro-1-naphthol.

photolysis of the azide, but may also cause UV-induced damage to biological compounds. The long-wave lamp is less likely to cause damage, but requires longer times of irradiation to cause photolysis. Irradiation at 302 nm has also been performed with a UV lightbox (Transilluminator, by Ultra-Violet Products Ltd) inverted directly over samples in small glass Petri plates or, for samples of less than 100 μl, in microcentrifuge tubes.

The procedure for the modification of L7/L12Cys89 with APTP (14) is as follows.

(i) All procedures are carried out under normal laboratory fluorescent lights unless stated otherwise, avoiding exposure of the photoreagents to sunlight. Resuspend lyophilized L7/L12Cys89 at 2 mg/ml in 50 mM sodium phosphate, pH 8.0. Determine the protein concentration by Bradford assay.

(ii) Completely reduce the protein by adding 10 μl of 2-mercaptoethanol per ml of protein solution (final concentration 0.14 M) and incubate at 30°C for 30 min. Remove excess reducing agent by gel filtration (de-salting) as described in Section 5.2.2 (i). Pool fractions containing protein, place the combined fractions on ice and determine the protein concentration. The final protein concentration should be about 0.6 mg/ml (~50 μM L7/L12Cys89); it should not exceed 80 μM for reasons detailed in (v) below.

(iii) Prepare a 10 mM stock solution of APTP by dissolving 1 mg of APTP in 340 μl of HPLC-grade acetonitrile. Determine the exact concentration of reagent from the A_{268} of a 500× dilution of the reagent into acetonitrile, using the molar extinction coefficient of 1.9×10^4 M^{-1} cm^{-1} (14).

(iv) Add 25 μl of HPLC-grade acetonitrile per ml of protein solution.

(v) Add 25 μl of 10 mM APTP in acetonitrile/ml of protein solution. The final APTP concentration is 0.25 mM, a 5-fold molar excess over protein. The final acetonitrile concentration is 5% (v/v). APTP (and many other reagents of this type) have limited solubilities (up to ~400 μM) in aqueous buffers that contain a small percentage of organic solvent. For protein solutions more concentrated than about 80 μM, modification must be carried out with either less than a 5-fold molar excess of reagent over protein, or with more organic solvent in the reaction mixture.

(vi) Allow the modification reaction to proceed for 2 h at 25°C.

(vii) Terminate the reaction by adding 1 volume of 6% (v/v) acetic acid, dialyse the protein against three changes of 500 volumes 6% acetic acid and lyophilize it.

(viii) Resuspend the lyophilized protein in 10 mM Tris–HCl, pH 7.4, to approximately 1 mg/ml, and determine the protein concentration. The extent of modification by APTP can be estimated from the absorbance of the modified protein at 268 nm compared with the unmodified protein. The molar extinction coefficient of modified protein at 268 nm (1.1×10^4 $M^{-1}cm^{-1}$) was determined with cytochrome *c* (14), and may vary with other proteins; hence, it only approximately measures the extent of incorporation of the photoreagent. If the photoreagent is radioactive, incorporation into protein can be monitored directly. Alternatively the extent of modification can be monitored by sulphydryl titration with [^{14}C]iodoacetamide, assaying for the loss of protein sulphydryls.

5.2.4 *Photo-cross-linking*

Because L7/L12 can be readily dissociated from and re-associated with the ribosome, a system composed of L7/L12 and L7/L12-deficient ribosomes is representative of the ligand–receptor complex described in *Figure 1*. L7/L12Cys89, modified at its single cysteine with APTP, has been re-bound to ribosomes lacking L7/L12 and irradiated with UV light to induce cross-link formation. The modified protein serves as a site-specific cross-linking probe of the binding site components for L7/L12 on the ribosome. 50S ribosomal subunits lacking only ribosomal protein L7/L12 are prepared by the method of Hamel *et al.* (37), with modifications to the procedure as reported by Tokimatsu *et al.* (38).

The photo-cross-linking procedure is given below.

(i) Incubate L7/L12-deficient 50S subunits in Buffer 1 (*Table 2*) at 37°C for 15 min. Remove excess reducing agent by gel filtration on Sephadex G-25 equilibrated in the same buffer without reducing agent. Determine the ribosome concentration by A_{260} (1 mg/ml = 15 A_{260}, equivalent to 0.33 mg/ml ribosomal protein).

(ii) Add an 8-fold molar excess of APT-L7/L12Cys89 in 10 mM Tris–HCl, pH 7.4, to the ribosomes (67 μg protein/mg ribosomes) and incubate at 37°C, 15 min.

(iii) Layer the reconstituted ribosomes over a 10% sucrose cushion in Buffer 1 without reducing agent (2 ml of ribosomes over 3 ml of cushion) in a thick-wall polycarbonate ultracentrifuge tube for a Beckman Ti 65 rotor (tube half full). Pellet the ribosomes at 58 000 r.p.m., 2 h, 4°C. Carefully remove the supernatant, which contains unbound APT-L7/L12Cys89.

(iv) Resuspend the ribosome pellet in Buffer 1 without reducing agent to approximately 10 mg/ml. Clarify the solution by centrifugation for 20 min, 7000 r.p.m., 4°C in a Sorvall SA-600 rotor. Keep the supernatant and discard the pellet. Determine the ribosome concentration in the supernatant and dilute the ribosomes to 3 mg/ml with Buffer 1 without mercaptoethanol.

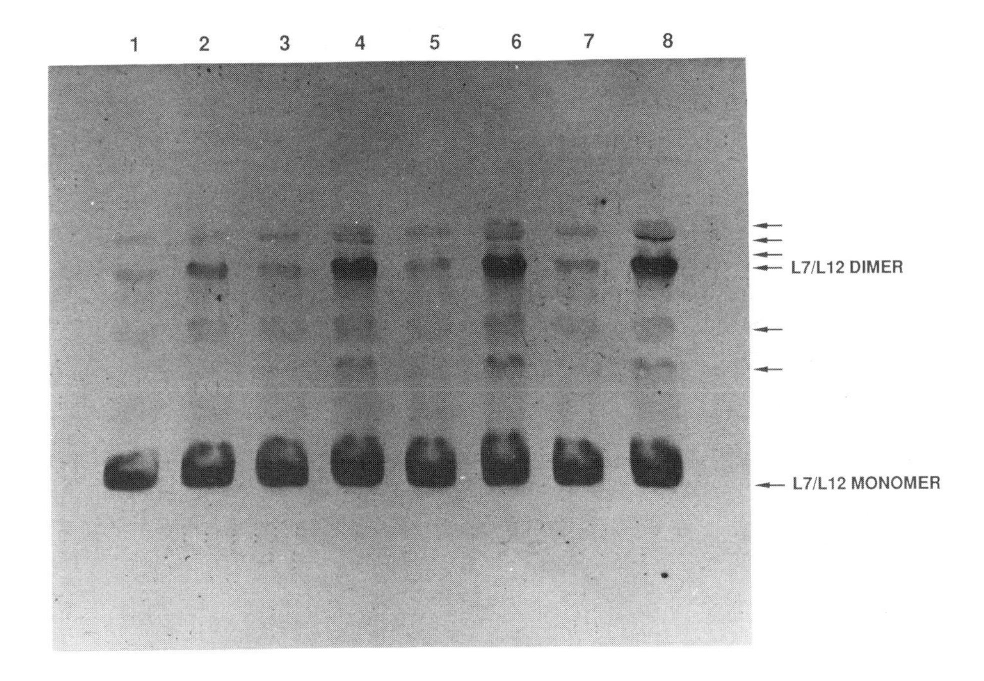

Figure 9. Immunoblot of a 15% polyacrylamide–SDS gel showing UV-induced cross-links formed between APT-L7/L12Cys89 and other ribosomal proteins. **Lanes 1, 3, 5** and **7** contain total 50S proteins from subunits with four copies of L7/L12w^{+} that were irradiated at 302 nm for 0, 5, 10 and 20 min, respectively. **Lanes 2, 4, 6** and **8** contain total 50S proteins from subunits with four copies of L7/L12Cys89 that were also irradiated at 302 nm for 0, 5, 10 and 20 min, respectively. The blot was probed with a monoclonal antibody to L7/L12 as described in *Figure 8B*. Arrows indicate the position of L7/L12, an L7/L12 cross-linked dimer and other cross-links between APT-L7/L12Cys89 and, as yet unidentified, ribosomal proteins.

(v) Place 4–8 μl samples of reconstituted ribosomes (24 μg) in 1.5-ml microcentrifuge tubes. Place the tubes, along with similar samples of control 50S subunits containing unmodified L7/L12, below the filter of a 302 nm light source (the filter dimensions are 15.5 × 36 cm). The tubes must be held tightly in a rack parallel to and directly beneath the lamp of the light source to ensure uniform irradiation of the samples. Irradiate the samples at 302 nm for 0, 5, 10 and 20 min, at a distance of 3.5 cm (the height of a microcentrifuge tube). Larger samples can be stirred with a magnetic stir bar in beakers and irradiated from above, or placed in quartz cuvettes and irradiated with a 'hand-held' light source (see Section 5.2.3) securely clamped to a stand with irradiation perpendicular to the clear surface of the cuvettes.

(vi) Add an equal volume of SDS–gel sample buffer containing 40 mM iodoacetamide (*Table 4*) and heat the samples at 65°C for 15 min. Analyse the proteins by SDS–PAGE in gels containing 15% acrylamide.

(i) *Detection of cross-links.* Cross-links between APT-L7/L12Cys89 and other ribosomal proteins are detected by SDS–PAGE and immunoblotting (Chapter 3) using monoclonal antibodies against L7/L12 (*Figure 9*). Several bands representing cross-links that contain L7/L12 and other ribosomal proteins, as well as the L7/L12 dimer, can be seen (lanes 2, 4, 6 and 8). Cross-link formation is nearly complete after 5 min of irradiation. None of these cross-links was formed in irradiated control ribosomes (lanes 1, 3, 5 and 7). Treatment of the UV-generated cross-links with SDS sample buffer containing 0.1 M dithiothreitol (DTT) (in place of iodoacetamide) at room temperature for 12 to 16 h (overnight), followed by incubation at 65°C for 15 min, completely reduced the cross-links. Diagonal electrophoresis, or the use of [^{125}I] APDP as outlined in *Figure 1*, can complete the identification of the binding site proteins.

6. ACKNOWLEDGEMENTS

Our sincere thanks to J. Haudenshield, Dr B. Perroud, Dr J. Gardner and Dr V. Williamson for helpful comments and suggestions. This work was supported by a grant from the US Public Health Service (GM17924).

7. REFERENCES

1. Sommer, A. and Traut, R. R. (1974) *Proc. Natl. Acad. Sci. USA,* **71**, 3946.
2. Kenny, J. W., Lambert, J. M. and Traut, R. R. (1979) In *Methods in Enzymology.* Hoblaue, K. and Grossman, L. (eds), Academic Press, New York, Vol. 59, p. 334.
3. Peters, K. and Richars, F. M. (1977) *Annu. Rev. Biochem.,* **46**, 523.
4. Das, M. and Fox, C. F. (1979) *Annu. Rev. Biophys. Bioeng.,* **8**, 165.
5. Ji, T. H. (1983) In *Methods in Enzymology.* Hirs, C. H. W. and Timasheff, S. N. (eds), Academic Press, New York, Vol. 91, p. 580.
6. Lundblad, R. V. and Noyes, C. M. (1983) *Chemical Modification of Proteins.* Vol. II, CRC Press.
7. Han, K.-K. Richard, C. and Delacourte, A. (1984) *Intl. J. Biochem.,* **16**, 129.
8. Wang, K. and Richards, F. M. (1974) *J. Biol. Chem.,* **249**, 8005.
9. Jue, R., Lambert, J. M., Pierce, L. R. and Traut, R. R. (1978) *Biochemistry,* **17**, 5395.
10. Vanin, E. F. and Ji, T. H. (1981) *Biochemistry,* **20**, 6754.
11. Kiehm, D. J. and Ji, T. H. (1977) *J. Biol. Chem.,* **252**, 8524.

12. Schwartz, M. D. (1985) *Anal. Biochem.*, **149**, 142.
13. Denny, J. B. and Blobel, G. (1984) *Proc. Natl. Acad. Sci. USA*, **81**, 5286.
14. Moreland, R. B., Smith, P. K., Fujimoto, E. K. and Dockter, M. E. (1982) *Anal. Biochem.*, **121**, 321.
15. Ultee, M. E. and Basch, R. J. (1985) *Anal. Biochem.*, **149**, 331.
16. Chong, P. C. S. and Hodges, R. S. (1981) *J. Biol. Chem.*, **256**, 5071.
17. Traut, R. R., Bollen, A., Sun, T. T., Hershey, J., Sundberg, J. and Pierce, L. R. (1973) *Biochemistry*, **12**, 3266.
18. Traut, R. R., Lambert, J. M., Boileau, G. and Kenny, J. W. (1980) In *Ribosomes: Structure, Function and Genetics*. Chambliss, G., Craven, G. R., Davies, J., Davis, K., Kahan, L. and Nomura, M. (eds), University Park Press, Baltimore, p. 89.
19. Traut, R. R., Tewari, D. S., Sommer, A., Gavino, G. R., Olson, H. M. and Glitz, D. G. (1985) In *Structure, Function and Genetics of Ribosomes*. Hardesty, B. and Kramer, G. (eds), Springer-Verlag, New York, p. 286.
20. Thomas, J. O. and Kornberg, R. D. (1975) *FEBS Lett.*, **58**, 353.
21. Cook, G. R., Yau, P., Yasuda, H., Traut, R. R. and Bradbury, E. M. (1986) *J. Biol. Chem.*, **261**, 16185.
22. Cover, J. A., Lamber, J. M., Norman, C. M. and Traut, R. R. (1981) *Biochemistry*, **20**, 2843.
23. Klotz, I. M. and Heiney, R. E. (1962) *Arch. Biochem. Biophys.*, **96**, 605.
24. White, F. H. Jr. (1972) In *Methods in Enzymology*. Hirs, C. H. W. and Timasheff, S. N. (eds), Academic Press, New York, Vol. 25, p. 541.
25. Carlsson, J., Drevin, H. and Axen, R. (1978) *Biochem. J.*, **173**, 723.
26. Perham, R. N. and Thomas, J. O. (1971) *J. Mol. Biol.*, **62**, 415.
27. King, T. P., Li, Y. and Kochoumian, L. (1978) *Biochemistry*, **17**, 149.
28. Lambert, J. M., Jue, R. and Traut, R. R. (1978) *Biochemistry*, **17**, 5406.
29. Howard, G. A. and Traut, R. R. (1974) In *Methods in Enzymology*. Moldave, K. and Grossman, L. (eds), Academic Press, New York, Vol. 30, p. 526.
30. Falke, J. J. and Koshland, D. E. (1987) *Science*, **237**, 1596.
31. Creighton, T. E. (1986) In *Methods in Enzymology*. Hirs, C. H. and Timasheff, S. N. (eds), Academic Press, New York, Vol. 131, p. 83.
32. Wells, J. A. and Powers (1986) *J. Biol. Chem.*, **261**, 6564.
33. Bradford, M. (1976) *Anal. Biochem.*, **72**, 248.
34. Ellman, G. L. (1958) *Arch. Biochem. Biophys.*, **74**, 443.
35. Chong, P. C. S. and Hodges, R. S. (1981) *J. Biol. Chem.*, **256**, 5064.
36. Hamel, E., Koka, M. and Nakamoto, T. (1972) *J. Biol. Chem.*, **247**, 805.
37. Tokimatsu, H., Strycharz, W. A. and Dahlberg, A. E. (1981) *J. Mol. Biol.*, **152**, 397.

CHAPTER 6

Determining the roles of subunits in protein function

EDWARD EISENSTEIN and HOWARD K. SCHACHMAN

1. INTRODUCTION

Why so many proteins are composed of subunits has proved to be a perplexing question that has stimulated hundreds of investigations aimed at delineating the structural and functional roles of subunits in oligomeric proteins. As a result of the pioneering ultracentrifuge studies of Svedberg and his colleagues, it was recognized 50 years ago that large numbers of proteins are composed of subunits (1). Indeed, research on oligomeric proteins has been so extensive that it is difficult to find a comprehensive discussion of the quaternary structure of proteins subsequent to the valuable review by Klotz *et al.* (2), which summarizes the stoichiometric constitution, the geometrical arrangements of subunits, the energetics of self-assembly, the communication between subunits and functional aspects of oligomeric proteins. Because the structure, stability and function of oligomeric proteins are determined by their constituent subunits and their interactions, investigations of the kinetics and thermodynamics of their inter-subunit interactions are crucial to our understanding of how the three-dimensional quaternary structures endow these proteins with their unique biological properties.

For many oligomeric enzymes composed of dissimilar polypeptide chains, such as lactose synthase and aspartate transcarbamoylase from *Escherichia coli*, we have clear answers to the question: why subunits? In such enzymes, and there are many, additional functions such as new activities or regulatory properties accrue as the result of the formation of the complexes from the individual proteins. With other oligomeric enzymes, however, containing identical polypeptide chains, such as aldolase and glyceraldehyde-3-phosphate dehydrogenase, unequivocal explanations for their existence as oligomers is lacking. One can postulate that the individual monomers fold into a compact tertiary structure having non-polar surfaces leading to insolubility and that the formation of tetramers in those cases leads to soluble functional enzymes. The formation of oligomers from monomers would, of course, lead to larger proteins thereby diminishing their permeability through membranes and also decreasing the osmotic pressure in various organelles containing large concentrations of protein. For many proteins, oligomer formation doubtless would confer enhanced stability relative to their monomeric constituents. In some cases the interactions between subunits in active oligomers may cause conformational changes in the monomers, which by themselves might not fold into active conformations. Recent investigations of some oligomeric enzymes have shown that the active sites are located at the interfaces between the polypeptide chains and that catalytic activity requires the joint participation of amino acid residues from each of the adjacent polypeptide chains. For such enzymes, the monomers are intrinsically inactive because the 'shared active site' is lacking, and the folded monomer even with the 'correct' tertiary structure can be considered as having a 'split active site'. If one considers the biosynthesis of proteins starting from DNA replication through transcription and translation to form the polypeptide chains which then fold and associate to yield functional proteins, it seems obvious that making oligomeric proteins from shorter polypeptide chains is more economical in the use of genetic information than making much longer polypeptide chains. Hence with oligomeric enzymes comprising multiple copies of the same chain or even of different polypeptide chains, there will be an increase in the fidelity of forming active proteins because of the reduction in the effects of errors in the replication, transcription and translation processes. When one recognizes that oligomeric proteins can disproportionate among themselves so that some molecules will be composed completely of fully competent chains, rather than mixtures of 'good' and 'bad' chains, another potential explanation of the large abundance of oligomeric proteins is at hand.

Multi-subunit proteins vary tremendously in complexity, ranging from dimers of identical chains, designated as α_2, and dodecamers of non-identical chains, described as an $\alpha_6\beta_6$ structure, to even more complex holo-enzymes such as H^+-ATPase which contains eight different types of polypeptide chains with the stoichiometry, $\alpha_3\beta_3\gamma\delta\chi\psi_2\omega_{10}$ (3). Determining the role of subunits in protein function is a difficult task because of the great diversity among oligomeric proteins. Hence we will focus on a variety of methods that have provided valuable information about the functional roles of the different subunits in aspartate transcarbamoylase (ATCase) of *E.coli*. ATCase can be considered an enzyme of intermediate complexity composed of six copies each of two types of polypeptide chains. The six catalytic chains, *c*, are organized in the holo-enzyme as two catalytic trimers, *C* or c_3, and the six regulatory chains are arranged as three regulatory dimers, *R* or r_2, to yield a structure, C_2R_3 or $(c_3)_2(r_2)_3$ having one 3-fold and three 2-fold axes of symmetry. ATCase serves as a useful model for investigations of the role of subunits in protein function because, at the holo-enzyme level, the effects of different types of subunits can be studied, and with the isolated *C* trimers or *R* dimers, attention can be focused on oligomers of identical chains. The questions we ask and the experimental approaches presented to answer them can be applied readily with suitable modification to other oligomeric proteins of lesser or greater complexity. Indeed, fruitful studies have been conducted on many other oligomeric enzymes, such as glutamine synthetase, pyruvate dehydrogenase and transcarboxylase, and the various aspects of the role of subunits in protein function illustrated here could easily have been drawn from investigations on other enzymes. Both because of convenience and because of our desire to present a coherent picture on a single enzyme, we have focused primarily on ATCase.

2. EXPERIMENTAL APPROACH

Most of the experimental approaches aimed at determining the functional roles of subunits in proteins are dependent upon the ability of the worker to dissociate the oligomers, separate and characterize the individual subunits, modify one or the other chemically, and then reconstitute complexes having structures similar to that of the native oligomeric protein. With many oligomeric enzymes these individual steps can be accomplished in high yield, but it must be emphasized that a procedure that is effective and efficient for one protein may be inappropriate for another. Therefore, a vast array of dissociating agents such as urea, guanidine hydrochloride (GdmCl), and NaCNS must be tried with each protein, and reconstitution experiments must be attempted with variations in ionic strength, temperature, pH and protein concentration.

The vast majority of oligomeric proteins containing two or more polypeptide chains are formed by non-covalent interactions between the individual subunits. In some cases, with antibodies such as IgG and IgM being the most notable examples, disulphide bonds are implicated as well as non-covalent attractive inter-chain interactions in the formation, stability and function of the intact protein. For such proteins, some of the procedures outlined below will need modification, since the mere rupture of non-covalent interactions will not suffice to produce the constituent subunits. Antibodies differ from most other oligomeric proteins in one other important regard. Oligomeric enzymes generally show a one-to-one correlation between the number of active sites and the number of polypeptide chains in the holo-enzyme whereas IgG, composed of two heavy and two light polypeptide chains, has only two binding sites for haptens or groups on the antigen.

For our purposes, the individual polypeptide chain represents the smallest subunit in an oligomeric protein and, in terms of nomenclature, it doesn't matter whether the chains associate to form oligomers only through non-covalent interactions, as in the great majority of multi-subunit proteins, or whether disulphide bonds are also implicated as 'cross-links'. In the case of oligomers containing different types of polypeptide chains, it is of use to consider the term, protomer, which was defined by Monod *et al.* (4) as the minimum-sized subunit that upon association with an integral number of identical subunits generates the quaternary structure of the oligomer. Hence, for a multi-subunit protein composed of identical polypeptide chains, the terms subunit, monomer and protomer are interchangeable. In contrast, a protomer in haemoglobin would be an $\alpha\beta$ combination whereas the subunits or monomers would be α and β polypeptide chains. Similarly for ATCase, the protomer would be a $c{:}r$ combination even though structurally the c chains form C or c_3 trimers and the r chains form R or r_2 dimers and no combination of individual c and r chains has as yet been detected.

2.1 **Basic considerations**

The ease with which a given oligomer can be dissociated into its constituent subunits depends upon both the strength of the interaction between the subunits and the architecture of the oligomer. For dimers of dissimilar chains we can represent the equilibrium between subunits and the complex as

$$A + B \rightleftharpoons A - B$$

which is characterized by the equilibrium constant for dissociation

$$K_d = [A][B]/[A - B]$$

Depending upon the magnitude of K_d and the protein concentration, we might expect to find primarily subunits and little complex or, alternatively, almost exclusively the complex and undetectable amounts of subunits. The former will generally be observed if K_d is as great as 10^{-3} M, whereas the latter obtains when K_d is 10^{-10} M or lower. These same considerations apply for oligomers composed of identical chains. Only dimers would be present in significant quantities at most reasonable protein concentrations if K_d is 10^{-10} M. In contrast, monomers would be the predominant species at most protein concentrations if K_d is 10^{-3} M. In practice, most workers consider protein concentrations in terms of mg/ml; obviously for the use of these elementary considerations the weight/volume concentrations must be translated into molar concentrations.

The dependence of the dissociation constant on experimental conditions such as temperature, pressure, ionic strength and the presence of co-factors will influence dramatically the ability to isolate the complex or the individual subunits. This complication has been illustrated most vividly during the isolation and purification of lactose synthase (5, 6). In the early stages of the purification, the $\alpha\beta$ complex was stable. However, as the purification proceeded and chromatography on Sephadex was employed, two separated species were obtained, neither of which was capable of catalysing the formation of lactose. Reconstitution of the complex by mixing the two independent proteins restored the desired catalytic activity, and a thorough study of the individual proteins permitted an unequivocal determination of the roles of the two subunits (5, 6).

These simple considerations become more complicated when one deals with trimers, tetramers or higher oligomers. With trimers, for example, the subunits must be arranged with cyclic symmetry, and the dissociation is thus dependent upon the rupture of two contact regions. Hence, if the intrinsic K_d for each inter-subunit 'bond' is 10^{-7} M the effective dissociation constant for the cyclic trimer would be very much smaller, and the trimer would be very stable. A dimer with the same energy of interaction between monomers would dissociate into monomers at comparable concentrations. For tetramers there are two possibilities. If the oligomer exhibits cyclic symmetry, C_4, the complex would be relatively very stable. But most tetramers exhibit dihedral symmetry, D_2, and they are, in effect, dimers of dimers. Hence, dissociation of the tetramers into dimers might be expected and, depending upon the interaction energy between monomers in the dimers, monomers might also be observed. These considerations apply to tetramers of identical chains such as aldolase and glyceraldehyde-3-phosphate dehydrogenase and to tetramers of non-identical chains such as haemoglobin.

There are additional considerations that apply to this equilibrium inside the cell. Numerous weak interactions are important for the stability of quaternary ensembles. Due to the high concentrations of protein commonly found under cellular conditions, excluded volume considerations are additional determinants of reactivity and may stabilize otherwise weakly associating interactions (7–11). In addition, post-translational processing by covalent modification, for example, or by proteolysis may affect the association of subunits. Some subunit assemblies may be more labile kinetically than thermodynamically. Thus, a polypeptide chain that associates with another to form a discrete quaternary arrangement may not be in a reversible equilibrium as described above. Rather, dissociation may be followed by unfolding instead of functional re-association. Hence, stability of this ensemble is limited by the kinetic lifetime of the complex, in contrast to the above thermodynamic treatment based on reversible interconversion.

2.2 **Questions**

The research worker with a purified oligomeric protein that has not yet been characterized is generally confronted with a series of questions worth answering. How many polypeptide chains are there in the oligomer? Are they identical? What is the stoichiometry? Can the different types of chains be separated from one another? Can the individual subunits be isolated by sufficiently mild means that some activities are preserved? Can the original oligomer be reconstituted from the independent polypeptide chains? During the assembly process, is the restoration of the original biological properties coincident with the formation of the oligomer? Are intermediates detected in the assembly process? Are they active? Is there a conformational change in the subunits associated with the assembly of the oligomer? What forces stabilize the oligomer? When ligands bind to one subunit, are the effects propagated to the other subunits leading to a global conformational change? How answers are obtained to some of these questions is illustrated by the procedures described below.

3. DETERMINATION OF SUBUNIT STRUCTURE OF OLIGOMERIC PROTEINS

Evaluating the architecture of multi-subunit proteins at first seems an overwhelming task when one contemplates comparing the results from relatively indirect approaches to the exquisite structural details resulting from X-ray diffraction studies. Yet our knowledge of the various levels of structure for the majority of proteins has been accomplished without the benefit of detailed crystallographic studies. Certainly in the case of H^+-ATPase the subunit composition and stoichiometry are known and will doubtless prove indispensable later for the construction of a three-dimensional model (3). Another striking example of a complex oligomeric enzyme, for which a tentative composition comprising about

20 polypeptide chains, 10 of them distinct from one another, is the DNA polymerase III holo-enzyme of *E.coli* (12). Although the hypothetical model proposed for this enzyme is likely to be altered as further evidence is accumulated, it serves as an invaluable basis for additional experimentation. Many methods including chemical cross-linking (Chapter 4) and hybridization experiments have been applied for the determination of complex arrangements of subunits in proteins. Of vital importance in a structural analysis of any multimeric protein is knowledge of the various classes of polypeptide chains in the protein. Only after such information is obtained would it be possible to assign functional roles for the individual subunits. ATCase is an excellent example illustrating this assignment of functional roles to individual subunits.

3.1 **Can the oligomeric protein be dissociated into smaller, stable subunits still possessing biological activities?**

For many oligomers, in which there are relatively weak attractive interactions between the various subunits, the dissociation into subunits can be accomplished readily by simple dilution. One such example was the discovery that lactose synthase was a complex composed of a galactosyl transferase and α-lactalbumin (5, 6). Similarly, the discovery of calmodulin represents a classical example of the dissociation of an important 'subunit' necessary for protein function. During the course of purifying cAMP phosphodiesterase, the workers observed a precipitous decrease in enzyme activity during fractionation by ion-exchange chromatography (13). Combining fractions led to a restoration of phosphodiesterase activity and subsequent research has shown elegantly that protein–protein interactions account for the marked stimulation in enzyme activity (14, 15). However, not all subunits in oligomeric enzymes can be separated in this way, and there are only a few examples of enzymes for which dissociation into dissimilar subunits has been demonstrated merely by dilution.

Chemical and physical treatments have been used with varying success to dissociate various oligomeric enzymes into subunits. Many procedures make use of the differential solubility of distinct subunits under specific conditions. Changes in temperature, pH, ionic strength, or the addition of denaturants or specific 'co-factors' have been used for the successful dissociation of many oligomeric proteins (2).

Both differential solubility at low pH and differential thermal stability have been used to isolate the α and β_2 subunits of tryptophan synthase (16). At pH 4.3, the β_2 subunit precipitates irreversibly and the α subunit remains soluble (17), whereas the α subunit is heat-sensitive at high temperature while the β_2 subunit is unaffected (18, 19). More recent improvements in the purification of the $\alpha_2\beta_2$ complex from enzyme-overproducing strains have led to more effective procedures for generating the purified subunits (20, 21).

Gerhart and Pardee (22) showed initially three methods to eliminate the inhibition of ATCase by CTP. These included heat treatment at 60°C in low ionic strength buffer, addition of 0.8 M urea and, finally, treatment of the enzyme with various mercurials. ATCase treated with mercurials exhibited an increased V_{max}

and apparent K_m, and the resulting enzyme exhibited hyperbolic saturation kinetics in contrast to the sigmoidal kinetics observed with the untreated enzyme. Subsequent experiments (23) demonstrated that the addition of certain mercurials causes the quantitative dissociation of the enzyme into two distinct subunits: catalytic subunits, which possess all the enzymic activity, and regulatory subunits, which bind nucleotide effectors. The ability of mercurial reagents, in particular *p*-mercuribenzoate and neohydrin, to displace a structural zinc atom from its tetrahedral coordination by four cysteine sulphydryl groups in the regulatory chains has permitted the subunits to be isolated readily by ion-exchange chromatography (24). Reconstitution of the holo-enzyme structure was accomplished simply by adding 2-mercaptoethanol to complex the mercurial, thereby regenerating the regulatory chain sulphydryl groups, followed by the addition of zinc ions for proper regulatory chain folding and then catalytic subunits. In this way, coupled with hybridization and cross-linking experiments (25–27), it was possible to demonstrate that ATCase is composed of two catalytic subunits and three regulatory subunits.

3.2 How can the number of polypeptide chains in an oligomer of identical chains be determined?

In principle the most direct method to answer this question involves the determination of the molecular weight of both the oligomer and the polypeptide chains produced by denaturants such as urea or sodium dodecyl sulphate (SDS). This procedure, employing sedimentation equilibrium for measuring the molecular weight of the oligomer and gel electrophoresis for the dissociated protein, is effective in many cases. However, there are proteins for which ambiguous results are obtained with this approach, and it is important to note that cross-linking followed by electrophoresis in polyacrylamide gels containing SDS and hybridization techniques have proved invaluable as simple, alternative techniques for determining the subunit composition of oligomers (Chapter 5).

Davies and Stark (28) used dimethyl suberimidate as a reagent to cross-link the various chains in an oligomeric protein and they showed that the pattern obtained by electrophoresis in polyacrylamide gels containing SDS yielded directly, and often unambiguously, the number of polypeptide chains in the oligomer. This technique has been used effectively to determine the number of polypeptide chains in aldolase, glyceraldehyde-3-dehydrogenase and the catalytic subunit of ATCase. As pointed out by Davies and Stark, cross-linking of the chains from different oligomers can also occur if the modification with bifunctional imidoester is performed at too high a protein concentration. A great many imidoesters are now available (29) with extensive variations in the separation of the two functional groups. Hence difficulties in forming a covalent cross-link between two polypeptide chains in an oligomer because of the lack of reactivity of some amino groups or an orientation of them which precludes reaction with a small reagent like suberimidate can be overcome with a cross-linker having a large spacer between the functional regions (Chapter 5).

Hybridization techniques of the type used originally in the study of isozymes have also been invaluable in determinations of the subunit composition of oligomers. This approach has the additional advantage that the hybrids that are formed and isolated can be used profitably for studying 'communication' between polypeptide chains and the formation of individual chains. Since variants of a given oligomeric enzyme often are not available, it is necessary to produce an altered protein by relatively specific chemical modification. The use of anhydrides, such as succinic anhydride, for modifying amino groups of lysine side-chains and introducing carboxylate residues has proved to be very valuable. The experimental procedure is presented in *Table 1*. Because of the conversion of positively charged residues to negatively charged groups, the modified protein invariably has a markedly altered electrophoretic mobility, and its chromatographic behaviour on DEAE–Sephadex is also changed substantially. Thus hybrids formed between the modified and native oligomers are readily detected and isolated. The formation of the hybrids requires a procedure for dissociating both the native oligomeric enzyme and the succinylated protein to yield single polypeptide chains. This is then followed by reconstitution of the oligomeric species through removal of the perturbant (or denaturant) that was used to produce the individual polypeptide chains.

Figure 1a illustrates the experimental procedure used by Meighen *et al.* (25) to demonstrate that the *C* subunits of ATCase were trimers of identical polypeptide

Figure 1. Determination of the number of polypeptide chains in the *C* subunit of ATCase by hybridization of the native protein with succinylated protein. The isolated subunits, C_{nnn} at a concentration of 2–10 mg/ml in 50 mM Tris–HCl buffer at pH 8.0 containing 2 mM 2-mercaptoethanol and 0.2 mM EDTA are succinylated with succinic anhydride (0.5 M in dioxane) at 4°C (*Table 1*). The molar ratio of the anhydride to protein lysine groups was about 3.5, and these conditions generally lead to the modification of about four lysine residues/polypeptide chain. Usually the succinylation is accomplished by the periodic addition of the anhydride in 20 μl portions at 4°C, followed by the addition of chilled 0.5 M NaOH to maintain the pH at 7.8–8.0. The extent of modification is readily determined by electrophoresis. Since the anhydride is hygroscopic and repeated opening of the bottle causes its conversion to succinic acid, it is important to test the anhydride by varying the amount added to the protein and analysing its effect on the electrophoretic mobility. Too much succinic anhydride causes extensive modification of the protein, leading to its dissociation into unfolded chains, and too little succinylation makes more difficult the electrophoretic resolution of the members in the hybrid set. This process, leading to the optimum extent of succinylation, is monitored ideally by electrophoresis to satisfy the criteria described in *Table 2*. C_{nnn} and the succinylated trimer, C_{sss}, are incubated together in either a denaturing medium like 5 M urea or a solvent that causes dissociation of the trimers with little unfolding of the chains, such as NaSCN. This process leads to the formation of monomers, which upon removal of the perturbant, urea or NaSCN, associate to yield the hybrid set shown schematically in (**a**). A typical hybridization experiment is illustrated in (**b**) which shows the kinetics of formation of hybrids between C_{nnn} and C_{sss} that were incubated together for various lengths of time in 1 M NaSCN. The mixture contained equal amounts (0.2 mg/ml) of C_{nnn} and C_{sss} at 25°C in 1 M NaSCN, 40 mM Tris–HCl, and 0.2 mM EDTA, pH 7.5. Aliquots (5 μl) of the mixture were removed at specific times and diluted with 45 μl of buffer containing no NaSCN. After 2 h, these diluted solutions were analysed by electrophoresis in polyacrylamide gels. The resulting patterns from *left* to *right* correspond to 0.25, 5, 20, 45 and 120 min, respectively, of incubation of C_{nnn} and C_{sss} in the 1 M NaSCN solution. (**c**) Effect of various salts on the dissociation of C_{nnn} and C_{sss} subunits. A solution containing ^{125}I-labelled C_{nnn} subunits at 0.066 mg/ml and unlabelled C_{sss} subunits at 1.32 mg/ml in 1 M salt, 40 mM Tris–HCl, and 0.2 mM EDTA, pH 7.5, was incubated for various times at 25°C. At specific times, indicated on the *abscissa*, aliquots (10 μl) were removed and diluted 10-fold in 40 mM Tris–HCl and 0.2 mM EDTA, pH 7.5. After 3 h, the diluted solutions were subjected to electrophoresis, the gels were stained and sliced, and the amount of radioactivity in each fraction was measured to determine the per cent of C_{nnn} trimers remaining. ◆, 1 M NaCl; ■, 1 M $NaNO_3$; ▲, 1 M $NaClO_4$; ●, 1 M NaSCN.

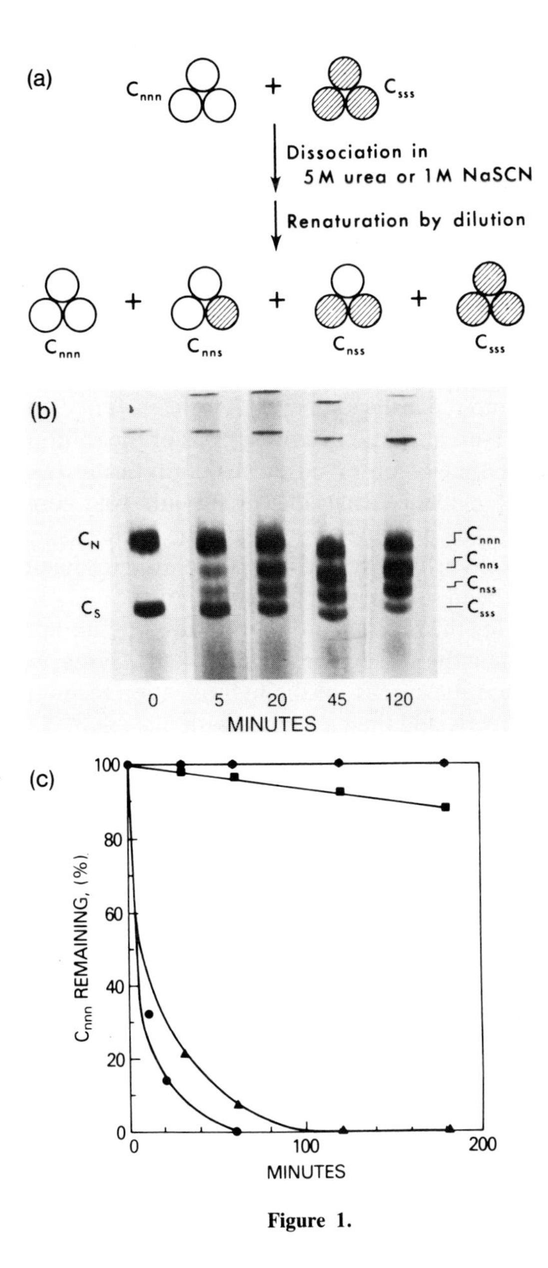

(a)
Cnnn
+
Csss
Dissociation in 5 M urea or 1 M NaSCN
Renaturation by dilution
Cnnn
+
Cnns
+
Cnss
+
Csss
(b)
CN
CS
Cnnn
Cnns
Cnss
Csss
0
5
20
45
120
MINUTES
(c)
100
80
60
40
20
0
Cnnn REMAINING, (%)
0
100
200
MINUTES

Figure 1.

Table 1. Succinylation of oligomeric proteins.

1. Prepare concentrated protein solutions (2–10 mg/ml) in 50 mM Tris–HCl, pH 8.0, 2 mM 2-mercapto-ethanol, 0.2 mM EDTA at 4°C.
2. Prepare 0.5 M succinic anhydride in dioxane.
3. For *C* subunit of ATCase, use 3.5 molar excess of succinic anhydride per protein lysine group, which yields approximately four modified lysines/monomer. The exact amount of succinic anhydride to be added should be determined empirically by titration with freshly prepared succinic anhydride stock solutions.
4. Add 20 μl aliquots of succinic anhydride to the protein solution, stirring briskly at 4°C. Maintain the pH at 7.8–8.0 by addition of chilled 0.5 M NaOH.
5. Evaluate the extent of modification, for example, by electrophoresis, and characterize the modified protein.

chains. The native (C_n) and succinylated (C_s) subunits were mixed in equal concentrations, dissociation was accomplished by the use of urea (4–6 M suffices for most oligomers), and reconstitution in high yield was effected either by a 1:10 dilution of the protein–urea solution or by dialysis of the solution to remove the urea. The number of species observed electrophoretically (in this case, four) provides unequivocal evidence that the *C* subunit was composed of three polypeptide chains. It was possible to separate the various species on DEAE–Sephadex, and the two hybrids, C_{nns} and C_{nss}, were used for functional studies.

This procedure has been shown to have general utility, and some of the criteria that must be satisfied for the succinylation procedure are summarized in *Table 2*. It should be noted that the dissociating agent, 4–6 M urea, yielded unfolded polypeptide chains, and the success of the hybridization technique requires that the chains re-fold correctly and then associate with one another randomly to give the four reconstituted trimeric species. In some cases, the dissociation can be achieved with 1 M NaSCN to yield folded monomers that, upon removal of the NaSCN, then associate to regenerate trimers (30, 31).

The effectiveness of NaSCN is shown in *Figure 1b*, which illustrates the kinetics of formation of the hybrids as revealed by electrophoresis in polyacrylamide gels. Also, a comparison of NaCl, $NaNO_3$, $NaClO_4$ and NaSCN in promoting the dissociation of the trimers and the formation of hybrids is shown in *Figure 1c*. For this experiment C_{sss} is used in large excess relative to ^{125}I-labelled C_{nnn} (a ratio of

Table 2. Criteria for satisfactory modification of residues in oligomers to produce a variant for use in hybridization experiments.

1. The modifying reagent must react specifically with side-chains of the protein to produce a relatively homogeneous derivative.
2. The modified protein must have a quaternary structure similar to the native protein and must have comparable energies of interaction between subunits.
3. The quaternary structure of the variant must be capable of being reconstituted after dissociation of the oligomers into subunits.
4. If hybrids are to be separated, the chemical modification must alter some physical property of the variant (e.g. electrophoretic mobility or chromatographic behavior).
5. The modified derivative need not be active; there are both advantages and disadvantages to inactivation.

20 to 1) and the per cent of C_{nnn} remaining is shown as a function of time. Of these reagents, NaSCN is seen readily to be the most effective.

Various perturbants have been employed to promote the dissociation of oligomers (2). These include co-factors and ligands, substrate analogues, pH, temperature, pressure, ionic strength and organic solvents. Preliminary studies that monitor some physical parameter, such as the absorbance in the UV or optical rotation, usually yield valuable information regarding the effectiveness of the dissociating agent and reversibility of the process.

4. EVALUATION OF THE STRENGTH OF INTERACTIONS BETWEEN SUBUNITS IN OLIGOMERS

Quantitative information on the strengths of the inter-chain interactions in oligomeric proteins is often imperative for a complete understanding of the quaternary structure of the protein and its function. Usually, however, the analysis of the interaction energies is hampered by the inherent strength and the multiplicity of the interactions. In ATCase, for example, there are two areas of contact between a *c* chain and the neighbouring adjacent *c* chains in the same trimer; in addition, each *c* chain has some area of contact with two *r* chains that helps to maintain the stability of the holo-enzyme. There are other 'bonds' as well, responsible for the structure of this enzyme. As a consequence, it has been virtually impossible to detect dissociation of ATCase by any physical chemical technique. Indeed, enzyme assays with ATCase at concentrations less than 1 nM show that the enzyme is largely undissociated, as might be expected from the multiplicity of inter-chain interactions. Although, in principle, dissociation might be expected if the protein concentration is decreased another 10^6-fold, there would be few physical chemical techniques sufficiently sensitive to detect the presence of subunits. Hence, for strongly associating systems, alternative methods are needed for the evaluation of the strength of the interactions and the effect of specific ligands and other factors on the relevant interaction energies. The hybridization technique described above, with appropriate modifications, is capable of providing valuable information for oligomeric proteins which, by physical chemical methods, show almost undetectable levels of dissociation into subunits. Determinations of the rates of subunit dissociation by subunit exchange leading to hybrid molecules can yield values for the relative interaction energies between subunits and especially of the factors that reduce these energies.

Subunit exchange is based on the principle that all contacts between chains are breaking and forming at certain rates and that a finite (but perhaps immeasurable) amount of subunit is always present in equilibrium with the oligomers. Hence, the presence of another oligomer of the same basic structure but containing chains that differ from those of the oligomer under investigation will lead to the progressive-formation of hybrids as a function of time. If, for example, K_d for a dimer is 10^{-8} M, representing a complex of high stability, and the second-order rate constant for dimer formation is $10^8\ M^{-1}\ sec^{-1}$, then dissociation of the dimers occurs with a first-order rate constant of $1\ sec^{-1}$. This slow dissociation is readily measured by subunit exchange.

As shown below, the exchange of subunits can take place in oligomers of identical chains or in oligomers containing different types of subunits. Both types of experiments require a variant of the oligomer being studied, which can be produced chemically or by genetic manipulation. As long as the two proteins differ in electrophoretic mobility, for example, the progress of formation of hybrids can be measured quantitatively to give rate constants for the dissociation process. Succinylation of proteins as described in *Figure 1* and *Table 1* and, according to the criteria summarized in *Table 2*, has yielded variants that have served for studies of this type with both the catalytic trimers of ATCase (30, 32) and with mutant forms of the holo-enzyme, where the goal was to measure the rate of exchange of one of the catalytic subunits in the enzyme (E. Eisenstein and H. K. Schachman, unpublished).

4.1 **Can mutational alteration affect the inter-chain interactions in oligomers of identical chains?**

Determining the rate of hybrid formation between ^{125}I-labelled mutant C trimers of ATCase and wild-type trimers modified by succinic anhydride enabled Yang and Schachman (32) to assess the effect of mutational alterations on the inter-chain interactions of the catalytic trimer. The mutant catalytic trimers were labelled with ^{125}I by the procedure of Syvanen *et al.* (33) and the wild-type trimers were succinylated as described in *Table 1*. After incubating mixtures of the two proteins for various lengths of time, aliquots were withdrawn and subjected to electrophoresis in polyacrylamide gels to separate the four species. The slowest moving protein, corresponding to the iodinated species, was excised from the gel and measurements of the radioactivity were performed with a Nuclear Chicago gamma counter. As seen in *Figure 2*, the fraction of the initial protein, C_{xxx}, decreased with time and, moreover, the initial rates for the three mutants and the wild-type trimers varied significantly. It is clear from the data in *Figure 2*, that C_{231}, a mutant in which the glycine residue at position 128 in the polypeptide chain is replaced by aspartic acid, is much more stable than C_{WT}. Conversely, C_{745}, containing two amino acid substitutions that as yet have not been identified, is much less stable than the wild-type trimers.

Thus far these preliminary experiments have not been used in attempts to determine the free energies of association of the polypeptide chains in the trimers, but it is clear that the subunit exchange technique is very sensitive and capable of yielding quantitative data. It should be emphasized that the method, as used, provided direct information about the inter-chain interactions and not the over-all stabilities of the different mutants. In this regard it is worth noting that C_{231}, the mutant with the stronger inter-chain interactions, is significantly less stable to heat denaturation with C_{WT}. Thus it is possible to differentiate between the stability of the quaternary structure in terms of inter-chain interactions and the over-all denaturation which destroys the tertiary and secondary structures of the mutant proteins.

This same technique is also of great value in determining the effects of ligands on the strength of the inter-chain interactions. When mixtures of native wild-type

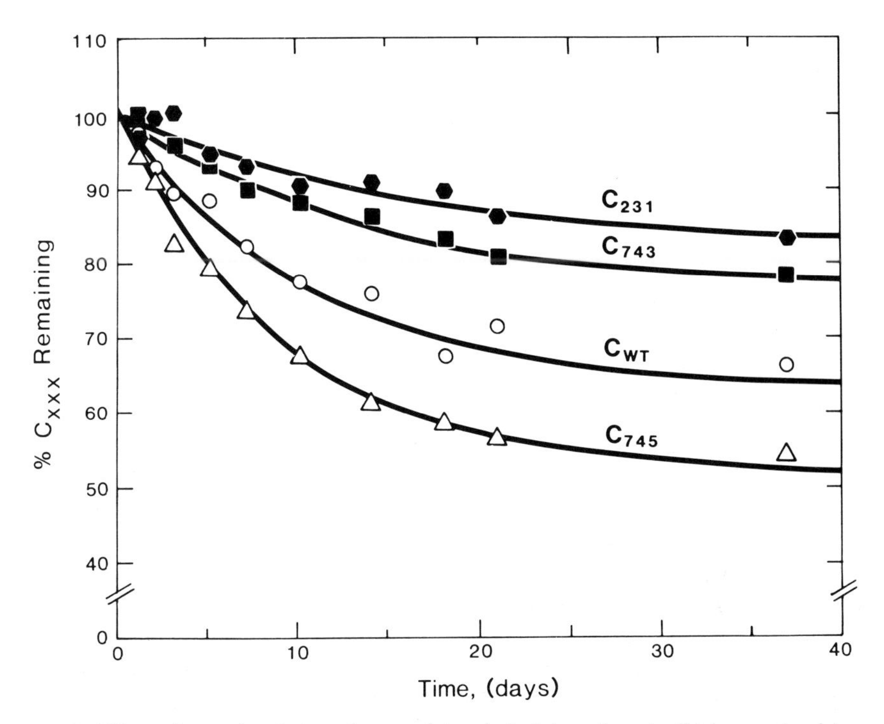

Figure 2. Effect of mutational alterations on inter-chain interactions in C trimers. A mixture of ^{125}I-labelled C_{xxx} (0.3 mg/ml; 10^5 c.p.m./μg) and C_{sss} (3 mg/ml) was incubated for various times at 4°C in 50 mM Tris–HCl buffer, pH 7.5, containing 2 mM 2-mercaptoethanol and 0.2 mM EDTA. Aliquots were removed at specific times and analysed by electrophoresis in polyacrylamide gels using a Hoefer Mighty Small slab gel electrophoresis and the discontinuous buffer system of Jovin *et al.* (34). The 7.5% gels were stained with Coomassie brilliant blue G-250 in 12.5% (w/v) trichloroacetic acid and de-stained with 5% methanol in 5% acetic acid. The various bands were cut from the gels and measurements of ^{125}I were made with a Nuclear Chicago gamma counter. The amount of radioactivity migrating as C_{xxx} was measured, and the results were plotted as per cent C_{xxx} remaining as a function of time. C_{123} (⬢), C_{743} (■), C_{WT} (○), and C_{745} (△).

protein, C_{nnn}, were incubated in the presence of the substrate, carbamoyl phosphate, or the bisubstrate analogue, *N*-(phosphonacetyl)-L-aspartate (PALA), no hybrids were detected after several days (35). In contrast, sodium pyrophosphate, ATP, CTP and UTP, which are competitive inhibitors of carbamoylphosphate, stimulated greatly the formation of the four-membered hybrid set. Thus these ligands, through their binding, cause a marked decrease in the strength of the inter-chain interactions. The discovery of the pronounced weakening of the inter-chain interactions by pyrophosphate has led to the best procedure now available for making hybrids of enzymically active C trimers of ATCase. Because it leads to little disruption of the tertiary structures of the individual chains, it is preferable to other procedures, such as the use of urea or NaSCN which invariably lead to lower yields of reconstituted trimers (32).

4.2 Can free subunits exchange with those in oligomers composed of non-identical chains?

Experiments on the rate of exchange between free subunits and those contained within holo-enzymes are particularly useful in studies of both the structure and function of multi-subunit proteins composed of different types of polypeptide chains. A low level of exchange is often an indication of the multiplicity and strength of the inter-chain interactions, and exchange kinetics can often provide invaluable information about those interactions and the effect on them of ligands or mutational alterations in the various polypeptide chains.

Creighton and Yanofsky (36) used the competition between wild-type and mutant subunits to measure the rate of dissociation of tryptophan synthase from *E.coli*. For these experiments they used a mutant α^* subunit that was enzymically inactive but still possessed an affinity for the β_2 subunit virtually identical to that of the wild-type α subunit. Because they had already demonstrated that the catalytic activity of wild-type α subunits was greatly enhanced when they were incorporated into the $\alpha_2\beta_2$ complex, Creighton and Yanofsky could use activity measurements to determine the dissociation rate of the complex due to subunit exchange. If α, α^* and β_2 subunits are incubated together, an equilibrium mixture results in which some of the α subunits exist in $\alpha_2\beta_2$ complexes and some are free because the potentially available β_2 subunits are occupied by α^* yielding $\alpha_2^*\beta_2$ complexes. Enzyme assays of the equilibrated mixture provide a measure of the fraction of α subunits within $\alpha_2\beta_2$ complexes, because each of the two α subunits within the complex are independent catalytically. By adding free α subunits to the mixture and measuring the rate of increase in activity, Creighton and Yanofsky were able to determine the dissociation rate of $\alpha_2^*\beta_2$, which provided the β_2 subunits for combination with the added α subunits. In this way they calculated a half-life for the $\alpha_2^*\beta_2$ complex of 2.4 min. Moreover, they showed that the effectors L-serine and pyridoxal 5′-phosphate caused a 30-fold increase in the stability of the complex.

This study of tryptophan synthase is of special interest because it was conducted at a time when no information was available about the arrangement of the subunits in the quaternary structure. Moreover, all the conclusions about the rate of dissociation and the effects of ligands were inferred from measurements of enzyme activity. There was no need to separate the holo-enzyme from the subunits in the kinetic studies. More recent experiments using stopped-flow kinetics to elucidate the mechanism of assembly of the $\alpha_2\beta_2$ complex have confirmed these effects of ligands on the association of the α and β_2 subunits and also have shed light on the structural changes in these subunits as a consequence of the assembly process (37).

The technique of subunit exchange using radioactive subunits is particularly valuable for studies of inter-chain interactions in holo-enzymes that are so stable that dissociation into their constituent subunits is hardly detectable by physical chemical procedures. For these studies it is necessary to remove aliquots at specific times and to separate the holo-enzyme from the subunits so as to measure the amount of radioactivity in the holo-enzyme and the subunits. Frequently this can

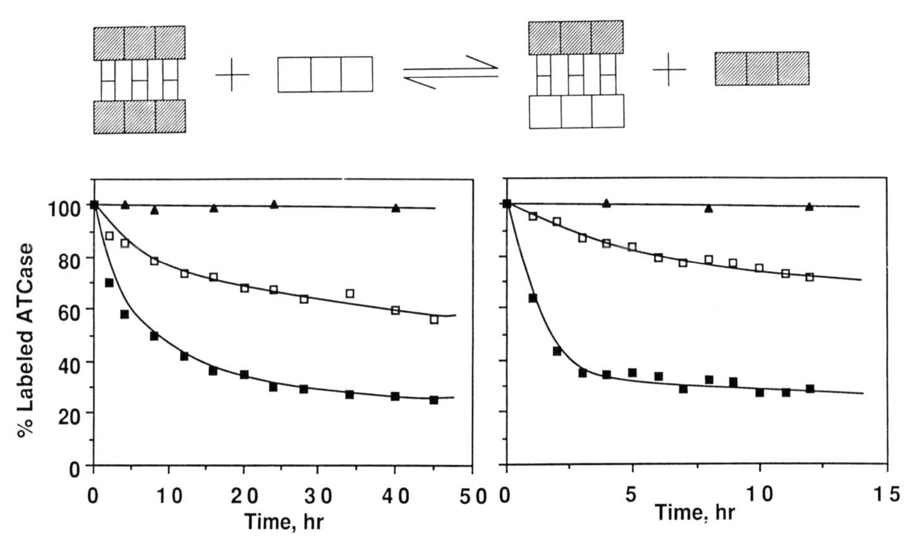

Figure 3. Subunit exchange in oligomers of dissimilar polypeptide chains. The technique is illustrated by the schematic model at the top, based on the known structure of ATCase containing two *C* trimers and three *R* dimers. Cross-hatched *C* subunits designate ^{125}I-labelled protein, and the technique is based on assembling holo-enzyme, C_2R_3, containing [^{125}I]*C* subunits and measuring the exchange with free unlabelled *C* subunits. The decrease in radioactivity of the holo-enzyme is measured as a function of time to give the plots shown for the two mutants. Mutant holo-enzymes, assembled from ^{125}I-labelled wild-type *C* subunits (65 μg, 6.2 × 10^4 c.p.m./μg) and purified mutant *R* subunits (35 μg), at a final concentration of 1 mg/ml, were incubated at 25°C with a 20-fold molar excess of unlabelled wild-type *C* subunits in 20 mM Tris–HCl, 5 mM 2-mercaptoethanol, 100 μM zinc acetate, 150 μM potassium phosphate, pH 7.15. At various times, aliquots were removed and subjected to electrophoresis and visualized as described in the legend to *Figure 2*. Lanes were cut into sections containing stained bands corresponding to an aggregate of ATCase formed in the assembly process, to the holo-enzyme, C_2R_3, to a regulatory subunit-deficient species, C_2R_2, to the *C* subunit, and to the dye front. The amount of radioactivity in these slices was measured in a Packard Multi-Prias gamma counter, and the results are plotted as the per cent of ^{125}I-labelled C_2R_3 remaining versus time. **Left**: wild-type ATCase + 130 μM PALA (▲); R142Glu → Asp ATCase (□); R142Glu → Asp ATCase + 130 μM PALA (■); **Right**: wild-type ATCase + 130 μM PALA (▲); R142Glu → Ala ATCase (□); R142Glu → Ala ATCase + 130 μM PALA (■).

be accomplished by electrophoresis or chromatography and in some cases by centrifugation. ATCase serves as a useful model for demonstrating this technique, which is illustrated in *Figure 3*. The holo-enzyme is dissociated into *C* and *R* subunits by treatment with a mercurial such as neohydrin, and the independent subunits are separated by chromatography on DEAE–Sephadex as described by Yang *et al.* (24). Purified *C* subunits are labelled with ^{125}I as described previously, and reconstituted ATCase is formed containing ^{125}I-labelled *C* subunits (shown hatched in *Figure 3*). Upon incubation of this labelled ATCase with unlabelled *C* subunits, some of the radioactivity appears as free *C* subunits and concomitantly there is a decrease in the extent of radioactivity of the holo-enzyme.

When the approach outlined above was used for wild-type ATCase containing ^{125}I-labelled wild-type *C* subunit and wild-type *R* subunits, no exchange was detected upon incubation of the reconstituted enzyme with unlabelled *C* subunits (*Figure 3*). However, mutant forms of the enzyme containing amino acid alterations in the regulatory chains at positions near or at the interface between the *c* and *r* chains were much less stable, and subunit exchange was readily measurable. As seen in *Figure 3*, the replacement of the glutamic acid residue at position 142 in the regulatory chain by either aspartic acid or alanine leads to a significant reduction in the strength of the inter-chain interactions. The extents of weakening the interactions differ for the two amino acid replacements, as is seen from the initial slopes in the kinetics presented in *Figure 3*. Moreover, the bi-substrate ligand PALA, which binds at the active site at a considerable distance from the interface between the *c* and *r* chains, has a marked effect in weakening the inter-chain interactions.

Subunit competition experiments of the type illustrated in *Figure 3* are easily performed and are very sensitive for measuring exchange with holo-enzymes that exhibit almost no dissociation. As with most techniques, there are pitfalls and certain criteria must be satisfied. First, the exchange process should be limited by the dissociation rate of the parental protein. If the product formed from the 'challenging' species is much less stable (thermodynamically) than the starting holo-enzyme, the extent of exchange might be vanishingly small and misleading results would be obtained. To avoid this potential difficulty, the experiments illustrated in *Figure 3* involved exchange-out of the ^{125}I-labelled *C* subunit rather than exchange-in. Conceptually it would be easier and certainly more direct to challenge unlabelled holo-enzyme with radioactively-labelled subunit. But such experiments could be criticized on the grounds that the native holo-enzyme was very stable and that the modified subunits were slightly damaged during the incorporation of the radioactive label. Under such circumstances thermodynamic considerations would disfavour exchange. It is of interest that exchange experiments of the type shown in *Figure 3* have been performed with ATCase mutants and a large excess of succinylated *C* subunits to give hybrid molecules containing one unmodified and one succinylated *C* subunit and unmodified *R* subunits. With ATCase both exchange-in and exchange-out yield similar results, but it cannot be taken for granted that this agreement will be observed with other holo-enzymes. A second criterion that must be satisfied, therefore, is that the product of the exchange process must have a stability comparable to the initial parental species. This consideration is particularly relevant in exchange studies involving mutant proteins. Third, since the holo-enzyme and the exchanging subunit must be separated physically, it is essential that the technique used for the separation, such as electrophoresis or chromatography, should not itself cause dissociation of the holo-enzyme. Centrifugation, for example, with the resulting pressure gradients in the tubes could cause dissociation, thereby leading to erroneous results. Finally, the separation technique must be sufficiently rapid that dissociation of the holo-enzyme during the electrophoresis or chromatography experiments can be neglected. These criteria are readily satisfied for those holo-enzymes, like ATCase, that exhibit a multiplicity of relatively strong inter-chain interactions.

5. DETERMINATION OF THE FUNCTIONAL ROLES OF DIFFERENT SUBUNITS IN COMPLEX OLIGOMERIC ENZYMES

Regulatory enzymes and multi-enzyme complexes generally display unique functional properties that are the result of interactions between different classes of subunits within the oligomers. With many of these complex proteins it has been possible to dissociate the oligomers into smaller units which, though lacking some of the functional properties of the intact complexes, still possess well-characterizable biological activities. Tryptophan synthase, for example, which is composed of two types of subunits, catalyses the reaction between indole 3-glycerol phosphate and L-serine to form L-tryptophan and D-glyceraldehyde 3-phosphate. Each of the isolated subunits, α and β_2, catalyses a partial reaction that is stimulated markedly when the subunits interact to form the $\alpha_2\beta_2$ multi-enzyme complex. Lactose synthase represents another type of complex formed from a galactosyl transferase and a 'specifier' protein, α-lactalbumin. Each of the subunits has been well characterized, and it is known that α-lactalbumin serves to inhibit the normal activity of the isolated galactosyl transfer and instead to alter its specificity, thereby catalysing the formation of lactose. In some respects ATCase is analogous to lactose synthetase. When ATCase is dissociated by treatment with mercurials like *p*-hydroxymercuribenzoate or neohydrin, the characteristic sigmoidal kinetics, as well as the inhibition by CTP and activation by ATP, are lost (23). This dissociation of the holo-enzyme yields different proteins, *C* subunits that exhibit Michaelian kinetics and *R* subunits that bind CTP and ATP. Upon removal of the mercurial from the *R* subunits and adding the free *C* subunits, Gerhart and Schachman were able to reconstitute the holo-enzyme having the characteristic allosteric properties of the original enzyme.

Clearly an understanding of oligomeric complexes comprising different types of subunits requires a detailed investigation of the numerous inter-chain interactions and this, in turn, is dependent on knowledge of the structure and function of the individual subunits that can be isolated from the complexes. Hence techniques are needed to dissociate the oligomers under mild conditions and then to separate the discrete subunits in pure form. With some systems, like lactose synthase, dissociation of the enzyme complex and isolation of the two constituent proteins occurs during chromatography in the late stages of the purification procedure; whereas with other oligomeric proteins, reagents like mercurials, urea at low concentrations, NaSCN and specific ligands are effective in causing dissociation. If the goal of the investigation is to determine the functional role of the subunits, it is essential that the different subunits that are isolated and studied be capable of associating with one another to yield reconstituted enzyme having the original biological properties. In many cases, such as those mentioned above, this goal has been attained.

6. RELATIONSHIP BETWEEN OLIGOMERIC STRUCTURE AND BIOLOGICAL ACTIVITY

For many oligomeric proteins containing identical polypeptide chains, we cannot answer unequivocally the question: are monomers active? In most cases the formation of monomers from the oligomer requires dissociating agents of sufficient strength that the tertiary structure of the monomers is disrupted along with the destruction of the quaternary structure. Only rarely are the inter-chain interactions sufficiently weak and different in kind, compared with the inter-chain interactions, that the dissociation of oligomers into folded monomers can be achieved. Even when it is possible to isolate monomers that are 'folded' according to physical chemical criteria, such as circular dichroism and hydrodynamic properties, it is not known whether the tertiary structure of such isolated folded monomers is the same as when they are incorporated into the native oligomers. Physical chemical criteria are not sufficiently sensitive, and as a result there is almost always ambiguity in knowing why activity is lost when oligomers are dissociated into monomers. Two alternative interpretations such as those illustrated in *Figure 4* must be entertained, and experimental procedures must be used to determine which is correct. On the one hand, no matter how mild the procedure used to dissociate the trimers depicted in *Figure 4*, there is a conformational change so subtle that it escapes detection by many physical chemical techniques. This change, depicted as a separation of residues comprising the active site, leads to inactivation. On the other hand, the active sites in the trimers may be at the interfaces between the polypeptide chains, and catalysis would be dependent on the joint participation of amino acid side chains from each

Conformational change

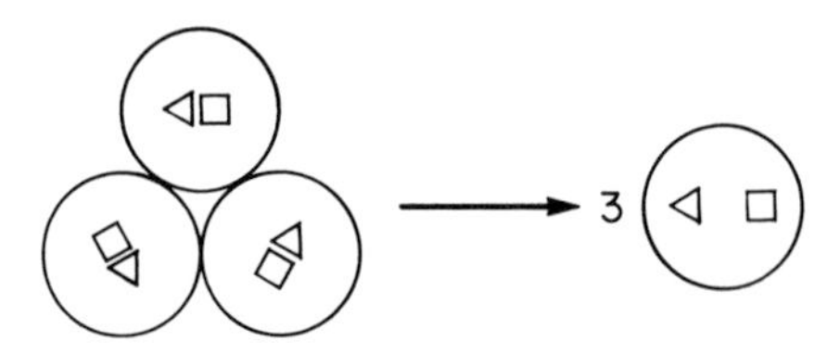

Shared active sites

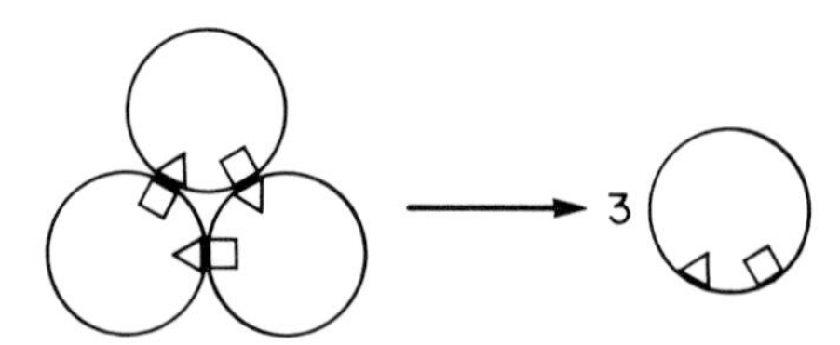

Figure 4. Hypothetical models illustrating loss of activity in dissociation of oligomers composed of identical chains. Individual monomers in the trimer are designated by large circles and the active sites requiring the accurate positioning of critical amino acid side chains are designated by ◁□. The upper diagram illustrates the loss of activity due to a conformational change in which the residues indicated by the ◁ and □ are no longer correctly positioned in the monomer. In the lower diagram of a hypothetical trimer, the three active sites are at the interfaces between adjacent polypeptide chains and critical amino acid residues from each of the two adjoining chains contribute to form a shared active site. For this model, dissociation into monomers yields three inactive subunits, each of which has a split active site.

of the adjacent chains. If the oligomers do have such shared active sites, then monomers are intrinsically inactive because they would have split active sites as shown in *Figure 4*. Differentiating between these alternatives is a difficult task, but the procedures described here can be effective in achieving that goal.

6.1 How can one determine whether monomers are active?

Three alternative procedures have been developed and used successfully to demonstrate whether active monomers can be produced by the dissociation of some oligomeric enzymes. Each technique has advantages and pitfalls. However, when used with caution they provide convincing evidence that in some instances monomers are active, whereas for other oligomeric enzymes monomers are intrinsically inactive. These techniques involve measuring the size of the enzymically-active species at great dilution where dissociation is favoured, assaying for active species during the reconstitution of active oligomers and measuring the activity of matrix-bound monomers that are so immobilized that re-association of oligomers cannot occur.

6.1.1 *Active enzyme centrifugation*

In principle all oligomeric proteins must dissociate into monomers at great dilution. However physical chemical techniques seldom are sufficiently sensitive that accurate data can be obtained at the very low concentrations required to produce monomers. In contrast, enzyme assays frequently can be performed with precision at much lower concentrations. As a consequence there are reports that seem at first glance to be paradoxical. For example, sedimentation velocity and sedimentation equilibrium studies of certain enzymes show association–dissociation behavior at concentrations of about 1 mg/ml. At the same time, enzyme assays indicate that the activity varies linearly with concentration even at 1 μg/ml. This could indicate that the dissociated oligomer (i.e. monomers if the dissociation is complete) has the same specific activity as the intact oligomer. But this conclusion may be unwarranted, because sedimentation studies are rarely performed in the presence of substrates, which may have a profound effect in stabilizing the oligomer. Hence the physical chemical measurements and the assays for detecting the active species must be conducted under identical conditions. This combined goal can be achieved by the technique of active enzyme centrifugation (38–40). The technique involves sedimentation of an enzyme at very low concentrations through a solution of substrates and makes use of the absorption optical system in the analytical ultracentrifuge to detect spectrophotometrically the formation of product as the enzymically-active species migrates through the cell. Hence the sedimentation coefficient of the active species is determined directly.

Active enzyme sedimentation has been used for a variety of systems, and only one example is illustrated here. Sedimentation velocity experiments had shown that the dimeric enzyme, carbamoyl phosphate synthetase from rat liver, dissociates readily as the concentration is reduced and that the substrates MgATP,

NH_4^+, HCO_3^- and acetylglutamate promote the dissociation to monomers (41). In view of these observations it was of interest to determine whether monomers or dimers (or both) were enzymically active. Since the synthesis of carbamoyl phosphate is accompanied by the release of a proton, the reaction could be followed conveniently by measuring the absorbance change at 560 nm due to binding of the protons to phenol red added to the solution. Lusty (41) was able to demonstrate by active enzyme sedimentation that the sedimentation coefficient of the active species was 7.65S, in agreement with the result by conventional sedimentation velocity experiments on dilute solutions of the enzyme. In this way she showed that both monomers and dimers are active (44).

As with most techniques, active enzyme sedimentation has its limitations. It is necessary that the concentration of active enzyme be adjusted so that it controls the extent of reaction. Hence the protein concentration range available for study is limited. For many systems, coupling the catalysed reaction to another is necessary so that an absorbance change can be measured at a suitable wavelength. To avoid experimental complications due to the absorbance of the protein, substrates and certain buffer constituents, it is advisable to couple the release of product to an auxiliary reaction that leads to a change in the visible region of the spectrum. Finally, there are problems in analysing the data for associating–dissociating systems in which different species are present across the corporate moving boundary.

6.1.2 *Kinetics of formation of active species during reconstitution of oligomers*

As has been demonstrated in many studies (42, 43), an analysis of the kinetics of re-activation, coupled with knowledge of the time course of oligomer formation, often provides evidence showing that monomers are active or, alternatively, that catalytic activity is dependent upon the acquisition of quaternary structure. If, during the reconstitution of active oligomers, activity is generated prior to the formation of the oligomer, we could conclude that monomers are active. Alternatively, the restoration of activity may be coincident with the formation of the oligomer. In such cases it is likely that the monomers are inactive. Caution is necessary in drawing conclusions, and it is important to analyse the kinetics to determine the nature of the rate-determining process. Reconstitution of oligomers from unfolded polypeptide chains produced by urea or GdmCl denaturation is generally a first-order process, because the slow step is the folding of the individual chains to yield monomers capable of rapid assembly into active oligomers. In contrast, the assembly of oligomers from folded monomers usually exhibits a concentration dependence characteristic of second-order kinetics. Hence it is important to determine the nature of the process and have available experimental techniques for analysing the different aspects of the assembly process.

The techniques used to study the assembly of the *C* trimers of ATCase are illustrated in *Figure 5*. For these studies (30) the dissociation of the trimers into folded monomers was accomplished by the use of 1 M NaSCN. This occurs rapidly, and sedimentation velocity and spectroscopic studies indicate that much of the secondary and tertiary structure of the active trimers still exists in the

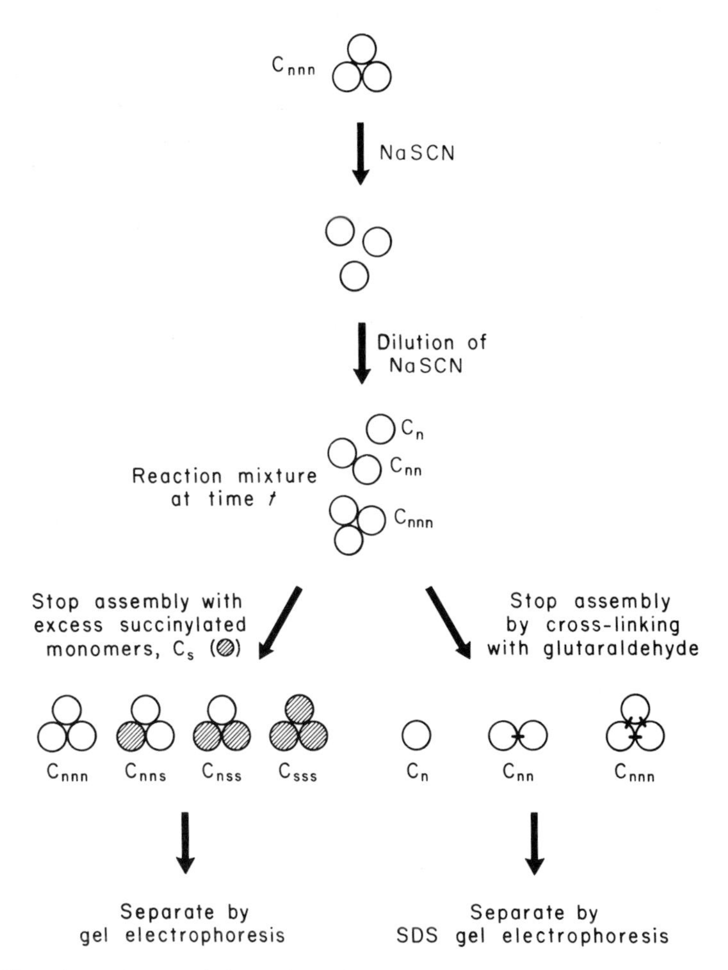

Figure 5. Experimental approach for analysis of assembly process and determination of the nature of the active species. Native C_{nnn} trimers were dissociated in 1.25 M NaSCN solutions, and assembly was initiated by the dilution (at least 20-fold) of the perturbant. At various times, aliquots of the assembly mixture, containing unreacted monomers, C_n, fully assembled trimers, C_{nnn}, and perhaps dimeric intermediates, C_{nn}, were removed and the assembly was stopped in two different ways. In one set of experiments (on the left), a large excess of C_s, produced by treating succinylated C_{sss} trimers with 1.25 M NaSCN, was added as a chase. These C_s monomers react with C_n monomers for form C_{nss} hybrids. Similarly, partially assembled C_{nn} dimers would react with C_s monomers to form C_{nns} hybrids. Fully assembled C_{nnn} trimers would not react with C_s monomers, and the excess C_s monomers would interact to form assembled C_{sss} trimers. Thus, this chase would lead to a four-membered hybrid set, and the individual species are separated by electrophoresis in polyacrylamide gels. In the alternative set of experiments (on the right), the assembly was stopped by cross-linking the various species with glutaraldehyde. The three putative species consisting of fully assembled cross-linked trimers, partially assembled cross-linked dimers and monomers are then separated by electrophoresis in polyacrylamide gels containing SDS.

monomers. Assembly is initiated by rapid dilution to form a hypothetical mixture containing monomers, dimers and trimers. The composition of this mixture in terms of the amounts of the three putative species can be determined as illustrated in *Figure 5*, either by the cross-linking procedure using glutaraldehyde described in *Table 3* or by the hybridization procedure using succinylated protein as described in *Table 1*. In these studies Burns and Schachman (30) obtained the data presented in *Figure 6*. As seen in *Figure 6a* the restoration of enzyme activity was both time-dependent and concentration-dependent. As expected, the formation of trimers was concentration-dependent (*Figure 6b*). When the data from independent measurements of trimer formation by both the hybridization and the cross-linking technique are combined with the restoration of enzyme activity (*Figure 6c*), it is seen that regeneration of activity is coincident with trimer formation. Thus this study shows clearly that monomers are not active. However, the data do not permit one to distinguish between the two alternative explanations for their inactivity illustrated in *Figure 4*.

It is of interest that an analysis of the kinetics showed the assembly to be a second-order process. The data could be fit satisfactorily by a scheme in which two monomers associate reversibly to form dimers and then another monomer reacts irreversibly and more slowly to form stable trimers. According to this mechanism, the concentration of dimers at any stage of the assembly must be very low. This was demonstrated to be the case experimentally by the cross-linking technique.

The analysis of the reconstitution of active oligomers from folded monomers provided convincing evidence that monomers are inactive, but often it is not possible to isolate folded monomers and studies, of necessity, must be conducted with unfolded polypeptide chains. Such chains produced by urea or GdmCl denaturation are frequently the starting point for reconstitution experiments (43). As mentioned above, the restoration of enzyme activity in the formation of *C* trimers of ATCase from disorganized polypeptide chains was first-order, thereby indicating that the rate-determining step was the folding of the chains into monomers capable of associating into active trimers (31). If the protein concentration for these reconstitution experiments had been much lower, then the

Table 3. Use of cross-linking by glutaraldehyde to determine species formed during reconstitution of oligomers.

1. Prepare the protein solution in suitable dissociating agent such as urea and dilute in 40 mM tri-ethanolamine–HCl, pH 7.5 at 25°C to initiate re-association. Because glutaraldehyde reacts with Tris and 2-mercaptoethanol, these components should be omitted from the reconstitution mixture.
2. At various times during the assembly process, remove 150 μl aliquots and add 12 μl of 0.2 M glutaraldehyde and 20 μl of 1 M borate at pH 8.9, yielding a final glutaraldehyde concentration of 14 mM and pH of 8.5.
3. The cross-linking reaction is stopped after 20 sec by the addition of 15 μl of 1 M $NaBH_4$; samples can be stored at 0°C until run on SDS–polyacrylamide gels (Chapter 3).
4. Because the glutaraldehyde reaction may not cross-link completely all the subunits in every species, control experiments measuring the extent of cross-linking of the intact, native oligomer should be performed. In this way a correction factor can be determined that will enable an accurate assessment of the intermediate and fully-assembled species during re-assembly.

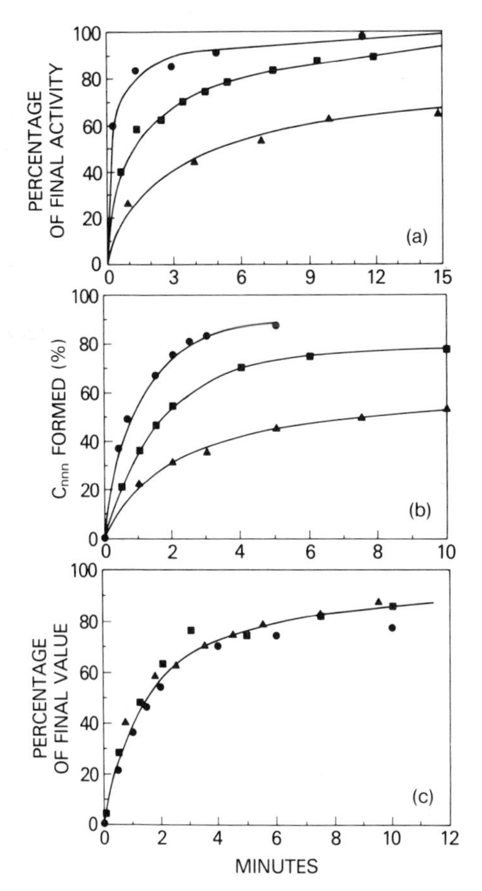

Figure 6. Regeneration of enzyme activity and re-assembly of catalytic trimers from folded monomers. (**a**) Restoration of enzyme activity. Native catalytic subunits at 0.2 mg/ml were dissociated into folded monomers at 25°C by incubation in 1 M NaSCN for 1 h. Re-assembly was initiated by diluting samples at least 20-fold into 40 mM Tris–HCl, pH 7.5, 2 mM 2-mercaptoethanol, 0.2 mM EDTA. At various times, aliquots (10–250 μl) were removed and diluted directly into assay mixtures containing 20 mM L-aspartate and 1 mM carbamoyl phosphate. The enzyme assays were of sufficiently short duration that, except for the earliest times, the bulk of the re-activation occurred prior to the addition of substrates. Assays were performed on all aliquots 24 h after the initiation of assembly to determine the maximum re-activation (the final activity was 90% that of the native subunits). Values are plotted as percentage of final activity versus time. (●) 10 μg/ml; (■) 1 μg/ml; (▲) 0.25 μg/ml. (**b**) Assembly of catalytic trimers (C_{nnn}) measured by hybridization. ^{125}I-Labelled catalytic trimers and succinylated catalytic subunits were converted separately to monomers by treatment with 1.25 M NaSCN at 0°C. At various times after the initiation of assembly of the ^{125}I-labelled monomers, aliquots were removed, and succinylated monomers were added rapidly in a molar ratio of at least 30 to 1. In this way, C_n monomers were trapped as C_{nss}, C_{nn} dimers were trapped as C_{nns}, and fully assembled trimers were unaffected. Thus, the individual species in the four-membered hybrid set could be separated by electrophoresis, and the amount of radioactivity corresponding to C_{nnn} trimers was determined. Because of the low levels of protein in the hybrids, carrier protein was added prior to electrophoresis for ease in visualization of the hybrid set. Results are expressed in terms of the per cent of radioactivity migrating as native C_{nnn} trimers versus time. Protein concentrations in the assembly mixture were 4 μg/ml (●), 1 μg/ml (■), and 0.4 μg/ml (▲). (**c**) Coincidence of the restoration of enzyme activity and trimer formation measured by the hybridization technique illustrated in *Figure 1* and by chemical cross-linking with glutaraldehyde. Summarized in *Table 3*. The assembly of trimers from ^{125}I-labelled monomers at 1 μg/ml was measured by regeneration of enzyme activity (▲), hybridization (●), and cross-linking with glutaraldehyde (■). Results are presented as the per cent of the final values observed for each of the independent techniques. The results from chemical cross-linking needed to be corrected by about 9% due to incomplete reaction of glutaraldehyde with native catalytic chains.

concentration-dependent association reactions would have been much slower and the half-time of the restoration of activity would have varied with concentration, since monomers were devoid of activity. By contrast, if the monomeric product of the folding process is active, first-order kinetics are observed even though the subsequent association reactions to form the oligomer may be slow. In a particularly interesting study of the assembly of muscle aldolase, Chan and his co-workers (44) found first-order kinetics for the reconstitution of active aldolase tetramers from chains produced by GdmCl denaturation of the enzyme. But by a careful search for transient intermediates in the assembly process, they showed that the active monomers produced by the folding process were sensitive to 2.3 M urea. Hence, following the reconstitution by adding urea at various stages of the over-all process enabled them to demonstrate not only that active monomers were produced but also that they were much less stable than the final reconstituted tetramers. All of this work was performed with enzyme assays only. Obviously, coupling such experiments with physical chemical data lends more confidence to the analysis of the process.

6.1.3 *Matrix-bound subunits*

By far the most general approach for determining the activity of individual monomers from oligomeric enzymes makes use of an insoluble, relatively rigid matrix onto which subunits are covalently attached at positions sufficiently remote from one another that association between the immobilized subunits cannot occur (45–47). In this way independent subunits can be assayed directly to determine whether they are active. The technique illustrated schematically in *Figure 7* requires chemical activation of the Sepharose or agarose by CNBr, followed by covalent coupling of the oligomeric enzyme. Experimental conditions are designed so that each oligomer is attached covalently via only one subunit and so that the amount of immobilized enzyme is sufficiently low that the bound molecules are far apart. Exhaustive washing of the matrix-bound oligomers with a denaturant like urea or GdmCl leads to the removal of all the non-covalently bound subunits and subsequent washing with buffer permits the re-folding of the immobilized monomers, which cannot re-associate because of the rigidity of the matrix.

It is essential that the initial covalent attachment involves only one subunit and that the enzyme activity will not be affected by the chemical reaction used for the immobilization. The matrix can be used either in the form of a column or an insoluble suspension, and it is possible to perform accurate quantitative measurements of the amount of oligomer bound as well as its activity. As a consequence, specific activities can be determined for the bound oligomer as well as the re-folded immobilized monomers. Also, as shown in *Figure 7*, the oligomers can be reconstituted on the matrix by adding soluble enzyme in the denaturant, and then washing the matrix to permit re-folding of the chains and association between the covalently-linked subunits and the free subunits. This control experiment is an important part of the procedure since the restoration of the initial

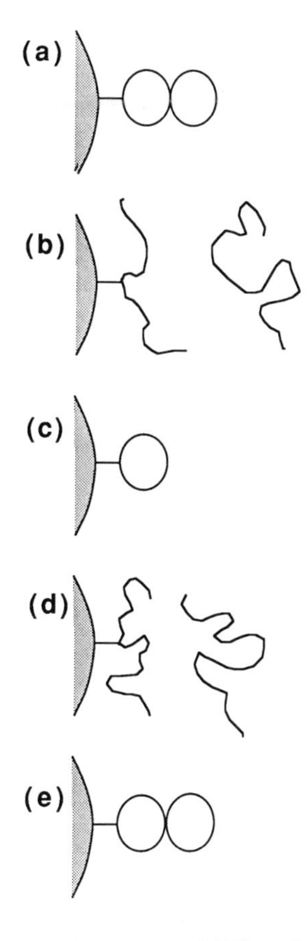

Figure 7. Assays for activity of matrix-bound monomers. (**a**) The soluble oligomer (illustrated here by a dimer) is covalently coupled to CNBr-activated Sepharose to yield the matrix-bound dimers linked through only one subunit. The amount of dimer bound to the column is determined by either nitrogen or protein determinations, and assays of enzyme activity are performed to permit a determination of the specific activity of the immobilized dimers. (**b**) The column is washed with a denaturant such as 8 M urea or 6 M GdmCl to cause dissociation of the dimers and unfolding of the covalently-bound chains. Protein determinations are conducted on both the material remaining and that in the eluate. (**c**) Buffer is passed through the column to remove the denaturant and permit re-folding of the bound chains to yield the folded monomers, which are than assayed for enzyme activity. Quantitative studies of both the amount of protein bound and the activity yield a value of the specific activity. (**d**) A control experiment is performed with the bound monomers by adding unfolded chains in the denaturant and then (**e**) the denaturant is removed so as to allow re-folding of the bound and free chains to determine whether assembly to dimers occurs. The amount of bound protein is measured to determine whether dimers are reconstituted, and assays are performed to measure the specific activity.

specific activity of the oligomer shows that the monomeric chains can re-fold to yield the correct tertiary structure required for oligomer formation.

Matrix-bound subunits from many oligomeric enzymes have been studied, and the technique has provided unequivocal evidence that the monomers from dimeric creatine kinase (48) and from tetrameric aldolase (45) are fully active. In contrast, the same technique showed clearly that the monomers obtained from dimeric phosphoglucose isomerase (49) are inactive. The technique has the added advantage that the stability of both the matrix-bound oligomers and active monomers can be compared with that of the oligomeric enzyme in solution. In that way, Chan (46) was able to demonstrate that the matrix-bound active monomers obtained from aldolase were less stable in dilute urea than the tetrameric enzyme. This result is in complete agreement with the conclusion inferred from the re-assembly studies, which indicated that a transiently-formed active but relatively unstable monomer is formed during the renaturation of urea-treated aldolase (44).

A thorough study by McCracken and Meighen (50, 51) on *E.coli* alkaline phosphatase showed some of the pitfalls of the technique. They demonstrated that the small amount of enzyme activity observed for the matrix-bound protein after the denaturation and re-folding was attributable not to monomers but to re-formed dimers, thus showing clearly that monomers were indeed inactive. The matrix was not completely rigid and neighbouring covalently-attached monomers re-associated on the matrix. Flexibility in the Sepharose matrix that permits subunit interaction between immobilized monomers had been identified previously by Green and Toms (52) to account for the increase of biotin complexed strongly to matrix-bound monomers of avidin at 37°C relative to that observed at 4°C. Cross-linking Sepharose with divinyl sulphone prior to avidin coupling decreased this subunit interaction, and the bound monomers displayed temperature–independent biotin-binding behaviour. These studies illustrate vividly potential problems with this approach, and the precautions and modifications of the technique they described serve as a model for workers planning to use this powerful method.

6.2 Are active sites within monomers, or are they shared between monomers and require the joint participation of amino acid residues from adjacent polypeptide chains?

This question is readily answered for those oligomeric enzymes which upon dissociation yield fully active monomers. For these enzymes, inter-subunit interactions and contacts are not needed for activity but instead serve other roles such as stabilization. However, many oligomers cannot be dissociated without the loss of activity and it is not clear which of the two schemes depicted in *Figure 4* provides the correct interpretation. Similarly, assembly kinetics showing the coincidence of restoration of activity and oligomer formation do not differentiate between the two possible mechanisms (the reverse of the dissociation processes in *Figure 4*). A method of potentially general applicability has been described recently that can demonstrate the existence of shared active sites and the requirement for participation of amino acid residues from adjacent polypeptide chains (53, 54). This technique is based on hybridization experiments whereby active hybrids are formed from two different inactive parental variants of the same oligomeric enzyme. The variants can be produced by chemical modification leading to an alteration in residues critical for catalysis (Chapters 9 and 10) or by genetic techniques involving random or site-directed mutagenesis (Chapter 11).

The rationale for using hybridization experiments to determine whether the oligomer contains active sites within the monomers or at the interfaces between adjacent chains is illustrated schematically in *Figure 8*. As in the earlier discussion of the nature of an active site, the amino acid residues indicated by ◁ and by □ are

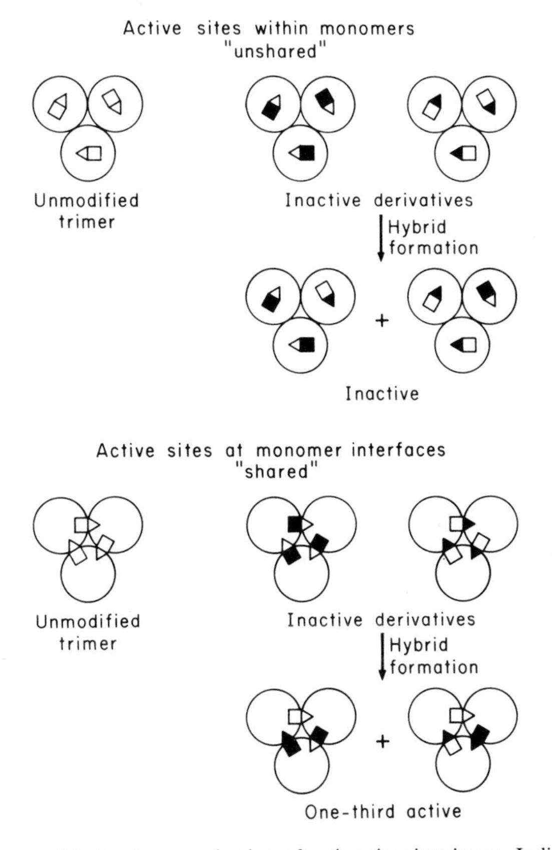

Figure 8. Alternative models for the organization of active sites in trimers. Individual polypeptide chains are represented by large circles and ◁ and □ denote active site residues in the unmodified or wild-type enzyme. Modification of ◁ or □ to yield either ◀ or ■ causes inactivation of the enzyme. A functional active site requires both ◁ and □ in proper proximity. (**Upper**) This model corresponds to active sites that are contained totally within a monomer. Hybridization of the two inactive parental proteins containing ◀ or ■ does not give rise to active enzyme because no monomer in either hybrid contains ◁ and □ in the proper orientation. (**Lower**) This model corresponds to active sites that are located at the interface between adjacent subunits. In this case, the adjoining essential functional residues are contributed from two different monomers. Hence, hybrids formed from two inactive parental proteins will yield species with one functional active site (◁□). Note that each hybrid contains one active site per trimer; thus the specific activity of each hybrid should be 33% that of the initial unmodified, wild-type protein.

considered essential and must be juxtaposed properly for catalytic activity. If either of them is altered chemically or by mutagenesis to give a ◀ or ■, enzyme activity is lost. Hence it is possible, in principle, to form two inactive variants or mutant proteins in which all of the sites are either ◀□ or ◁■. When hybrids are formed from the two inactive parental oligomers, dramatically different results are seen depending on whether the active sites in the original or wild-type protein are within monomers or at the interfaces between adjoining polypeptide chains. As seen in *Figure 8*, enzyme activity is generated from the inactive parental proteins only for those oligomers containing shared sites in which the mutations involve sites from both parts of the shared site. Indeed, according to the scheme illustrated in *Figure 8* for trimers, each of the two hybrids would have one functional active site, and their specific activity would be 33% that of the unmodified trimer. Thus the activation through hybrid formation would be substantial. If the hybridization procedure involved a random assortment of chains yielding one unit of each inactive parental protein and three units of each of the two hybrids, the total activity generated would equal 25% of that of the corresponding amount of native wild-type trimers. It is important to note that the schemes illustrated in *Figure 8* are readily modified for dimeric and tetrameric enzymes. The basic conclusions are the same. If the active sites of these oligomers are contained within the monomers, no activity is formed in the hybrids. For any oligomer containing shared sites, however, very active hybrids can, in principle, be formed from the appropriate inactive parental mutants.

The approach described here was used initially by Robey and Schachman (53) and by Wente and Schachman (54) to determine whether the active sites in the *C* trimers of ATCase were shared as had been inferred from crystallographic and chemical modification experiments. In the studies of Robey and Schachman (53) using a chemically-modified derivative and a mutant produced by site-directed mutagenesis, they were able to measure about a 100-fold increase in enzyme activity merely by mixing the two *C* trimers together in a dilute buffer containing a ligand known to promote hybrid formation. Wente and Schachman (53) showed that incubating two virtually inactive *C* trimers in 0.8 M urea led to a dramatic increase in activity approaching a limiting value about 10^5 times that of the initial solution. Moreover, the final activity was almost equal to that calculated for a binomial distribution resulting from the random re-assortment of polypeptide chains. Variations and extensions of this method have been reported recently for a proteolytic variant of the B2 protein of *E.coli* ribonucleotide reductase (55) and for *in vivo* experiments with two inactive mutants of *Rhodospirillum rubrum* ribulose bisphosphate carboxylase-oxygenase (56, 57).

For many oligomeric enzymes, insufficient information exists to warrant attributing catalytic roles to individual amino acid residues; hence it is difficult to know which residues should be altered so as to provide the variants needed for the aforementioned hybridization experiments. Chemical modification experiments often lead to relatively homogeneous derivatives devoid of activity, as in the inactivation of the *C* trimers of ATCase by treatment with pyridoxal 5′-phosphate and $NaBH_4$ (58). After the altered residue Lys 84 had been identified in this way, it was profitable to construct mutants in which Lys 84 was replaced by Gln or Arg residues (59). These mutants proved to be virtually devoid of activity and, like the chemically-modified trimers, ideal for the hybridization experiments described above. Target residues for modification by site-directed mutagenesis frequently can be selected on the basis of X-ray diffraction studies of complexes between enzymes and substrate analogues.

Increases in enzyme activity as a result of hybrid formation from two relatively inactive variants are not, in themselves, sufficient to warrant conclusions that the active sites are shared between adjacent polypeptide chains. Indeed, slight re-activation, though significant, may not be indicative of shared sites. If one or both of the mutants used for the hybridization experiment is inactive because of an amino acid substitution that interferes with the proper folding of the chains, then re-activation may be attributable to conformational corrections in the hybrid. It is likely that the extent of re-activation resulting from such conformational corrections will not approach that obtained by re-forming a wild-type active site from two inactive mutants. Nonetheless, caution should be used in interpreting the results. This ambiguity in interpreting the results can be overcome by an additional hybridization experiment devised by Wente and Schachman (54).

As seen in *Figure 9*, for trimers containing shared active sites the hybrid set formed from wild-type trimers and a double mutant with ◀ and ■ in place of ◁ and □ leads to four species, two of which are the parental proteins. One hybrid trimer, composed of two wild-type chains and one chain containing the two mutational alterations, has only one active site. The other hybrid containing one wild-type chain and two chains with the double mutant has no active sites. This

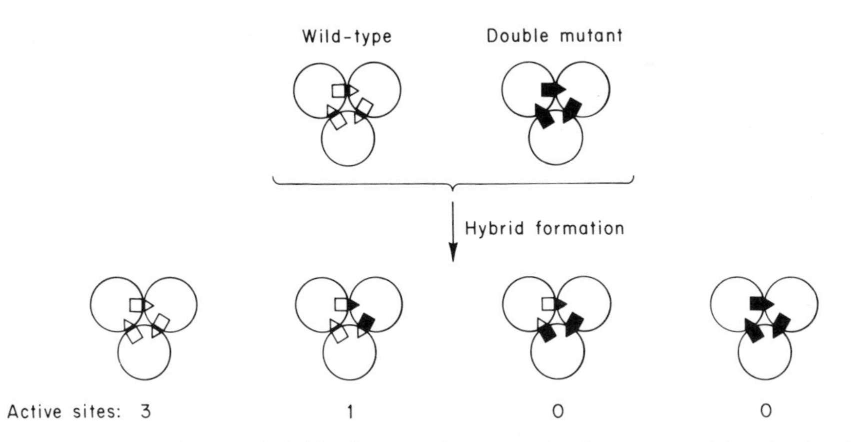

Figure 9. Schematic diagram of hybridization procedure to confirm the presence of shared active sites by inactivation of wild-type trimers. The wild-type enzyme contains essential residues designated by ◁ and □, and the inactive protein has a double mutation represented by ◀ and ■. The inactivation of wild-type trimer by hybridization with the double mutant is illustrated by the four-membered hybrid set that is formed between them. Because the double mutant is missing essential residues on both sides of the subunit interface, only the hybrid with two wild-type chains is active, and it contains only one site. The other hybrid, containing one wild-type chain and two chains with the double mutation, is totally inactive. Thus by the re-arrangement of the chains to form the hybrid set, the total enzyme activity is decreased.

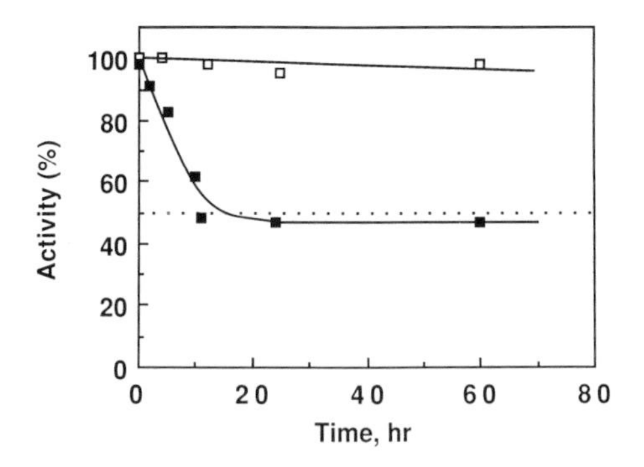

Figure 10. Inactivation of wild-type catalytic subunit by trimers containing double mutations. Mixtures of equal amounts of wild-type *C* trimers and inactive double mutants (Lys 84 → Gln and His 134 → Ala) at 2 mg/ml were incubated at 4°C in 50 mM Tris acetate, pH 8.3, 2 mM 2-mercaptoethanol, 0.2 mM EDTA in the absence (□) or presence (■) of 10 mM sodium pyrophosphate. Aliquots were removed at various times and diluted 1:1000 into 40 mM potassium phosphate, pH 7.0, containing 4 mM carbamoyl phosphate and 10 mM L-aspartate. Activity was measured spectrophotometrically as described by Foote (60). The results are plotted as activity in per cent of the initial value for the mixture versus time. The progressive decrease in activity to 46% of the initial value is in excellent agreement with the predicted 50% decrease in activity based on random hybridization of active and inactive parental species to form a binomial distribution of species in the hybrid set.

negative complementation experiment leading to inactivation of wild-type polypeptide chains as a result of its association into trimers with the mutant chains can be demonstrated readily by incubating the two proteins together under conditions that promote hybrid formation. Moreover, the extent of inactivation can be calculated for a random reorganization of the chains in the same manner as that used for forming active hybrids from two inactive parental mutants. The results of such an experiment with wild-type *C* trimers of ATCase and a double mutant in which Lys 84 is replaced by Gln and His 134 is replaced by Ala are shown in *Figure 10*. The gradual loss of activity promoted by 10 mM pyrophosphate is readily apparent, and the final activity is close to that expected for a binomial distribution of the various species in the 4-membered hybrid set (S. R. Wente and H. K. Schachman, unpublished).

6.3 **What is the specific activity of hybrid molecules composed of polypeptide chains from two different mutants?**

In the previous section we showed that the formation of active hybrids from two different inactive parental proteins constituted strong evidence for the existence of shared active sites in the wild-type oligomer. However, that conclusion was based strictly on the observed increase in activity when the two defective proteins were incubated together under conditions favouring hybridization. Clearly the isolation of each hybrid followed by quantitative assays of the specific activity would constitute more convincing evidence than the measured increase in activity of the incubation mixture and its correlation with the value expected on the basis of a binomial distribution. A similar conclusion pertains to the negative complementation experiment illustrated by *Figures 9* and *10*. The isolation of the relevant hybrids in pure form is highly desirable. This is generally not a simple task because the parental proteins often do not differ significantly from one another in electrophoretic or chromatographic behaviour. In principle the scheme illustrated in *Figure 1*, where one protein is succinylated, would suffice. But succinylation is essentially irreversible and the modification itself frequently causes inactivation. Hence what is needed is a reagent that can be used to modify proteins reversibly. The goal is to introduce charged groups on one of the two parental proteins, form the hybrids by some dissociation and reconstitution process, separate the members of the hybrid set by ion-exchange chromatography and then remove the charged groups to yield the individual hybrids as well as the recovered parental proteins.

A general method for the reversible modification of proteins and its application for the isolation of hybrid molecules was described by Gibbons and Schachman (61) and is illustrated in *Figure 11*. The acylation with 3,4,5,6-tetrahydrophthalic anhydride (THPA) leads principally to modification of amino groups, thereby producing a net change of −2 in the charge on the protein for each group acylated. Depending on the protein, dissociation of the oligomers can be promoted by urea, NaSCN or specific ligands like 10 mM pyrophosphate, which is extremely effective for the *C* trimers of ATCase. When the perturbant is removed from a mixture of the two proteins, the reconstitution, if random, leads to a hybrid set composed of four species. Because of the introduced charges, these species differ in their chromatographic behaviour and, following their separation, the proteins are incubated at pH 6.0 for about 25 h, which leads to virtually complete deacylation and restoration of the original amino groups.

A protocol for the use of THPA is given in *Table 4*. The extent of reaction with

Table 4. Reversible tetrahydrophthaloylation of oligomeric proteins.

1. Prepare the protein solution at approximately 5 mg/ml in 50 mM phosphate buffer, 2 mM 2-mercaptoethanol, 0.2 mM EDTA, pH 8.2, at 4°C.
2. Prepare 0.5 M 3,4,5,6-tetrahydrophthalic anhydride (THPA) in acetone or freshly-distilled dioxane (ε_{255} = 4470 M^{-1} cm^{-1}).
3. For the *C* subunit of ATCase, addition of 0.6 equivalents of THPA per amino group (15 amino groups per polypeptide chain) yields a fairly homogeneous derivative with approximately four acylated lysine residues per chain. With other proteins, this value should be determined empirically in titration experiments by varying the ratio of THPA to protein.
4. Add the necessary amount of THPA all at once with stirring at 4°C. Maintain the pH between 7.8 and 8.2 with the addition of ice-cold 0.5 M NaOH. Allow the reaction to proceed 10 min once the pH stops decreasing.
5. The extent of modification can be assessed quantitatively by determining the remaining amino groups with either ninhydrin as described by Moore and Stein (62) as modified by Fraenkel-Conrat (67) or with trinitrobenzene sulphonate (TNBS) using the method of Habeeb (64). Alternatively, the extent of reaction can be measured qualitatively by the increase in absorbance at 250 nm.
6. Dialyse modified protein solution at pH of 8 or above to inhibit undesired deacylation. The tetrahydrophthaloylated oligomer is then used in hybridization experiments.
7. Removal of the tetrahydrophthaloyl groups from modified lysine residues is accomplished by dialysis of the acylated derivative at pH 6 for 25–40 h, which yields complete deacylation and restoration of functional amino groups.

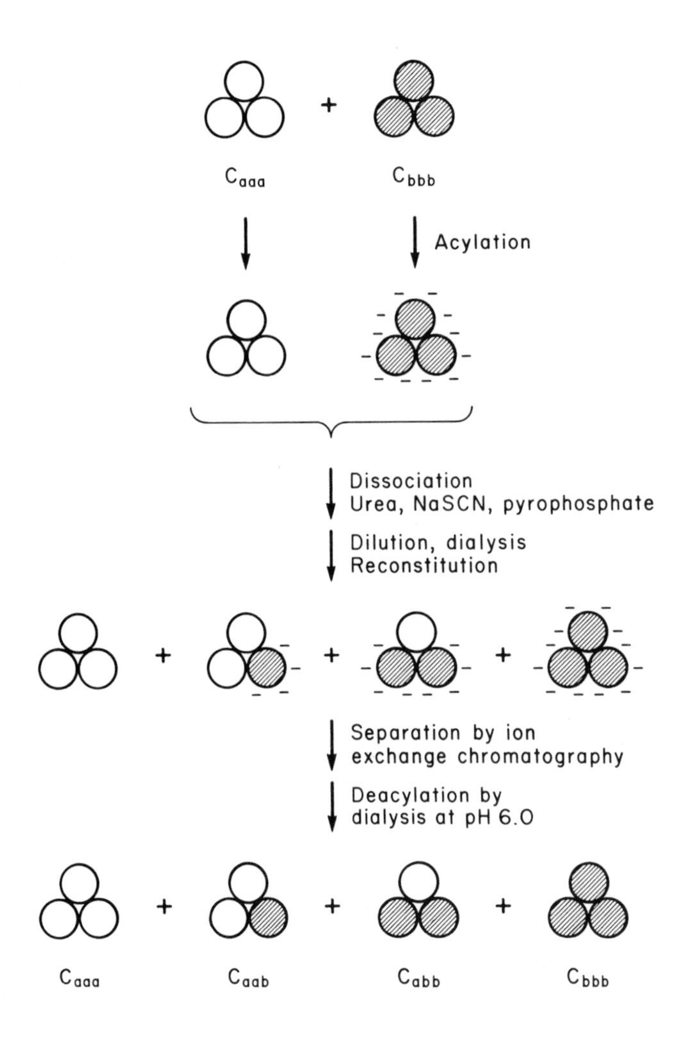

Figure 11. Protocol for the construction and purification of individual hybrid *C* trimers. (**Step 1**) Choose two parental oligomers to hybridize, for example one whose subunits are designated by open circles, C_{aaa}, and one whose subunits are represented by the cross-hatched circles, C_{bbb}. (**Step 2**) One of the parental proteins, C_{bbb}, is acylated with THPA (*Table 4*), yielding a derivative that is altered substantially in its net charge. (**Step 3**) The parental proteins are mixed in identical amounts and dissociation is achieved by the addition of an appropriate reagent, such as urea, NaSCN or, in the case of *C* trimers of ATCase, 10 mM pyrophosphate. (**Step 4**) Reconstitute hybrid enzymes by removing the denaturant to allow random subunit association which yields, in this case, a four-membered hybrid set. (**Step 5**) Hybrid species are fractionated using ion-exchange chromatography, yielding the purified species. (**Step 6**) Deacylation is effected by dialysis at pH 6.0 for 24–48 h at room temperature, yielding individual hybrid enzymes, C_{aab} and C_{abb}, as well as the two parental proteins, C_{aaa} and C_{bbb}.

THPA with ε-amino groups can be determined qualitatively by the increase in absorbance at 250 nm; generally the slight modifications required to introduce the desired number of charged groups will produce only a small increase in the absorbance of the protein at 280 nm. Since there are many ε-amino groups in proteins and they differ somewhat in reactivity, a heterogeneous population of modified molecules results from the reaction of oligomeric proteins with THPA. Too much modification often leads to dissociation of the oligomers, perhaps because the attendant net increase in negative charge causes too much electrostatic repulsion. Obviously this must be avoided. So also should underacylation be avoided, because this leads to greater heterogeneity and little change in net charge, so that separation of members of the hybrid set is difficult. There clearly is an optimum extent of acylation and this will doubtless vary for different proteins. But it is not difficult to determine this optimum empirically by titrating the oligomer with varying amounts of THPA.

The acylated derivative is stable at slightly alkaline pH (~8.0) even at room temperature, yet the acylation is fully reversible under slightly acid conditions (~ pH 6.0) so mild as not to destabilize the hybrid oligomers. Gibbons and Schachman (61) showed that the deacylation rate was first-order and increased with decreasing pH. The half-time for deacylation of amino groups at room temperature is 8 h at pH 6 and greater than 800 h at pH 8. Moreover, after limited modification of both aldolase and the *C* subunit of ATCase, they were able by deacylation to recover fully active enzymes with electrophoretic mobilities indistinguishable from the unmodified proteins.

Limitations in the use of THPA as a reversible acylating group include those expressed for the succinylation of proteins described in *Table 2*. Because some deacylation occurs in neutral or even slightly alkaline solutions, the fractionation of the members of the hybrid set must be conducted as rapidly as possible. In addition the deacylation procedure requires that the hybrid oligomers be stable in dilute solutions at room temperature and pH 6 for about 24 h.

This procedure for isolating trimers from different mutants of ATCase has been used successfully in demonstrating that the active sites in the enzyme are shared and require the joint participation of amino acid residues from adjacent polypeptide chains (54). *Table 5* summarizes the data illustrating that the active hybrids formed from two defective mutants contain one active site per trimer. Also, the results from a hybridization experiment involving the wild-type trimers and an inactive double mutant show that the hybrid containing two wild-type chains and one chain with the double mutant has one active site and that the hybrid containing one wild-type chain and two chains with the double mutant is inactive. This is precisely the result predicted in the negative complementation experiment illustrated in *Figure 9*. The widespread application of this recently-developed hybridization technique doubtless will depend on the ability of research workers to dissociate other oligomeric proteins reversibly, so that the reconstituted hybrid oligomers can be isolated and studied in terms of their functional properties.

The complementation technique described here can probably be adapted to form active hybrid oligomers from two defective proteins, one of which is first

Table 5. Specific activity of purified hybrid catalytic trimers of ATCase.

Trimer	K_m *(aspartate)* *(mM)*	V_{max} *(μmol/h/μg)*	*% Activity relative to* C_{WWW}
Experiment 1a[a]			
C_{WWW}	7.2	34	100
C_{LLL}	—[c]	0.003	0.008
C_{LLS}	4.7	11.6	34
C_{LSS}	7.9	10.8	32
C_{SSS}	—	0.001	0.003
Experiment 2[b]			
C_{WWW}	7.2	33	100
C_{WWD}	6.6	11	33
C_{WDD}	—	0.005	0.015
C_{DDD}	—	<0.003	<0.009

Trimers were purified as described in the text. Enzyme assays at various aspartate concentrations were conducted at 30°C in 50 mM imidazole acetate, pH 7.0, 0.2 mM EDTA, 4.8 mM carbamoyl phosphate, using the method of Prescott and Jones (66) as modified by Pastra-Landis *et al.* (66). Data were fit to the Michaelis–Menten equation. *C* trimers are designated by subscripts representing the type of each of the polypeptide chains in the trimers. The subscript W corresponds to a wild-type polypeptide chain; L represents a mutant chain in which Lys 84 is replaced by Gln; *S* corresponds to a mutant in which Ser 52 is replaced by His; and D represents a chain with a double mutation in which Lys 84 is replaced by Gln and Ser 52 by His.

[a] Results from Wente and Schachman (54).

[b] Results from Raumann, Wente and Schachman (unpublished).

[c] Because of the low values of V_{max}, accurate determinations of K_m were not possible.

immobilized covalently as a monomer to a Sephadex matrix. By adding the other protein in a denaturant according to the procedure outlined in *Figure 7*, a pure hybrid of known composition could be produced. If the protein originally immobilized on the matrix is linked by a cleavable bond, pure hybrids could be obtained in solution without the need for the fractionation procedure described above. It is worth noting that complementation experiments of a slightly different form have been performed using immobilized enzymes. When either dimeric phosphoglucose isomerase (49) or dimeric glycogen phosphorylase (67) are coupled covalently to a Sepharose matrix and a denaturant is used to disrupt quaternary interactions between subunits, the resulting immobilized monomers are inactive. By mixing these folded, inactive monomers with inactive polypeptides produced either by chemical modification with pyridoxal 5′-phosphate as in the case of phosphoglucose isomerase (49) or by using inactive pyridoxal 5′-phosphate derivatives for glycogen phosphorylase (67), enzyme activity could be reconstituted. Thus active oligomeric enzymes were formed from a potentially active subunit and an intrinsically inactive subunit. This result could be the consequence of forming the requisite shared site. However, the findings in themselves do not preclude the alternative interpretation that a conformational change occurred as a consequence of oligomer formation. Distinguishing between these alternative views requires complementation between two intrinsically inactive monomers.

7. EVALUATION OF CONFORMATIONAL CHANGES IN OLIGOMERIC PROTEINS

Conformational changes in proteins resulting from their interaction with substrates, analogues and regulatory effectors often have been invoked to account for their functional behaviour. Yet in spite of the enormous number of reports, rarely is it possible to answer many fundamental questions. Does a biologically significant change in a protein resulting from an interaction with a ligand involve only a few or many atoms? Do these atoms undergo small or large displacements from their equilibrium positions in the unliganded protein? Are effects stemming from ligand binding to one chain propagated to other chains to produce conformational changes in the unliganded chains? Do conformational changes occur rapidly (within milliseconds) or slowly (over seconds)? Answers to some of these questions can be obtained but we will concentrate on only those that have been of special interest in studies of the regulatory properties of ATCase.

7.1 **Do conformational changes occur in subunits upon their association to form an oligomer?**

As indicated previously, the isolated C subunits of ATCase exhibit Michaelian kinetics and upon their interaction with R subunits to form the holo-enzyme, sigmoidal kinetics are observed and the resulting enzyme is inhibited by CTP and activated by ATP (23). In order to determine whether structural changes occur in the C trimers as a result of the assembly of ATCase, Kirschner and Schachman (68, 69) used C trimers that were modified with tetranitromethane. The resulting nitrated C trimers (C_{nit}), containing about 0.8 nitrotyrosyl groups per polypeptide chain, retained about 90% of the original catalytic activity and were fully competent to combine with R subunits to yield ATCase-like molecules exhibiting the sigmoidal kinetics and feedback inhibition characteristic of the native enzyme. By measuring the absorbance at 430 nm (due to the nitrotyrosyl chromophore) they showed that there was a substantial decrease in the absorbance as a result of the assembly process. This result showed clearly that the environment of the nitrotyrosyl groups in the holo-enzyme differed from that of the free C_{nit} subunits. Moreover, spectral titration of the C_{nit} trimers and $(C_{nit})_2R_3$ as a function of pH demonstrated that a significant change in the pK of the nitrotyrosyl residues had occurred. It was clear, therefore, that assembly of the subunits into the holo-enzyme had caused a conformational change affecting the tertiary structure of the chains containing the nitrotyrosyl groups. But from this experiment alone the functional significance of this spectral perturbation was not evident. Hence, additional studies were required to determine the effect of active-site ligands on the environment of the chromophore.

7.2 Do global conformational changes occur in the quaternary structure upon the binding of ligands?

Since the studies with the isolated C_{nit} trimers showed that the addition of succinate, an analogue of the substrate aspartate, in the presence of carbamoyl phosphate, caused a decrease in the absorbance at 430 nm, it was possible to use the chromophore as a spectral probe for analysing the local changes in the reconstituted allosteric derivative $(C_{nit})_2R_3$. As seen in *Figure 12*, there was a pronounced change in the absorbance as a function of the succinate concentration. Thus the spectral probe provided an indication of the occupancy of the active sites by succinate. In an effort to estimate the extent of the linkage between the global allosteric transition and site occupancy, Kirschner and Schachman (68) also measured the decrease in the sedimentation coefficient of the protein. The disparity between the two titration curves shown in *Figure 12* is striking. A significant fraction of the enzyme molecules was converted to the swollen conformation before much succinate was bound. These results indicate that even low succinate concentrations are sufficient to release the quaternary constraint of the low-activity state of ATCase. Moreover, the global transition to the high-activity state is complete even though the active-sites are not fully saturated.

Although this observation is completely consistent with a concerted transition from the low-activity to the high-activity state of ATCase, it also seemed possible that the spectral change could be attributed to two distinct phenomena. On the one hand, a change in absorbance due to the binding of succinate to a polypeptide chain containing a nitrotyrosyl group certainly must occur. The results with C_{nit} trimers demonstrated this type of conformational change. On the other hand, was it conceivable that the global conformational change revealed by the decrease in the sedimentation coefficient was also accompanied by a spectral change even in unliganded polypeptide chains? Could this putative indirect spectral change be in the opposite direction to the clearly measurable direct change observed with C_{nit}

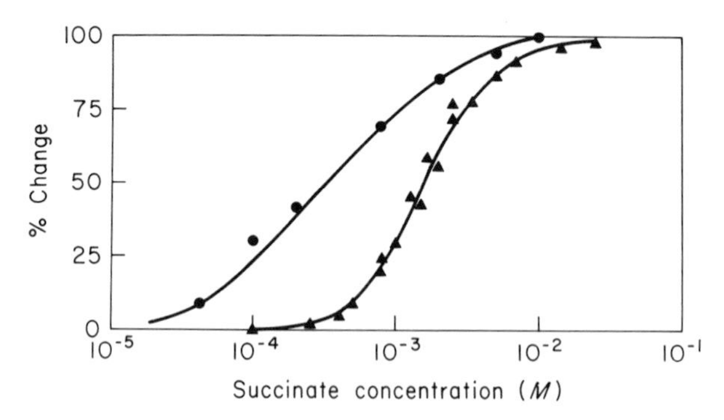

Figure 12. Relationship between local and global conformational changes in $(C_{nit})_2R_3$. Local changes due to occupancy of the active sites by succinate in the presence of carbamoyl phosphate (2 mM) were measured by the per cent decrease in the absorbance of the nitrotyrosyl groups at 430 nm. The titration of this spectral change as a function of succinate concentration is indicated by ▲. Global changes as a function of succinate concentration in the presence of carbamoyl phosphate (2 mM) were measured by the per cent decrease in the sedimentation coefficient (●), which provides a measure of the swelling of the enzyme molecules (68).

trimers? Answering these questions required the construction of another hybrid described in the next section.

7.3 Is there cross-talk or communication between polypeptide chains in an oligomer?

To establish inter-subunit communication between liganded and unliganded polypeptide chains in $(C_{nit})_2R_3$, it was necessary to separate the direct contribution of ligand binding on the spectrum of the chromophore from the possible indirect effect on the nitrotyrosyl groups stemming from the global conformational change indicated by the decrease in the sedimentation coefficient. Therefore, Yang and Schachman (70) constructed an ATCase-like hybrid with normal unmodified R subunits, one native C subunit which, of course, bound ligands, and a second C subunit that was modified chemically so as to contain the sensitive nitrotyrosyl chromophores and to be inactive and incapable of binding ligands. The hybrid, to be useful for determining whether there was cross-talk between subunits, also had to exhibit allosteric kinetic properties and the ligand-promoted swelling indicated by the decrease in the sedimentation coefficient.

The hybrid designated, $C_NC_{nit,P}R_3$, where C_N refers to the native C subunit, $C_{nit,P}$ designates the trimer in which each chain is nitrated (69) and pyridoxylated according to the procedure of Greenwell *et al.* (58) to render the trimers inactive, and R_3 refers to the three R subunits, was constructed by a series of chemical modifications summarized in *Figure 13*. The nitration of C_N was performed first to give C_{nit}, which was characterized as described in the previous sections. Then C_{nit} was treated with pyridoxal 5′-phosphate and $NaBH_4$ to yield the inactive doubly-modified derivative $C_{nit,P}$ that no longer bound substrate analogues. Thus, unlike C_{nit}, the pyridoxylated and nitrated C trimers showed no change in spectrum upon the addition of succinate in the presence of carbamoyl phosphate. Subsequent steps in the protocol illustrated in *Figure 13* involved acylation of $C_{nit,P}$ with THPA to introduce negative charges as shown in *Figure 11*, reconstitution of ATCase-like molecules by mixing equal amounts of C_N and $C_{nit,P,T}$ and an excess of R subunits, fractionation of the three-membered hybrid set, and then deacylation of the hybrid to give $C_NC_{nit,P}R_3$.

The inter-subunit hybrid $C_{nit,P}R_3$ exhibited sigmoidal kinetics with a V_{max} almost exactly 50% that of the wild-type enzyme, consistent with the fact that the hybrid contained only three functionally active sites compared with the six in wild-type ATCase. In addition, the hybrid was inhibited by CTP and activated by ATP. When succinate was added to $C_NC_{nit,P}R_3$ in the presence of carbamoyl phosphate, there was a marked change in the visible absorption spectrum indicating a perturbation in the environment of the chromophores on the unliganded polypeptide chains. The difference spectrum caused by the binding of succinate is presented in *Figure 14a*. It is striking that there was an *increase* in absorbance at 430 nm, in marked contrast to that observed with $(C_{nit})_2R_3$ or C_{nit}, where the direct effect of succinate binding to the chains containing the chromophore leads to a *decrease* in the absorbance. An additional control was performed with a hybrid prepared by a procedure similar to that illustrated in *Figure 13*. In the hybrid,

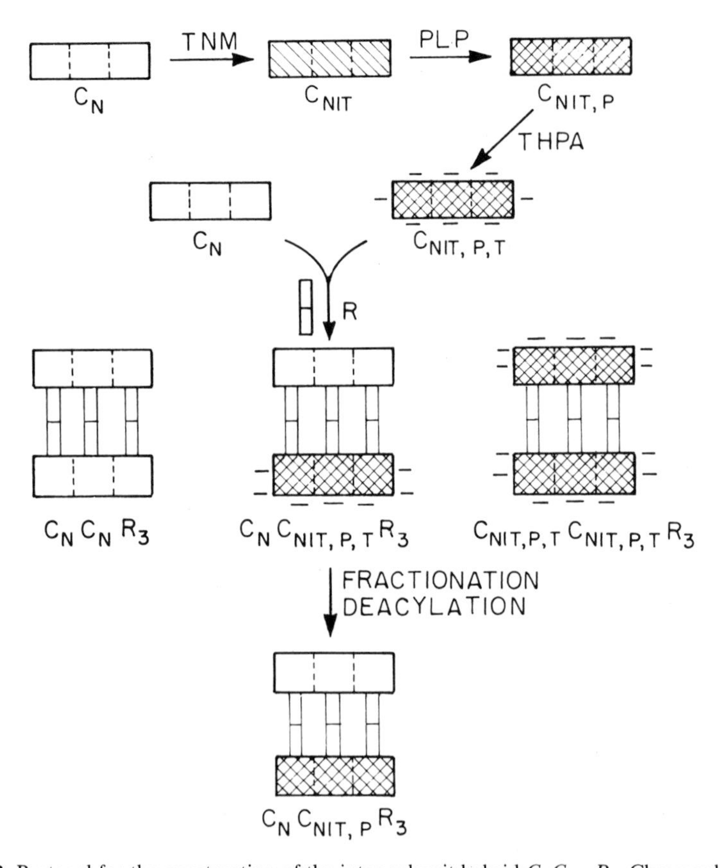

Figure 13. Protocol for the construction of the inter-subunit hybrid $C_NC_{nit,P}R_3$. Chromophores were introduced by treating C_N subunits (horizontal rectangle divided into three parts to indicate their trimeric structure) at 3 mg/ml with 20 mM tetranitromethane (TNM) in the presence of 20 mM carbamoyl phosphate and 50 mM succinate for 75 min at room temperature (69). The resulting derivative, C_{nit}, with 0.6–0.8 nitrotyrosyl residues per *c* chain, exhibited 85–90% of the enzyme activity of C_N. C_{nit} was inactivated by specific modification of a single lysyl residue with pyridoxal 5′-phosphate (PLP), followed by reduction of the Schiff base with $NaBH_4$ (58, 71). The doubly-modified derivative, $C_{nit,P}$, contained 0.98 residue of pyridoxamine 5′-phosphate per polypeptide chain and exhibited 2.4% the activity of C_N. All subsequent operations were performed with aluminium foil in a partially-darkened laboratory to minimize photo-reactivation due to the light-catalysed splitting of the pyridoxamine 5′-phosphate moiety from the protein (79). The net negative charge of $C_{nit,P}$ was increased by the reaction with THPA (*Table 3*) at a molar ratio of 0.6 THPA per protein lysine groups (61). The negative charges due to the acyl groups are illustrated by minus signs in the triply-modified protein, $C_{nit,P,T}$. ATCase-like molecules were reconstituted by mixing C_N and $C_{nit,P,T}$ and adding excess native regulatory subunits (vertical rectangles divided into two sections to indicate their dimeric structure). The final concentrations in the reconstitution mixtures were 2 mg/ml for C_N and $C_{nit,P,T}$ and 3 mg/ml for regulatory subunits. The three complexes in the hybrid set, $C_NC_NR_3$, $C_NC_{nit,P,T}R_3$ and $C_{nit,P,T}C_{nit,P,T}R_3$ differed in charge as shown schematically; they were separated by chromatography on DEAE–Sephadex by the procedure of Meighen *et al.* (25) as modified by Gibbons *et al.* (71). The pooled fraction containing $C_NC_{nit,P,T}R_3$ was then incubated for 1–2 days at room temperature in 40 mM potassium phosphate, pH 6.1, 2 mM 2-mercaptoethanol, 0.2 mM EDTA. This procedure led to the formation of the hybrid $C_NC_{nit,P}R_3$ with the active chains in one subunit, C_N, and the spectral probes on inactive chains in $C_{nit,P}$.

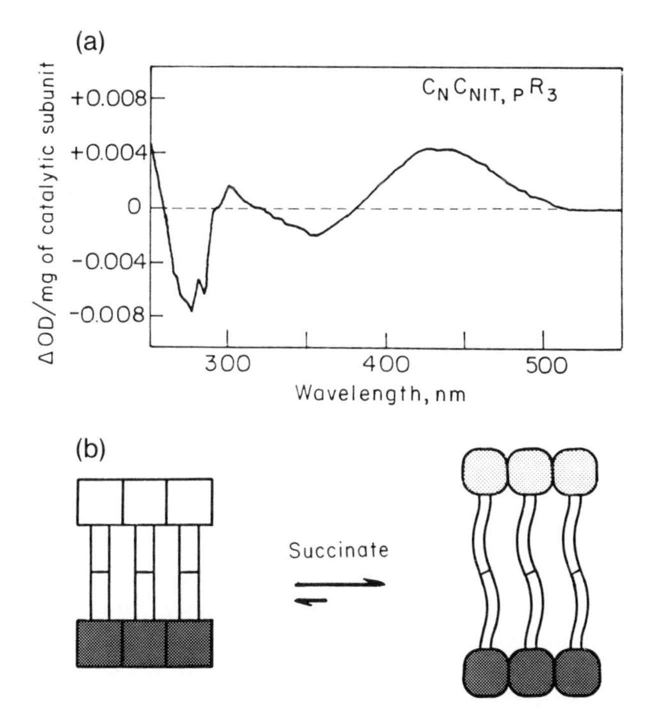

Figure 14. Global nature of succinate-promoted conformational change in the inter-subunit hybrid $C_NC_{nit,P}R_3$ (**a**) Communication difference spectrum resulting from succinate binding to the inter-subunit hybrid $C_NC_{nit,P}R_3$. Identical protein solutions were placed in the two cuvettes for determination of a base line. Then 2 mM carbamoyl phosphate was added to one cell and an equivalent volume of buffer was added to the other. Difference spectra caused by the addition of carbamoyl phosphate were measured and showed a broad positive peak at 440–460 nm due to the increased absorbance of the nitrotyrosyl residues. Succinate was then added to the cell containing carbamoyl phosphate, and an equal volume of buffer was added to the reference cell. Spectra were again measured, and the values for the perturbation due to carbamoyl phosphate were subtracted from those caused by the addition of succinate (5 mM) to give the results presented. Carbamoyl phosphate binds to the pyridoxylated chains and causes a marked change in the spectrum at 330 nm. In addition, carbamoyl phosphate causes an increase in the absorbance in the nitrotyrosyl residues; the positive peak in the difference spectrum was at about 490 nm. Succinate binds only to the C_N subunit, and the large positive peak at 450 nm corresponds to changes in the tertiary structure of the unliganded chains in $C_{nit,P}$. (**b**) Diagrammatic illustration of the effect of succinate binding on the global conformation of the inter-subunit hybrid $C_NC_{nit,P}R_3$. The unliganded inter-subunit hybrid, illustrated on the left side of the equilibrium, is composed of one native catalytic subunit, C_N (horizontal rectangle divided into three parts to indicate its trimeric structure), one catalytic subunit that had been previously nitrated and pyridoxylated as described in the legends to *Figure 12* and *Figure 13* (shaded horizontal rectangle), and three regulatory subunits (vertical rectangles divided in two to indicate their dimeric structure). Upon binding the inhibitor succinate to the C_N subunits in the hybrid in the presence of carbamoyl phosphate, indicated by the light shading on the right side of the equilibrium, the tertiary and quaternary structure of the enzyme changes. The conversion of the enzyme to the 'swollen', more active conformation is coincident with a change in the tertiary structure of the C_N, R and $C_{nit,P}$ subunits, indicated by a rounding of the subunits. The absorbance change of the nitrotyrosyl chromophores at 450 nm reflects the change in the global structure of the enzyme, even though their location is on the unliganded catalytic chains.

$C_{nit}C_PR_3$, the chains containing the nitrotyrosyl groups were active and therefore bound succinate when carbamoyl phosphate was present, while the other subunit, C_P, was inactive. This hybrid, which also exhibited allosteric kinetics, permitted an evaluation of the direct effect of ligand binding. The succinate difference spectrum was the opposite of that shown in *Figure 14a*; that is, there was a decrease in absorbance at about 450 nm when succinate was bound.

Thus the communication difference spectrum observed for $C_NC_{nit,P}R_3$ demonstrates clearly the global nature of the ligand-promoted conformational change, which is illustrated diagrammatically in *Figure 14b*. Ligands binding to the active chains cause a change in the entire molecule with the result that the unliganded chains undergo an alteration in structure affecting the environment of the nitrotyrosyl groups. It is of interest that this indirect effect on the absorption spectrum is in the opposite direction to the direct effect. Hence in the titration of $(C_{nit})_2R_3$ there are two opposing spectral changes as succinate is bound. One, the decrease in absorbance, is linked to the occupancy of the active sites and the other, the increase, is a response to the global swelling of the molecules. These opposing spectral perturbations doubtless account for the lack of a spectral effect of succinate in the region of 10^{-4} M in *Figure 12*. It is worth noting that the titration of $C_NC_{nit,P}R_3$ with succinate as measured by the communication difference spectrum, the decrease in the sedimentation coefficient and the enhanced reactivity of the sulphydryl groups on the *r* chains yielded superimposable curves. All three methods provide a direct indication of the global conformational change associated with the allosteric transition whereby the low-activity form of the enzyme is converted to the high-activity state.

8. SUMMARY

The various roles for subunits in the architecture, stability and biological function of oligomeric proteins can be established by the application of a wide assortment of biochemical techniques, some of which have been discussed generally in this chapter. Using the simple but powerful approach of combining chemical cross-linking and subunit hybridization will permit the determination of the subunit arrangement and stoichiometry of oligomers, supplying important structural information needed to develop mechanisms to account for the functional behaviour of proteins. Determining the strength of the interactions between both similar and dissimilar polypeptides in multi-subunit proteins can provide the thermodynamic rationale to explain the effects of substrates, inhibitors and effector ligands on the biological activity of oligomers. Moreover, the impact of mutational alterations that affect inter-subunit interactions can be studied quantitatively using the convenience of subunit exchange techniques.

Since the over-all goal of numerous studies concerning multi-subunit proteins is to relate the inter-dependence of oligomer structure to biological activity, the question raised earlier, 'Why subunits?' rests upon, in many cases, evaluating whether isolated, folded monomers are in themselves active, or whether the formation of active protein is coincident with the acquisition of quaternary structure. Methods described here for the preparation, fractionation and isolation of purified hybrid oligomers comprise a powerful strategy advantageous to resolving this issue by eliminating the ambiguities inherent in many other techniques. Hybrid trimers of the catalytic subunit of ATCase, composed of various amino acid replacements, were proven extremely valuable in establishing that, perhaps for many enzymes, activity is dependent upon oligomer formation because active sites require the joint participation of amino acid residues from adjacent chains, rather than some conformational adjustment in the active site upon assembly of the oligomer. The use of hybrid ATCase-like molecules composed of native and chemically-modified subunits have, in addition, been valuable in evaluating conformational changes in subunits as a consequence of their association and in demonstrating changes in the global structure of hybrid molecules, stemming from ligand binding to active subunits and communicated to the inactive subunits that are incapable of binding ligand.

9. ACKNOWLEDGEMENTS

This work was supported by United States Public Health Service grant GM 12159 from the National Institute of General Medical Sciences and by National Science Foundation Research grant DMB 85-02131 (to H.K.S.), and by a National Research Service Award GM 11067 (to E.E.).

10. REFERENCES

1. Svedberg, T. and Pedersen, K. O. (1940) *The Ultracentrifuge*. Clarendon Press, Oxford.
2. Klotz, I. M., Darnall, D. W. and Langerman, N. R. (1975) In *The Proteins, Third Edition*. Neurath, H. and Hill, R. L. (eds), Academic Press, New York, Vol. 1, p. 293.
3. Foster, D. L. and Fillingame, R. H. (1982) *J. Biol. Chem.*, **257**, 2009.
4. Monod, J., Wyman, J. and Changeaux, J.-P. (1965) *J. Mol. Biol.*, **12**, 88.
5. Ebner, K. E. and Magee, S. C. (1975) In *Subunit Enzymes: Biochemistry and Functions*. Ebner, K. E. (ed.), Marcel Dekker, Inc., New York, p. 137.
6. Hill, R. L. and Brew, K. (1975) *Adv. Enzymol.* **43**, 411.
7. Minton, A. P. (1981) *Biopolymers,* **20**, 2093.
8. Minton, A. P. (1983) *Mol. Cell. Biochem.*, **55**, 119.
9. Fulton, A. B. (1982) *Cell,* **30**, 345.
10. Srere, P. A. (1980) *Trends Biochem. Sci.*, **5**, 120.
11. Srere, P. A. (1981) *Trends Biochem. Sci.*, **6**, 4.
12. Kornberg, A. (1988) *J. Biol. Chem.*, **263**, 1.
13. Cheung, W. Y. (1971) *J. Biol. Chem.*, **246**, 2859.
14. Cheung, W. Y. (1980) *Science,* **207**, 19.
15. Klee, C. B. and Vanaman, T. C. (1982) *Adv. Protein Chem.*, **35**, 213.
16. Yanofsky, C. and Crawford, I. P. (1972) In *The Enzymes, Third Edition*. Boyer, P. D. (ed.), Academic Press, New York, Vol. VII, p. 1.
17. Henning, U., Helinski, D. R., Chao, F. C. and Yanofsky, C. (1962) *J. Biol. Chem.*, **237**, 1523.
18. Wilson, D. A. and Crawford, I. P. (1965) *J. Biol. Chem.*, **240**, 4801.
19. Höberg-Raibaud, A. and Goldberg, M. E. (1977) *Biochemistry,* **16**, 4014.
20. Miles, E. W. (1979) *Adv. Enzymol.* **49**, 127.
21. Miles, E. W., Bauerle, R. and Ahmed, S. A. (1987) In *Methods in Enzymology*. Kaufman, S. (ed.), Academic Press, New York, Vol. 142, p. 398.
22. Gerhart, J. C. and Pardee, A. B. (1962) *J. Biol. Chem.*, **237**, 891.
23. Gerhart, J. C. and Schachman, H. K. (1965) *Biochemistry,* **4**, 1054.
24. Yang, Y. R., Kirschner, M. W. and Schachman, H., K. (1978) In *Methods in Enzymology*. Hoffee, P. A. and Jones, M. E. (eds), Academic Press, New York, Vol. 51, p. 35.
25. Meighen, E. A., Pigiet, V. and Schachman, H. K. (1970) *Proc. Natl. Acad. Sci. USA,* **65**, 234.

26. Nagel, G. M. and Schachman, H. K. (1975) *Biochemistry,* **14**, 3195.
27. Cohlberg, J. A., Pigiet, Jr., V. P. and Schachman, H. K. (1972) *Biochemistry,* **11**, 3396.
28. Davies, G. E. and Stark, G. R. (1970) *Proc. Natl. Acad. Sci. USA,* **66**, 651.
29. Peters, K. and Richards, F. M. (1977) *Ann. Rev. Biochem.,* **46**, 523.
30. Burns, D. L. and Schachman, H. K. (1982) *J. Biol. Chem.,* **257**, 8638.
31. Burns, D. L. and Schachman, H. K. (1982) *J. Biol. Chem.,* **257**, 8648.
32. Yang, Y. R. and Schachman, H. K. (1987) *Anal. Biochem.,* **163**, 188.
33. Syvanen, J. M., Yang, Y. R. and Kirschner, M. W. (1973) *J. Biol. Chem.,* **248**, 3762.
34. Jovin, T., Chrambach, A. and Naughton, M. A. (1964) *Anal. Biochem.* **9**, 351.
35. Burns, D. L. and Schachman, H. K. (1982) *J. Biol. Chem.,* **257**, 12214.
36. Creighton, T. E. and Yanofsky, C. (1966) *J. Biol. Chem.,* **241**, 980.
37. Lane, A. N., Paul, C. H. and Kirschner, K. (1984) *EMBO J.,* **3**, 279.
38. Cohen, R. and Mire, M. (1971) *Eur. J. Biochem.,* **23**, 267.
39. Cohen, R. and Mire, M. (1971) *Eur. J. Biochem.,* **23**, 276.
40. Kemper, D. L. and Everse, J. (1973) In *Methods in Enzymology*. Hirs, C. H. W. and Timasheff, S. N. (eds), Academic Press, New York, Vol. 27, p. 67.
41. Lusty, C. J. (1981) *Biochemistry,* **20**, 3665.
42. Jäenicke, R. (1984) *Angew. Chem Intl. Ed. Engl.,* **23**, 395.
43. Jäenicke, R. (1987) *Prog. Biophys. Mol. Biol.,* **49**, 117.
44. Chan, W. W.-C., Mort, J. S., Chong, D. K. K. and Macdonald, P. D. M. (1973) *J. Biol. Chem.,* **248**, 2778.
45. Chan, W. W.-C. (1970) *Biochem. Biophys. Res. Commun.,* **41**, 1198.
46. Chan, W. W.-C. (1976) In *Methods in Enzymology*. Mosbach, K. (ed.), Academic Press, New York, Vol. 44, p. 491.
47. Bickerstaff, G. F. (1980) *Intl. J. Biochem.,* **11**, 201.
48. Bickerstaff, G. F. and Price, N., C. (1978) *Biochem. J.,* **173**, 85.
49. Bruch, P., Schnackerz, K. D. and Gracy, R. W. (1976) *Eur. J. Biochem.,* **68**, 153.
50. McCracken, S. and Meighen, E. (1980) *J. Biol. Chem.,* **255**, 2396.
51. McCracken, S. and Meighen, E. (1987) In *Methods in Enzymology*. Mosbach, K. (ed.), Academic Press, New York, Vol. 135, p. 492.
52. Green, N. M. and Toms, E. J. (1973) *Biochem. J.,* **133**, 687.
53. Robey, E. A. and Schachman, H. K. (1985) *Proc. Natl. Acad. Sci. USA,* **82**, 361.
54. Wente, S. R. and Schachman, H. K. (1987) *Proc. Natl. Acad. Sci. USA,* **84**, 31.
55. Sjöberg, B.-M., Karlsson, M. and Jörnvall, H. (1987) *J. Biol. Chem.,* **262**, 9736.
56. Larimer, F. W., Lee, E. H., Mural, R. J., Soper, T. S. and Hartman, F. C. (1987) *J. Biol. Chem.,* **262**, 15327.
57. Larimer, F. W., Lee, E. H., Mural, R. J., Soper, T. S. and Hartman, F. C. (1988) *J. Biol. Chem.,* **263**, 2575.
58. Greenwell, P., Jewett, S. L. and Stark, G. R. (1973) *J. Biol. Chem.,* **248**, 5994.
59. Robey, E. A., Wente, S. R., Markby, D. W., Flint, A., Yang, Y. R. and Schachman, H. K. (1986) *Proc. Natl. Acad. Sci. USA,* **83**, 5934.
60. Foote, J. (1983) *Anal. Biochem.,* **134**, 489.
61. Gibbons, I. and Schachman, H. K. (1976) *Biochemistry,* **15**, 52.
62. Moore, S. and Stein, W. H. (1948) *J. Biol. Chem.,* **176**, 367.
63. Fraenkel-Conrat, H. (1957) In *Methods in Enzymology*. Colowick, S. P. and Kaplan, N. O. (eds), Academic Press, New York, Vol. 4, p. 247.
64. Habeeb, A. F. S. A. (1966) *Anal. Biochem.,* **14**, 328.
65. Prescott, L. M. and Jones, M. E. (1969) *Anal. Biochem.,* **32**, 408.
66. Pastra-Landis, S. C., Foote, J. and Kantrowitz, E. R. (1981) *Anal. Biochem.,* **118**, 358.
67. Feldman, K., Zeisel, H. J. and Helmreich, E. J. M. (1976) *Eur. J. Biochem.,* **65**, 285.
68. Kirschner, M. W. and Schachman, H. K. (1973) *Biochemistry,* **12**, 2997.
69. Kirschner, M. W. and Schachman, H. K. (1973) *Biochemistry,* **12**, 2987.
70. Yang, Y. R. and Schachman, H. K. (1980) *Proc. Natl. Acad. Sci. USA,* **77**, 5187.
71. Gibbons, I., Yang, Y. R. and Schachman, H. K. (1974) *Proc. Natl. Acad. Sci. USA,* **71**, 4452.
72. Ritchey, J. M., Gibbons, I. and Schachman, H. K. (1977) *Biochemistry,* **16**, 4584.

CHAPTER 7

Analysis of sequence-specific DNA-binding proteins

DANIELA RHODES

1. INTRODUCTION

The development of two extremely sensitive electrophoretic assay systems, the 'bandshift' and 'footprinting' assays described in this chapter, have permitted the identification and characterization of a growing number of very scarce proteins involved in cellular processes such as transcription and replication. Most of these proteins function by binding to DNA, and it is precisely this ability to bind to DNA in a sequence-specific manner and form fairly strong complexes that is exploited in both bandshift and footprinting assay systems. In these systems the identification or analysis of a protein is based on studying the effects of protein binding on the DNA. The use of radioactively labelled DNA fragments permits the detection of extremely small amounts of protein (in the region of 10^{-20}–10^{-15} mol). The principal applications for the bandshift and footprinting assays are for obtaining information on the DNA-binding site of a protein and for detecting and monitoring the purification of sequence-specific DNA-binding proteins from crude cell extracts.

The use of the two electrophoretic methods requires prior knowledge of the approximate region of DNA containing the binding site for the protein or proteins of interest. This information is obtained from gene deletion and functional analyses and is beyond the scope of this chapter (for an example, see reference 1). The bandshift, or preferably the footprinting assay, is then used to map precisely the binding site of a protein, which in turn permits the application of DNA affinity chromatography and purification of the cognate protein as described in Chapter 8.

Each system is introduced by a discussion on the basis of the assay and a description of a generally applicable procedure, but because the physical properties of specific DNA-binding proteins vary significantly, no one protocol may be effective in all cases. For this reason the application of the two electrophoretic systems is illustrated with results from two extensively studied transcription factors that have somewhat different requirements; the yeast heat shock factor (HSF) and transcription factor IIIA (TFIIIA) from *Xenopus*. They provide a good starting point for finding optimal assay conditions for other sequence-specific DNA-binding proteins, either in crude cell extracts or purified.

The electrophoretic systems used in the two assays described in this chapter have been adopted from those developed for the separation of double- and single-stranded DNA fragments; consequently, most biochemistry laboratories should possess both the equipment and expertise to carry out bandshift and footprinting studies.

2. SOME GENERAL CONSIDERATIONS

When studying sequence-specific DNA-binding proteins it is helpful to keep in mind a few simple facts.

(i) The binding constants of sequence-specific DNA-binding proteins vary over a wide range (between 10^9 and 10^{14} M^{-1}). Consequently the molarity of protein or binding-site DNA required for complex formation is dependent on the binding constant and is likely to be different for different DNA-binding proteins.

(ii) The concentrations of different DNA-binding proteins in a cell extract might be both variable and extremely low. Therefore, when setting up a preliminary binding experiment, it is advisable to use a relatively high concentration of binding-site DNA (10^{-9}–10^{-8} M) to facilitate protein–DNA complex formation, even if protein is limiting.

(iii) Cell extracts contain many DNA-binding proteins that are capable of interacting with DNA in a sequence-independent manner and thus may compete with the protein of interest for binding-site DNA. This problem can be minimized by the inclusion of competition or carrier DNA in binding reactions.

(iv) When attempting to define a DNA-binding activity or target DNA, consider the possibility that one protein may bind specifically to more than one structurally related sequence and, conversely, that a sequence may be recognized by more than one protein.

(v) The conditions required for optimal complex formation and stability for different proteins may vary considerably in terms of pH, ionic strength and metal ion content. For example, proteins containing the 'zinc-finger' motif, such as TFIIIA, require zinc as an essential co-factor, and their binding activity may be inhibited by the presence of ethylenediamine tetraacetic acid (EDTA) in buffers.

(vi) Differences in the size, aggregation state and pI of protein–DNA complexes affect the choice of conditions used for the bandshift assay. Consider that the process of electrophoretic separation may destabilize protein–DNA complexes. The ionic strength and composition of gel loading and electrophoresis buffers should therefore be adjusted in accordance with the known or suspected stability of the protein–DNA complex.

(vii) Some chemical reagents, such as dimethylsulphate (DMS) and hydroxyl radical used in the footprinting assay for probing the DNA in a protein–DNA complex, may react with or damage the protein. The integrity of the protein–DNA complex in the footprinting buffer conditions should be checked in bandshift gels. Also note that nucleases are themselves DNA-binding proteins and may affect the stability of the complex studied, if used at high concentrations.

3. FORMATION OF SEQUENCE-SPECIFIC PROTEIN–DNA COMPLEXES

A practical and often successful approach is to titrate a fixed amount of a radioactively-labelled DNA restriction fragment, containing the (known or supposed) protein binding site, with increasing amounts of crude cell extracts, partly-fractionated cell extracts or purified protein. As explained in Section 2, the use of a high concentration of binding-site DNA favours complex formation even if the amount of protein is limiting. Note that a protein at different stages of purification is likely to have somewhat different binding condition requirements. The often necessary inclusion in binding reactions of carrier protein (e.g. bovine serum albumin, BSA) and competitor DNA is discussed below.

Complex formation is often first tested on simple and quick bandshift gels, then the most promising conditions are analysed in the more laborious but more informative footprinting gels. In some cases, however, complex formation may only be detected in the less disruptive conditions of the footprinting assay. Bandshift gels permit the detection of variable amounts of formed complexes, but detection in footprinting gels requires most of the binding site DNA to be bound by protein. Conditions for protein–DNA complex formation are given for both HSF from yeast (Section 3.5) and TFIIIA from *Xenopus* (Section 3.6). The former is used for illustration of the bandshift assay (Section 4) and the latter the footprinting assay (Section 5).

3.1 **Preparation of DNA fragments containing the binding site for a protein**

3.1.1 *General considerations*

Success in interpretation of results from either the bandshift or footprinting assay depends critically on the choice of a suitable DNA fragment containing the binding site for the protein of interest. Some general considerations regarding the DNA to be used in the two electrophoretic assay systems are given below. The details of how protein–DNA complex formation is detected are given in the sections describing the bandshift (Section 4) and footprint (Section 5) assays, respectively.

(i) DNA fragments from the region of DNA believed to be involved in binding of one or more proteins (for instance the promoter region of a gene) are produced by cleavage with restriction endonucleases, fractionated in non-denaturing polyacrylamide gels and recovered as described in detail in the volume *Gel Electrophoresis of Nucleic Acids: A Practical Approach* (ref. 2) in this series.

(ii) If a DNA fragment contains multiple binding sites, the specific interaction of several proteins is likely to give rise to complex results. (As discussed in

Section 4.8, several sites of protein binding can be resolved in footprints and may result in the appearance of several bands in bandshift gels.) For simplicity the following discussion presupposes the availability of a DNA fragment containing the binding site for a single protein.

(iii) The length of a DNA fragment containing the binding site for a protein affects both the size of the change in mobility in bandshift gels and the desired single-base resolution within and around a footprint. As the resolution in denaturing polyacrylamide gels is limited to about 300 nucleotides, a suitable fragment length for use in both assay systems is between 50 and 300 bp.

(iv) For the bandshift assay, the DNA fragment may be labelled continuously or at the ends. In contrast, the footprinting analysis requires the DNA fragment to be radioactively labelled exclusively at one of the four ends of a double-stranded fragment.

(v) The distance between the radioactive label and protein binding site (as well as fragment length, iii) affect the location of the footprint, so it is practical to choose a fragment that contains the protein binding site at approximately its centre. The optimal distance between protein binding site and label is about 20–100 nucleotides.

(vi) For successful formation and analysis of protein–DNA complexes, the purity and integrity of the binding-site DNA must be high. In particular, it is important that the binding-site DNA be unnicked and free of gel contaminants and salt. The stringency of these requirements varies with the purity of the protein; for example, purified proteins are more sensitive to chemical and oxidative damage than proteins in crude mixtures.

3.1.2 *End-labelling of DNA fragments*

A detailed description of the end-labelling reaction is given in ref. 2.

The label used for labelling DNA fragments in the experiments described in this chapter is ^{32}P, but the use of the recently available [^{35}S]nucleotides (Amersham International plc) is advised because ^{35}S has a longer half-life than ^{32}P and produces sharper bands in the autoradiographs. A label is introduced at either of the two ends of a DNA strand by replacement of the 5′-phosphate with T4 polynucleotide kinase (Boehringer Corporation Ltd) and [γ-^{32}P]ATP (300–3000 Ci/mmol; Amersham International plc) or by filling in at the 3′ end with the appropriate [α-^{32}P]deoxynucleotide (300–3000 Ci/mmol) and reverse transcriptase (Anglian Biotechnology) or the Klenow fragment of *Escherichia coli* DNA polymerase (Boehringer Corporation Ltd). Because 5′-labelling introduces label into both strands, the unwanted label must be removed by re-cutting with a suitable restriction enzyme, followed by re-isolation of the labelled DNA fragment. After the binding site of the relevant protein has been identified, short (20–50 bp) synthetic oligonucleotides of a length somewhat longer than the binding site may also be used in the detection and analysis of protein binding.

3.2 Competitor DNA

When working with crude or partially purified cell extracts it is essential to include unlabelled DNA in binding reactions in substantial excess over binding-site DNA to prevent interference of sequence-specific binding by non-specific interactions with DNA. The most commonly used competitor or carrier DNA is poly[d(I-C)] (Boehringer Corporation Ltd), but mixed sequence or plasmid DNA and poly[d(A-T)] can also be used. The amount and composition of competitor DNA required to maximize sequence-specific binding is dependent on both the purity of the protein (pure proteins require little or no competitor DNA, compare Sections 3.5 and 3.6) and the specificity of its interaction with DNA, but is typically of the order of 100- to 1000-fold (w/w) excess over binding-site DNA. The amount of competitor DNA required to abolish binding gives a rough estimate of the difference between the sequence-specific and non-specific binding constant of a protein.

3.3 Carrier protein

Binding reactions are carried out at very low concentrations of DNA and protein, and DNA-binding proteins have a propensity to stick to the walls of reaction vials, so it is essential when working with purified protein to add from 50 μg up to 1 mg/ml of BSA (special quality for molecular biology, Boehringer Corporation Ltd).

3.4 Preparation of crude cell extracts and purified DNA-binding proteins

A detailed description of the preparation of cell-free extracts and purification of HSF used in the binding experiment described below is given in Chapter 8. In brief, HSF is prepared from whole-cell extracts of a protease-deficient strain of the budding yeast *Saccharomyces cerevisiae*. It is purified by a generally applicable purification procedure involving three columns:

(i) heparin–agarose;
(ii) mixed sequence DNA–Sepharose;
(iii) binding-site DNA–Sepharose.

The purification protocol for TFIIIA is not given because it is unlikely to be applicable to other transcription factors (3).

3.5 Formation of the HSF–DNA complex

The HSF from yeast is given as an example of sequence-specific protein–DNA complex formation using crude or partially-fractionated cell extracts (4, 5). The HSF protein (150 kd) is found in two states, phosphorylated and non-phosphorylated, resulting in complexes with different mobilities in bandshift gels (*Figure 1*). Note that sequence-specific complex formation in crude protein mixtures requires the inclusion of competitor DNA in the binding reactions (Section 3.2). Complex formation is described for extracts (fractionated on heparin-agarose and calf thymus DNA-Sepharose) and a ^{32}P-labelled 62-bp DNA fragment containing the HSF binding site.

(i) Binding reactions are carried out in 0.5–1.5 ml siliconized snap-cap Eppendorf tubes.
(ii) The binding buffer contains 20 mM Hepes (pH 7.9), 60 mM KCl, 1 mM $MgCl_2$, 12% (v/v) glycerol, 0.1% Nonidet P-40 and 1 mM dithiothreitol (DTT).
(iii) A 20-μl binding reaction contains binding buffer, 0.5 ng of ^{32}P-labelled binding-site DNA, 1 μg of supercoiled pUC1g DNA, 1 μg of poly[d(I-C)] and up to 40 μg of partly-fractionated protein extract.
(iv) Incubate the binding mixture at room temperature for 30 min.
(v) Analyse complex formation in non-denaturing polyacrylamide gels as described in Section 4.

3.6 **Formation of the TFIIIA–DNA complex**

TFIIIA is a member of a large family of proteins that contain the 'zinc-finger' motif as their DNA-binding domains (3). Because removal of zinc inactivates these proteins, the binding buffer is supplemented with Zn^{2+}, and care must be taken to omit EDTA from all other buffers. For the binding reaction described below, pure TFIIIA (40 kd) and a ^{32}P-labelled 236-bp restriction fragment containing the TFIIIA binding site were used (10).

(i) Binding reactions are carried out in 0.5–1.5 ml siliconized snap-cap Eppendorf tubes.
(ii) The binding buffer contains 20 mM Tris–HCl (pH 7.4), 70 mM KCl, 2 mM $MgCl_2$, 20 μM zinc acetate, 0.2 mM DTT, 6% (v/v) glycerol, 0.1% Nonidet P-40 and 50–100 μg/ml BSA.
(iii) A typical 30-μl reaction mix contains binding buffer and 0.3 pmol (10 nM) of binding-site DNA fragment labelled at the 3′ or 5′ end of one of the two strands.
(iv) Dispense 0.3–3 pmol of protein directly into this mixture and mix gently by tapping the tube.
(v) Incubate the binding mixture at room temperature for 30 min.
(vi) Analyse complex formation in non-denaturing polyacrylamide gels or agarose gels as described in Section 4.

4. BANDSHIFT ASSAY

The basis of the bandshift assay, sometimes called the gel retardation assay, is that protein–DNA complexes are surprisingly stable to electrophoretic fractionation in gels and migrate as distinct bands more slowly than the free DNA fragment. The assay is simple and quick, and the use of radioactive binding-site DNA makes it highly sensitive. Its primary use is in the detection of DNA-binding proteins in crude cell extracts (6), but it can also be used to study the binding activity (complex formation) of purified DNA-binding proteins, both highly sequence-specific (7, 8) and non-specific, such as histones (9, 10). Thus (provided non-specific binding has been eliminated by the use of competitor DNA, Section 3.2), the appearance of one or more slowly migrating bands may be taken to indicate that one or more proteins present in the extract bind in a sequence-specific manner to the binding-site DNA fragment.

The fractionation system most frequently used is non-denaturing polyacrylamide gels, but agarose gels can also be used. The size of a protein–DNA complex (or complexes) may vary considerably, and is often unknown at the beginning of a study, so both gel systems should be tried to test which gives the better separation. In general, protein–DNA complexes of high molecular weight are best fractionated in agarose gels, whereas for analysis of smaller or closely-related complexes polyacrylamide gels give the better resolution. In addition, the two gel matrices may affect the stabilities of various complexes differently.

4.1 **Technical aspects of non-denaturing gels**

The electrophoretic separation systems used for bandshift and footprint gels are adapted from those developed for the separation of double- and single-stranded DNA fragments. The construction of various types of apparatus, preparation of acrylamide and agarose gels, electrophoresis conditions, radioactive labelling of DNA, detection of radioactive DNA and recovery of DNA from gels have been described in great detail in ref. 2, and will not be repeated here. Instead, we will describe the application of these standard techniques to the fractionation and identification of protein–DNA complexes.

4.1.1 *Electrophoresis buffers*

Tris–borate–EDTA [TBE, 90 mM Tris base, 90 mM boric acid, 2 mM EDTA (pH 8.3)] is the electrophoresis buffer used in both acrylamide and agarose gels. In cases where EDTA must be omitted, Tris–borate (TB) is used alone.

4.1.2 *Sample buffers*

Protein–DNA complexes are loaded on to non-denaturing gels (acrylamide or agarose) supplemented with 1/5 volume of 20% glycerol, 0.1% (w/v) bromphenol blue. Careful application of the sample is required to prevent dilution.

4.2 **Non-denaturing agarose gels**

Agarose gels [0.7–1.5% (w/v)] are cast from high-temperature gelling agarose (BRL Ultrapure) and 0.25–1 × TBE or TB buffer (Section 4.1.1). The use of low-temperature gelling agarose may at times be convenient. The size of the gel can vary, but commercially available flat bed apparates usually have a bed size of about 10 × 10 cm or the larger 20 × 25 cm. The gel is cast approximately 0.5 cm in height (which would require either 50 ml or 200 ml of agarose). The smaller gels, called 'mini-gels' are most frequently used to assay complex formation because they require shorter electrophoretic times. The larger bed size is used when greater resolution in the separation is required. Both gel and electrophoretic buffers are half the concentration of TBE or TB. To prevent heating and the destabilization of protein–DNA complexes, apply a lower electric current than that used for the separation of DNA fragments: 20–30 mA for mini-gels and

50–60 mA for larger gels. Electrophoresis is carried out at room temperature or in the cold-room. It is customary to stop electrophoresis when the bromphenol dye has reached the bottom of the gel, but the electrophoresis time has to be optimized for the complex studied and the separation required. Following electrophoresis, the gel is placed on Whatman D52 paper and dried under vacuum at 60°C. Gels may also be exposed wet for autoradiography if elution of the DNA or protein–DNA complex is required.

4.3 Non-denaturing polyacrylamide gels

Non-denaturing polyacrylamide gels can be 3.5–6% (w/v, with an acrylamide to bisacrylamide weight ratio of 19:1 or 39:1) and the electrophoresis buffer 0.25–1 × TBE or TB (Section 4.1.1). Optimal compositions for gels and running buffers should be established empirically, as they may dramatically affect the quality of separation. 20 × 20 × 0.3 cm gels are convenient and can be run at 5–15 V/cm. It is customary to stop electrophoresis when the bromphenol dye has reached the bottom of the gel, but considerably longer electrophoresis times may at times be required for good separation. Such long running times (at 5 V/cm to prevent uneven heating) are essential in the resolution of the phosphorylated and non-phosphorylated forms of HSF (*Figure 1b*). Following electrophoresis, the gels are placed on Whatman 3MM paper without fixing and dried under vacuum at 80°C. Gels may also be exposed wet for autoradiography in cases where subsequent elution of the DNA or protein–DNA complex is desired.

4.4 Autoradiography

After gels have been dried, they are exposed to X-ray film (such as Kodak X-AR and Fuji RX) with an intensifying screen at −70°C (HI speed X), or without a screen at room temperature. The intensifying screen reduces the exposure time by a factor of 5–10.

4.5 Electrophoretic fractionation of the HSF–DNA complex

Figure 1 shows the result of the analysis in a non-denaturing polyacrylamide gel of complex formation between the 150-kd HSF present in crude or partially-purified yeast extracts and a DNA fragment containing the HSF binding site (4, 5). The DNA fragment not bound by protein has migrated to the bottom of the gel, whereas that bound by protein is retarded and found at the top of the gel.

Figure 1b shows that the bandshift assay can further be used to detect modifications in DNA-binding proteins. In this case, the heat shock-induced phosphorylation of HSF causes further retardation in the migration of the HSF–DNA complex, resulting in complexes with distinctly different electrophoretic mobilities, depending upon whether the protein is phosphorylated or not.

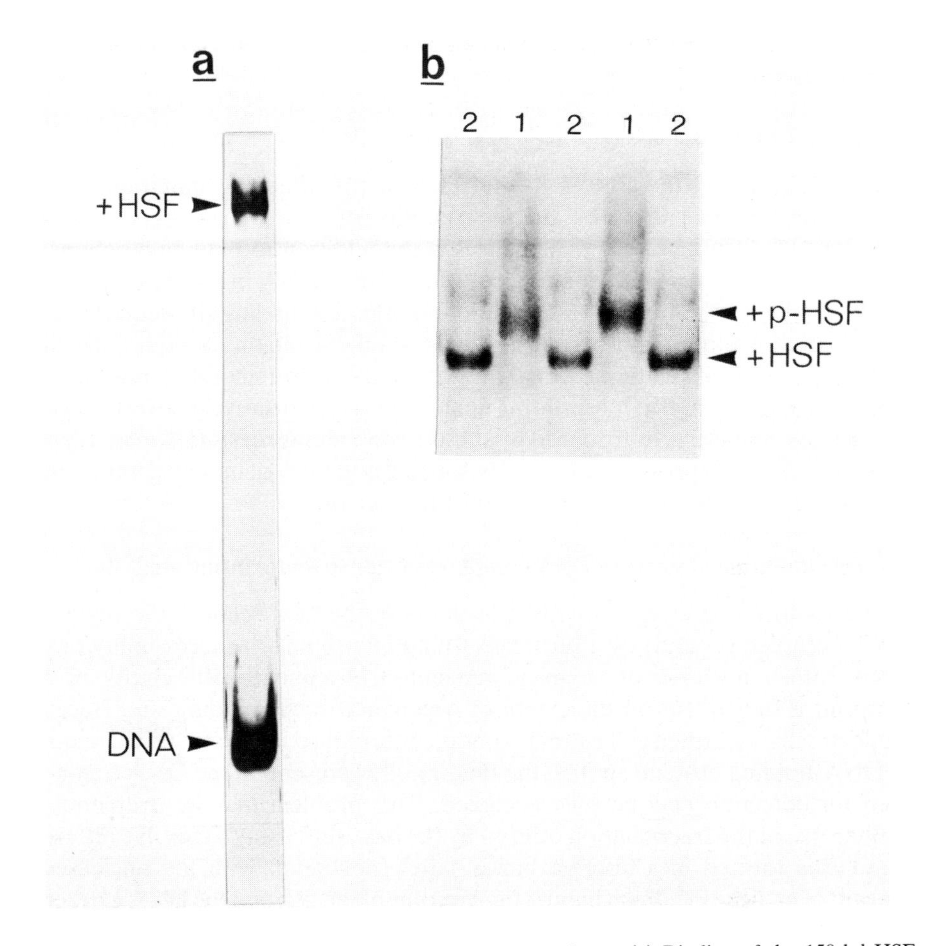

Figure 1. Electrophoretic fractionation of HSF–DNA complexes. (**a**) Binding of the 150 kd HSF protein causes the radioactive 62-bp DNA fragment containing the HSF binding site to migrate as a distinct band, found at the top of the gel, much more slowly than the unbound DNA in the band at the bottom of the gel. (**b**) Shows the effect of heat-shock-induced phosphorylation of HSF on the mobility of the HSF–DNA complex. The complex formed with phosphorylated protein (**lane 1**) migrates more slowly than the complex formed with non-phosphorylated protein (**lane 2**). Fractionation was carried out in a 4% non-denaturing polyacrylamide gel run in 22.5 mM Tris–borate, 0.6 mM EDTA, pH 8.3. To improve the fractionation of complexes containing modified and unmodified protein, electrophoresis in gel **b** was for 4 h rather than the 1 h of gel **a**. (Courtesy of P. Sorger and H. Pelham.)

4.6 The use of the bandshift assay to study the effects on complex formation of modification and mutations on protein or DNA

Several sequence-specific DNA-binding proteins undergo modifications, such as the phosphorylation of the HSF (see *Figure 1*). The bandshift assay may be used to study the effects of modification on the DNA-binding activity of a protein and/or the mobility of a protein–DNA complex. Modifications may induce changes in the protein structure or may allow other proteins to bind to an already formed protein–DNA complex.

4.7 **Cross-competition assay**

Bandshift gels lend themselves to the study of the effects of mutations in the protein or binding-site DNA on the stability of the cognate protein–DNA complex (11). The basis for this analysis is the addition to binding reactions of increasing amounts of, for example, cold competitor DNA in the form of a short synthetic oligonucleotide corresponding to a mutated binding site sequence (5). The cross-competition approach using short, synthetic oligonucleotides may also be very useful in dissecting overlapping or complex protein–DNA interactions used in combination with footprinting analysis (12). Alternatively, effects of gene deletions or mutations [introduced by site-directed mutagenesis (Chapter 11) and *in vitro* or *in vivo* expression] on the DNA-binding activity of mutant proteins may be tested. For an example of this type of study, see ref. 13.

4.8 **Fractionation of protein–DNA complexes prior to footprinting analysis**

In the footprinting assay, described in detail in the next section, the region of DNA occupied (bound) by a protein is found by probing the accessibility of the DNA with a nuclease or chemical reagent. Consequently the clarity of the footprint is dependent on the extent of occupancy of the binding site. Because crude or partly-purified cell extracts contain different concentrations of a number of DNA-binding proteins, not all the binding sites present on the DNA fragment used for detection may be fully occupied. This problem may be overcome by making use of the fractionation offered by the bandshift assay to resolve different complexes formed into discrete bands, after incubation with the nuclease or reagent of choice (14). Each band is then cut out of the gel, and the DNA extracted and analysed in denaturing polyacrylamide gels. Thus, the fractionation offered by bandshift gels, in combination with footprinting analysis, permits identification of the regions of DNA bound by protein in different complexes.

5. FOOTPRINTING ASSAY

The physical principle underlying the footprinting or protein protection assay is explained in *Figure 2*. First a protein – DNA complex is formed using protein and end-labelled binding-site DNA, as described in Section 3. Then a sample containing the protein–DNA complex and another containing naked DNA are subjected to mild enzymatic or chemical cleavage and the resulting fragments analysed in a denaturing polyacrylamide gel. The position of a band in the gel corresponds to the distance between the label and the point at which the DNA has been cleaved. Because the protein protects the DNA of its binding-site from cleavage, resulting in much slower cleavage rates for DNA bound by protein than for naked DNA, there will be an almost complete absence of fragments arising from within the protein binding site (15). The absence of fragments gives rise to a gap in the electrophoretic pattern of bands (or digestion pattern). This gap is the 'footprint'

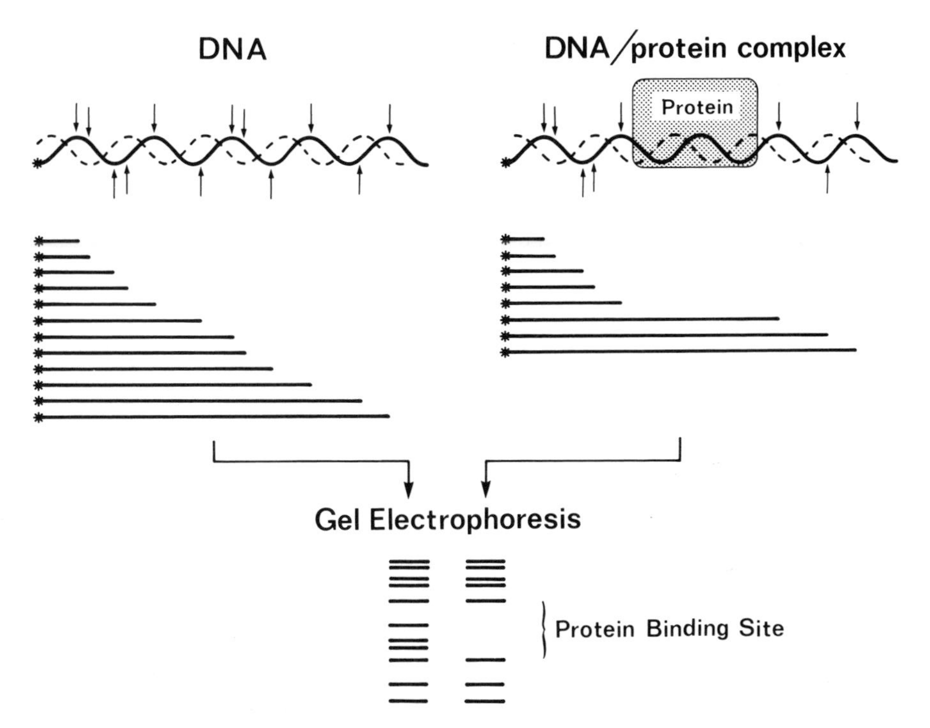

Figure 2. Schematic representation of the footprinting assay. The left side of the illustration shows a naked DNA fragment and the right side the same DNA fragment containing bound protein. Arrows represent the points of cleavage by a DNA cleavage reagent or nuclease. The basis of the footprinting assay is that the protein protects the DNA backbone from cleavage by whatever reagent is used: in contrast the naked DNA is cleaved along its entire length. Analysis of the resulting DNA fragments by denaturing gel electrophoresis (sequencing gels) reveals the differences between the pattern of bands (cutting patterns) of the naked DNA and that of the protein–DNA complex. Only fragments containing the unique radioactive label (indicated by *) are visible in the analysis. The absence of fragments arising from the region of DNA bound by protein gives rise to an absence of bands, or gap, in the cutting pattern of the protein–DNA complex, as shown in the right-hand lane.

left by the protein on the DNA (16). The location of the footprint in the sequence is found by including sequencing marker tracks in the analysis. It should be remembered that although the appearance of a footprint is clearly a good indication that a protein binds to the specific region of DNA, a footprint reflects the effects of protein binding on the cleavage rate of the reagent used. Therefore, an area of protection should not automatically be equated with direct protein-to-DNA contact.

The footprinting assay is superior to the bandshift assay because it gives information not only about the presence of a DNA-binding activity studied in solution (rather than in bandshift gels), but also about the size and sequence of the DNA-binding site of the as yet uncharacterized protein.

5.1 Factors that affect the clarity of a footprint

(i) Clean footprints require full occupancy of the binding site by protein. The amount of protein required is found by titrating binding-site DNA with protein (or extracts) and the occupancy is measured in bandshift gels. It is often practical to add an excess of protein to maximize complex formation. Alternatively, if only partial occupancy is obtained, the DNA bound by protein can be separated from unbound DNA in non-denaturing gels of the bandshift assay (Section 4) after incubation with the footprinting reagent, so that only the DNA bound by protein is analysed for protection.

(ii) The integrity of a protein–DNA complex should be checked on bandshift gels after incubation with footprinting reagents or nucleases, as these may damage or destabilize the pre-formed complex (see Section 2).

(iii) Labelled DNA fragments used for the detection of protein binding should be free from 'nicks', because the resulting fragments may obscure the digestion pattern obtained in the footprinting analysis. The aim is to introduce, on average, about one cut (or modification) per DNA strand. The number of cuts per strand can be calculated from the equation given in Section 6.

(iv) Care should be taken in the working up of the DNA after nuclease digestion or chemical attack, as the purity of the DNA affects the electrophoretic separation. Ensure that all protein is extracted and that DNA pellets contain a minimum of salt and reagents.

(v) Identification of a gap, or binding site of a protein, relies on comparing the digestion pattern of the protein–DNA complex with that of the naked DNA, so it is essential that the two samples are digested in the same buffer conditions, contain comparable amounts of radioactivity and are digested to a similar extent. This can be achieved most simply by digesting each sample for various times.

(vi) The electrophoretic resolution of a footprint affects the ability of the eye to discern differences between the digestion pattern of the naked DNA and that of the protein–DNA complex. If in doubt, change the distance between radioactive label and binding site, or analyse the footprint on the other DNA strand.

5.2 How to choose a footprinting reagent

The several available footprinting reagents are most frequently used to probe the DNA in a pre-formed protein–DNA complex. In other words the reagent is used to measure the 'protection' of the DNA provided by the protein. A different approach is the 'interference method' in which the DNA is methylated (on guanine and adenine bases) or ethylated (on phosphate groups) prior to protein binding. This latter method relies upon the separation of the protein–DNA complex from uncomplexed DNA (as a result of the modification) to identify the regions of the DNA that cannot be modified without interfering with complex formation. 'Interference footprinting' is not described in this chapter but references are given in *Table 1* together with a list of reagents used in 'protection footprinting'.

There are two main factors that affect the appearance or characteristics of a footprint: the sequence specificity of the reagent and its size. Clearly a large

Table 1. List of commonly used footprinting reagents.

Reagent	*Point of attack*	*Reference*
Protection		
DNase I	Phosphate	(17)
		This chapter
DNase II	Phosphate	(17)
		(10)
Dimethylsulphate	N-7 of guanines	(18)
	(N-3 of adenines)	This chapter
Methidium propyl-EDTA-Iron(II)	Sugar	(19)
		(20)
Cu-phenanthroline	Sugar	(21)
		(22)
Hydroxyl radical	Sugar	(23)
		This chapter
Interference		
Dimethylsulphate	N-7 of guanines	(18)
	(N-3 of adenines)	(24)
Ethylnitrosourea	Phosphate	(25)
		(24)

For each reagent the first reference refers to the chemistry of the reagent and the second gives one example of its use in footprinting. To permit direct comparison between the results obtained with different reagents most references given are for the their use in the study of the TFIIIA–5S RNA gene complex.

reagent such as DNase I cannot cleave the DNA in the immediate vicinity of bound protein because of steric hindrance. In general, all reagents used to probe the accessibility of the DNA in a protein–DNA complex exhibit some degree of sequence specificity in their rates of cleavage of naked DNA. Different reagents recognize different sequence-dependent structural variations of the DNA double helix (17). So, depending upon the sequence of the binding site, different reagents give somewhat different protection patterns or footprints. The limitations imposed by the sequence specificity, in addition to the size of each reagent or nuclease, may be overcome by using a set of reagents to give a more complete sampling, and hence more reliable and precise information about the binding site of the protein studied.

The footprinting method is illustrated with three different footprinting reagents, DNase I, hydroxyl radical and DMS, because they give protection patterns that look very different, but complement each other in the information given. The methylation reaction gives information about protection on the 'inside' of the DNA double helix, whilst DNase I and hydroxyl radical, both of which cleave the ribose–phosphate backbone, give information about protection by protein on the 'outside' of the DNA helix.

Each reagent is introduced by a description of its individual advantages and disadvantages and its mode of attack on the DNA. To permit the reader to evaluate each footprinting reagent, I show and describe the use of the above-mentioned reagents to probe the accessibility of the DNA in one specific complex, that formed between purified TFIIIA and its binding site on the 5S RNA gene.

5.3 **DNase I as a probe**

DNase I is the classical endonuclease of footprinting (15, 16). Like most nucleases, it is large, about 40 Å in diameter (26). It binds to the minor groove of the DNA double helix and introduces cuts in the backbone of each DNA strand independently. Because the rate of DNase I cleavage is dependent upon sequence-determined variation in DNA structure, not all backbone positions can be cleaved. One advantage of using a bulky probe, however, is that the protection offered by bound protein is amplified, resulting in very clear gaps in the cutting pattern of the DNA in a complex (*Figure 3a*). The main factor that affects the clarity of DNase I footprints is the extent of digestion. If a large excess of protein is used for complex formation, or if competitor DNA is included, considerably higher concentrations of DNase I or longer digestion times than those quoted below will be required to obtain a suitable extent of digestion.

5.3.1 *Digestion conditions* (10)

(i) Place 30 μl of TFIIIA/DNA complex (0.3 pmol of DNA + 0.9 pmol of protein) prepared as described in Section 3.6 and the control naked binding-site DNA (0.3 pmol in 30 μl of binding buffer) in the bottom of two siliconized snap-cap Eppendorf tubes.
(ii) Add 2 μl of DNase I (Type II-S, Sigma Chemical Company Ltd) at 2 U/ml to each mixture; mix gently by tapping and incubate at 20°C for times ranging from a few seconds to several minutes.
(iii) Take out 10 μl aliquots at, for example, 10, 30 and 90 sec and stop the digestion by dispensing into 10–20 μl of 5 mM EDTA, 1% SDS and 100 μg/ml of sonicated carrier DNA.
(iv) Prepare the samples for analysis in sequencing gels as described in Section 5.6.

5.4 **Hydroxyl radical as a probe**

This footprinting method makes use of hydroxyl radical (23), generated by the reduction of hydrogen peroxide by iron (II), to cleave the DNA backbone. The major advantages of this probe are that it is very small, about the size of a water

Figure 3. Footprints of the TFIIIA protein bound to its binding site on the 5S RNA gene. The three footprinting reagents: (**a**) DNase I, (**b**) hydroxyl radical and (**c**) DMS (methylation protection) show that the area of DNA protected by protein from cleavage or chemical modification is the same and located between nucleotides 45 and 97 of the 5S RNA gene. Note that the hydroxyl radical footprint runs in the opposite direction to the other two because in this experiment the ^{32}P label is located at the 5′ end of the non-coding strand, whereas in the other two it is at the 3′ end. The three different footprinting reagents give protection patterns that in detail look very different as a consequence of their rather different sizes and modes of attack. Electrophoretic fractionation was carried out in 8% sequencing gels containing 8 M urea and TBE buffer. Lanes labelled G show positions of guanine nucleotides within the sequence cut by DMS/piperidine. (Hydroxyl radical footprint, courtesy of M. Churchill.)

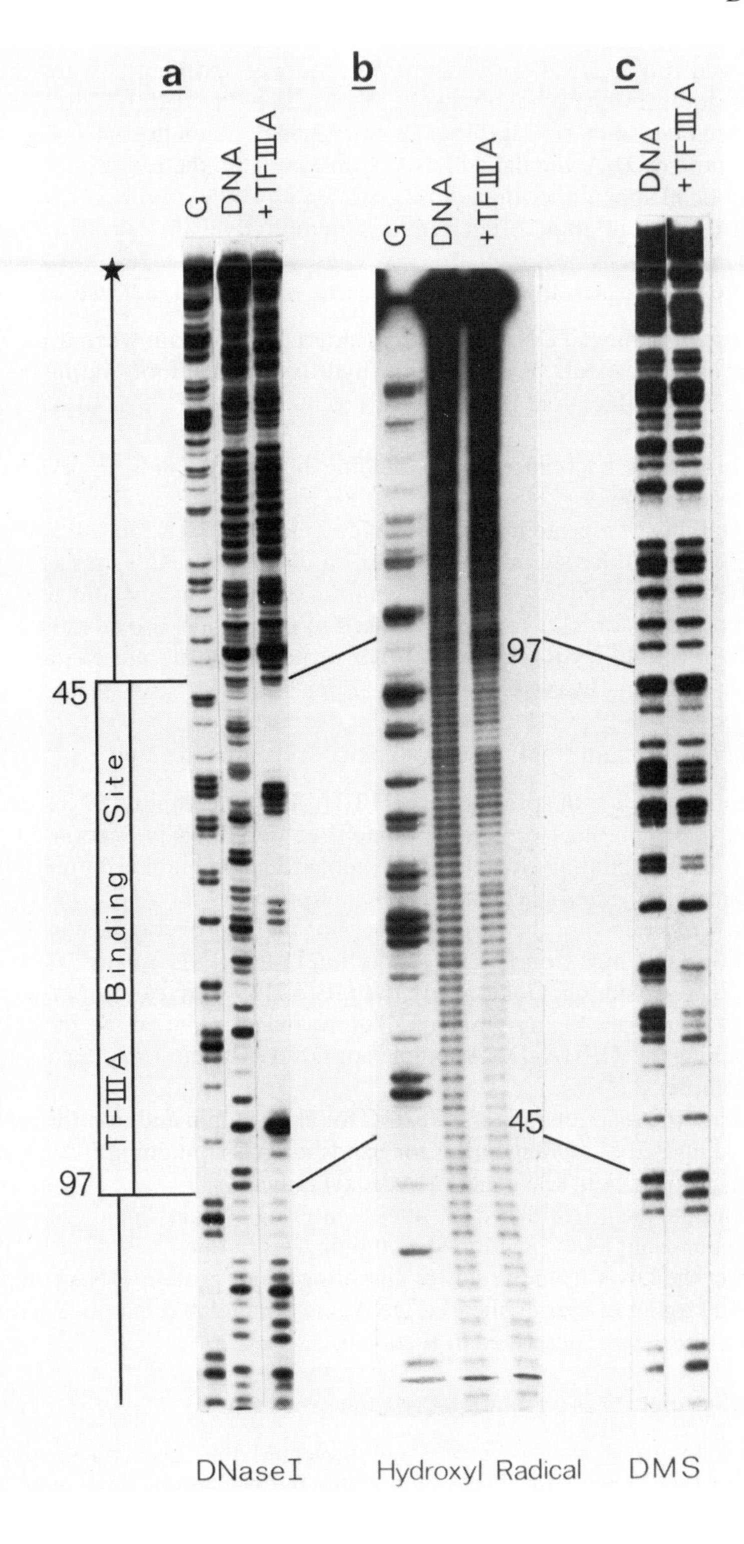
a
b
c
G
DNA
+TFIIIA
G
DNA
+TFIIIA
DNA
+TFIIIA
45
97
TFIIIA Binding Site
97
45
DNaseI
Hydroxyl Radical
DMS

molecule and it cuts DNA with almost no sequence specificity, so that a high resolution footprint can be obtained. A disadvantage of this probe, as a consequence of its small size, is the small difference between the rates of cleavage of protein-bound DNA and naked DNA. Consequently, the footprint generated is more subtle in appearance than those generated by other probes (*Figure 3*) and an accurate interpretation often requires quantitative analysis like the one described in Section 6.

There are three factors that bear on the clarity of hydroxyl radical footprints.

(i) The use of unnicked DNA fragments, although important when using other footprinting reagents, is essential for hydroxyl radical footprinting because this cleavage reaction is inefficient, usually leaving 75% of the DNA uncleaved.
(ii) Glycerol must be omitted from binding buffers, as it is a scavenger of hydroxyl radical.
(iii) The stability of a protein–DNA complex in the hydroxyl radical-generating reaction conditions should be tested on bandshift gels. As is the case for the TFIIIA–DNA complex, the stability of the complex may be affected by the concentrations of EDTA and H_2O_2 used to generate hydroxyl radical. This problem is unlikely to arise with other proteins that do not contain metal co-factors bound by cysteine residues (23).

5.4.1 *Reaction conditions* (M. Churchill, personal communication)

(i) Dispense 30 μl of pre-formed TFIIIA–DNA complex (0.3 pmol of DNA + 0.9 pmol of protein), prepared as described in Section 3.6, and control naked binding-site DNA (0.3 pmol in 30 μl of binding buffer) into the bottom of individual siliconized snap-cap Eppendorf tubes.
(ii) Place 4 μl of each of the three solutions of the hydroxyl radical-generating reaction (a) 1 mM iron II-EDTA solution [1 mM $(NH_4)_2Fe(SO_4)_2 6H_2O$, 2 mM EDTA, Aldrich, Gold Label], (b) 0.03% H_2O_2 and (c) 10 mM L-ascorbic acid (sodium salt, Sigma) on the wall of the reaction vial, above the solution containing the TFIIIA–DNA complex or DNA. Mix these and then mix with the complex or DNA.
(iii) Incubate the reaction mixtures at 20°C for about 1 min and stop the reaction by adding 5 μl of loading buffer for bandshift gels containing 50% glycerol, which is an efficient scavenger of hydroxyl radical.
(iv) Fractionate the mixture on a 6% non-denaturing polyacrylamide gel (as used in the bandshift assay, Section 4) containing TB buffer.
(v) Extract the DNA from the slowly-migrating band (protein–DNA complex); add 100 μg/ml of carrier sheared DNA and ethanol precipitate. Analyse in sequencing gels as described in Section 6.

5.5 **Dimethylsulphate as a probe (methylation protection)**

DMS, like hydroxyl radical, is a small molecule that gives more detailed information than a nuclease. DMS methylates the N-7 atoms of guanine bases located in the major groove of double-helical DNA. The N-3 atoms of adenine bases, located in the minor groove, are also methylated but at a much slower rate; consequently, the adenine reaction is not generally useful for measuring protection. The positions of modification are identified by eliminating the methylated base and subsequent cleavage of the DNA backbone by heating in piperidine. This reaction is taken from Maxam and Gilbert guanine-specific sequencing (18). Reduced cleavage of a guanine nucleotide in DNA bound by protein, compared with the cleavage of the same guanine in the naked DNA, is taken to indicate that the protein binds to the major groove and is in close proximity to that nucleotide. An obvious limitation of this reagent is that effects of protein binding can only be studied where there are guanine residues in the sequence. The clarity of methylation protection footprints is primarily dependent on two factors.

(i) DMS also methylates the protein, but presumably not at the interface with DNA. It is prudent, however, after incubation with DMS, to check the integrity of the protein–DNA complex on bandshift gels.
(ii) The concentration of DMS required to obtain a suitable extent of reaction depends upon the concentration of DNA and protein present and may have to be increased over the amount quoted below when probing complexes prepared in the presence of large excesses of competitor DNA or protein.

5.5.1 *Reaction conditions* (27)

(i) Methylation by DMS of the TFIIIA–DNA complex (0.3 pmol of DNA + 0.9 pmol of protein) prepared as described in Section 3.6, is carried out in parallel with control naked binding-site DNA (0.3 pmol in binding buffer) in volumes of 30 μl in siliconized screw-cap plastic tubes.
(ii) Dispense 3 μl of freshly diluted 2% (v/v) DMS (reagent grade, BDH Chemicals Ltd) directly into the complex and DNA mixtures, mix gently by tapping the tube and incubate at 20°C for up to 3 min.
(iii) Take out 10 μl aliquots at, for example, 40 sec, 1 min and 3 min and add to a screw cap tube containing 5 μl of 'stop solution' (1.5 M sodium acetate, pH 7.0, 1.0 M mercaptoethanol, 15 mM EDTA, 100 μg/ml sheared DNA). The role of EDTA in this and subsequent buffers is to chelate Zn^{2+} and Mg^{2+} present in the binding buffer, because these ions cause depurination when DNA is heated.
(iv) Add 40 μl of 95% ethanol (−20°C) and precipitate the DNA by placing the samples at −70°C (dry ice–ethanol bath) for 5 min; centrifuge at about 10 000 r.p.m. for 5 min; discard the supernatant.
(v) To remove all traces of reagent and salt re-dissolve the pellet in 20 μl of 0.3 M sodium acetate, 10 mM EDTA and precipitate the DNA with 50 μl of ethanol and centrifuge as in (iv).
(vi) Wash the pellet with 40 μl of 95% ethanol (−20°C), centrifuge, discard the supernatant and dry the DNA pellet under vacuum for a few minutes.
(vii) Cleave at the methylated bases by re-dissolving the pellet in 20 μl of 10%

(v/v) freshly-diluted piperidine (reagent grade, BDH Chemicals Ltd, re-distilled) and 10 mM EDTA. Wrap Teflon tape around the screw-on cap to improve the seal and place the tube for 30 min in a water bath at 90°C.

(viii) At the end of the incubation, remove the tubes from the water bath, centrifuge for a few seconds, punch holes in the caps and freeze-dry to remove the piperidine. Remove remaining traces of reagents, which interfere with the subsequent electrophoretic separation, by washing the DNA pellet with 70% (v/v) ethanol (−20°C).

(ix) Dry the DNA pellet and analyse on a sequencing gel as described below in Section 5.6.

5.6 Visualization of the footprint on denaturing polyacrylamide gels

Visualization of the site of protein-binding on the DNA, the footprint, requires fractionation of single-stranded fragments resulting from one or other of the footprinting reagents described above. Electrophoretic fractionation is carried out in denaturing polyacrylamide gels of the type used in DNA sequencing, followed by autoradiography.

5.6.1 *Preparation of DNA for analysis in denaturing polyacrylamide gels*

(i) At the end of an incubation with a footprinting reagent, extract the protein–DNA complex twice with phenol/chloroform (1/1 v/v) saturated with 50 mM Tris–HCl, pH 7.5.

(ii) Precipitate the aqueous phase containing the DNA with 2.5 volumes of ethanol, centrifuge, and remove remaining traces of salt by washing the pellet with one volume of 70% (v/v) ethanol (−20°C).

(iii) Dry the pellets and dissolve them in 5 μl of 'loading buffer': 90% (v/v) formamide (re-distilled), 10 mM NaOH, 3 mM EDTA, 0.1% (w/v) xylenecyanol and 0.1% (w/v) bromphenol blue.

(iv) Prior to electrophoresis, heat the samples for 1 min in a boiling water bath.

5.6.2 *Denaturing polyacrylamide gels*

For a detailed description of how sequencing gels are cast and run, see ref. 2.

Denaturing polyacrylamide gels, or 'sequencing gels', contain from 5 to 12% (w/v) polyacrylamide (acrylamide to bisacrylamide weight ratio of 19:1) and 7 or 8 M urea. Sequencing gels are about 20 cm wide, 40–50 cm long and 0.3 mm thick. Electrophoresis is carried out at 30 mA, and the resolution of the fractionation is improved by pre-running the gel for 30 min at 30 mA prior to loading samples. The buffer in the gel and electrophoretic reservoir is 1 × TBE buffer (Section 4.1). Uneven migration of fragments can be avoided by clamping an aluminium plate to the front glass plate. This produces even distribution of the heat that builds up during electrophoresis. It is customary to run the fast-migrating marker dye to the bottom of the gel, but longer electrophoresis times may be required to obtain single-bond resolution in the area of the footprint. Following electrophoresis the gel is fixed in methanol (10% v/v), acetic acid (10% v/v), transferred to a sheet of Whatman 3MM paper, and dried under vacuum at 80°C.

5.6.3 *Autoradiography*

After fixation, gels are exposed to X-ray film with an intensifying screen (HI-speed X) at −70°C. The intensifying screen reduces the exposure time but at the same time reduces the resolution of the bands in the autoradiograph. It is often advisable to expose a gel for various lengths of time, as the clarity of the footprint may depend on the intensity of the bands in the autoradiograph. Films used for the quantitative analysis described in Section 6 should be pre-flashed to an optical density of 0.15 absorbance units. The response of commonly used X-ray films (such as Kodak X-AR or Fuji RX) is linear only in the range of 0.15–1.5 absorbance units; consequently, films should not be over-exposed for this type of analysis.

5.7 **Assignment of bands in the autoradiograph to DNA sequence**

In the methylation protection experiments, the bands in the cutting pattern correspond to positions where there are guanine nucleotides in the sequence, so assignment of a band to a nucleotide position (strictly, bond positions) in the known sequence is straightforward. When a reagent such as DNase I is used, the bands in the cutting pattern cannot be aligned directly with sequence, but alignment is achieved by comparison with marker G or G + A tracks prepared by making use of the DMS/piperidine reaction (Section 5.5). In this alignment of cutting positions with bonds in the sequence, it is necessary to consider the chemistry of the various cleavage reactions. Note that the DMS/piperidine reaction results in the destruction of guanine bases, so 5′-end-labelled fragments end at the bond *before* a guanine and 3′-end-labelled fragments at the bond *after* a guanine in the sequence.

5.8 **Results: different reagents reveal different details of the footprint**

Figure 3 shows the results of footprinting TFIIIA bound to its binding site on the 5S RNA gene using the three different reagents described above: DNase I (Section 5.3), hydroxyl radical (Section 5.4) and DMS (Section 5.5). After incubation with the footprinting reagent, the resulting DNA fragments were fractionated in denaturing polyacrylamide gels. The footprint, that is the length of DNA affected by protein binding, is for each reagent about 50 bp and is located between nucleotides 45 and 95 of the non-coding strand. (Similar results were obtained for the coding strand, but are not shown here.) In detail, however, each footprint is strikingly different and is a consequence of the difference in size and mode of attack of the three different reagents used. These details not only contain information about where the protein is bound but, together with the quantitative analysis described in the next section, may be used to extend the footprinting assay to ask how a protein binds to DNA (27, 28).

6. QUANTITATIVE ANALYSIS OF FOOTPRINTING RESULTS

The process of finding evidence for sequence-specific binding by a protein involves looking for differences between the digestion pattern of the protein–DNA complex and that of the naked DNA. This involves scanning by eye hundreds of data points represented by bands in the digestion pattern seen in autoradiographs. A more accurate and objective method is to subtract the digestion pattern of the naked DNA from that of the protein–DNA complex so that the effects of protein binding can be seen more clearly. This involves calculating the probability of cleavage at each bond, which is related to the amount of radioactivity (or intensity) in the corresponding band in the cutting pattern (29). This analysis is particularly useful when the effects of protein binding on the cutting rate of a reagent, such as hydroxyl radical (*Figure 3b*), are not easily discernible by eye, or when partial protection is obtained, as is the case in the methylation protection experiment shown in *Figure 3c*. In addition, the combination of quantitative analysis with the use of several different reagents gives more reliable information about protein contacts than about the other effects of protein binding. (See ref. 27 for an example of such a study.) The disadvantage of quantitative analysis is that it can be time consuming, if one does not have access to computer-aided densitometry and calculations.

6.1 **Densitometry and probability calculations**

Lanes of the naked DNA and of the TFIIIA–DNA complex from the same autoradiograph that match closely, both in terms of band intensity and extent of digestion, are chosen for densitometry. The extent of digestion is judged from both the amount of radioactivity left in the undigested fragment and the radioactivity of the bands in the region of DNA outside the protein binding site. The amount of radioactivity refers here to the intensity of a band in the autoradiograph as measured by densitometry. Because quantification of band intensity is not possible for bands with an absorbance greater than 2.0, it is essential not to use over-exposed films. For calculation of the difference probability plot shown in *Figure 4* the autoradiograph was scanned manually in a Joyce-Loebl microdensitometer, and the area under each band measured using a computer-interfaced planimeter. The probability of cutting at bond *n* is calculated using the equation shown below; that is by dividing the integrated (A_n) of band *n* by the sum of the integrated area (A_m) representing uncut DNA plus all the fragments (bands) longer than and including band *n* (15).

$$P_n = A_n \Big/ \sum_{m=n}^{m=n_{max}} A_m$$

This calculation includes a correction for the decreasing probability of having only a single cut at increasing distances from the radioactive label. Summation of all probabilities of cutting in a lane gives the average number of cuts in a DNA strand. More recently, automated laser densitometers linked to a computer have become available (from for instance Bio-Rad, LKB-Pharmacia or IBI Ltd), making this type of analysis less time-consuming.

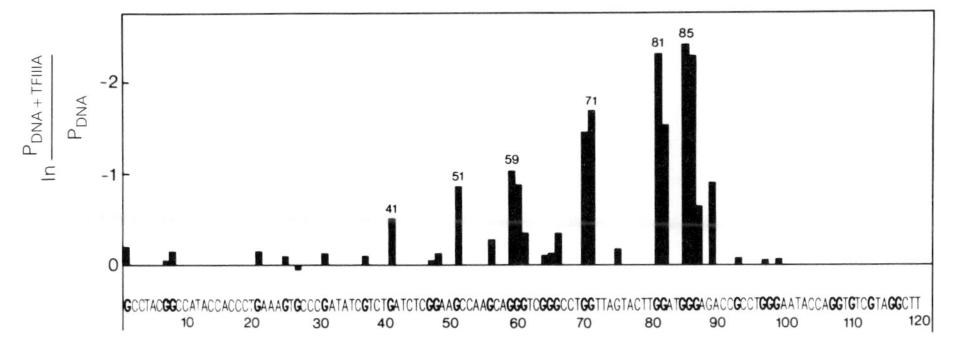

Figure 4. Methylation difference probability plot for the TFIIIA–5S RNA gene complex. The plot was calculated from densitometry of the DNA and TFIIIA–DNA lanes shown in *Figure 3c*. The difference in the extent of methylation at each guanine is represented by a bar and was calculated by subtracting ln (probability of cleavage) of each guanine residue in the naked DNA from the same guanine in the protein–DNA complex. Negative values show guanine residues that are protected, and values close to zero indicate a reactivity close to that of the naked DNA. This type of plot reveals the site of protein binding much more clearly and reliably than can be done by comparison by eye of the cutting patterns shown in *Figure 3c*.

6.2 Calculation of difference probability plots

The probabilities of cleavage are used to calculate a difference probability from the function of the kind $\ln(P_n) - \ln(F_n)$ along the entire length of the binding site and surrounding DNA (29). P_n is the probability of cutting at band n in the protein–DNA complex, while F_n is the probability of cutting at band n in the naked DNA. *Figure 4* shows the methylation difference probability plot for the TFIIIA–5S RNA gene complex calculated from the data for the non-coding strand shown in *Figure 3c*. The regions of DNA that do not bind the transcription factor have values close to zero, and negative values reflect the extent of protection. The advantages of quantitative analysis are, I think, obvious. At a glance one can see the region of the DNA double helix that is protected by protein. In addition, interesting details are revealed, such as the protection of guanine residues located 10 nucleotides apart, that were not obvious by looking at the raw data of the footprint of *Figure 3c*.

This type of quantitative analysis of footprinting results, plus plotting of the points of protection on to a model of the DNA double helix, are required to understand if, for example, two proteins bind to the same or opposite sides of the DNA double helix. Such data may even be used to deduce a model for the interaction between a sequence-specific DNA-binding protein and its binding site DNA (27).

7. OTHER APPLICATIONS FOR BANDSHIFT AND FOOTPRINTING GELS

In this chapter, I have described the application of bandshift and footprinting gels to assay complex formation between DNA and sequence-specific DNA-binding proteins. These two electrophoretic systems have also played an important role in the study of chromatin assembly, including the binding and sequence-determined positioning of histone octamers on DNA restriction fragments (29) and structural analysis of a triple complex formed between the histone octamer, a *Xenopus* gene for 5S RNA, and TFIIIA (10). One use of the fractionation offered by bandshift gels that has thus far not been generally exploited (but see ref. 10), and which may require radioactively-labelled protein, is for characterization of the protein or proteins present in protein–DNA complexes.

8. ACKNOWLEDGEMENTS

I thank my colleagues for much advice, in particular Louise Fairall, Peter Sorger and Mair Churchill.

9. REFERENCES

1. Pelham, H. R. B. (1982) *Cell,* **30**, 517.
2. Rickwood D. and Hames, B. D. (eds) (1982) *Gel Electrophoresis of Nucleic Acids: A Practical Approach,* IRL Press, Oxford.
3. Miller, J., McLachlan, A. D. and Klug, A. (1985) *EMBO J.,* **4**, 1609.
4. Sorger, P. K. and Pelham, H. R. B. (1987) *EMBO J.,* **6**, 3035.
5. Sorger, P. K., Lewis, M. J. and Pelham, H. R. B. (1987) *Nature,* **329**, 81.
6. Strauss, F. and Varshavsky, A. (1984) *Cell,* **37**, 889.
7. Garner, M. M. and Revzin, A. (1981) *Nucleic Acids Res.,* **9**, 3047.
8. Fried, M. G. and Crothers, D. M. (1981) *Nucleic Acids Res.,* **9**, 6505.
9. Varshavsky, A., Bakayev, V. V. and Georgiev, G. P. (1976) *Nucleic Acids Res.,* **3**, 477.
10. Rhodes, D. (1985) *EMBO J.,* **4**, 3473.
11. Hill, D. E., Hope, I. A., Macke, J. P. and Struhl, K. (1986) *Science,* **234**, 451.
12. Pietle, J. and Yaniv, M. (1987) *EMBO J.,* **6**, 1331.
13. Hope, I. A. and Struhl, K. (1986) *Cell,* **46**, 885.
14. Topol, J., Ruden, D. M. and Parker, C . S. (1985) *Cell,* **42**, 527.
15. Lutter, L. C. (1978) *J. Mol. Biol.,* **124**, 391.
16. Galas, D. and Schmitz, A. (1978) *Nucleic Acids Res.,* **5**, 3157.
17. Drew, H. R. (1984) *J. Mol. Biol.,* **176**, 535.
18. Maxam, A. M. and Gilbert, W. (1980) In *Methods in Enzymology.* Grossman, L. and Moldave, K. (eds), Academic Press, New York, Vol. 65, p. 499.
19. Hertzberg, K. P. and Dervan, P. B. (1984) *Biochemistry,* **23**, 3934.
20. Hatfull, G. F., Noble, S. M. and Grindley, N. D. F. (1987) *Cell,* **49**, 103.
21. Pope, L. E. and Sigman, D. S. (1984) *Proc. Natl. Acad. Sci. USA,* **81**, 3.
22. Kuwabara, M., Yoon, C., Goyne, T., Thederah, T. and Sigman, D. S. (1986) *Biochemistry,* **25**, 7401.
23. Tullius, T. M., Dombroski, B. A., Churchill, M. E. A. and Kam, L. (1987) In *Methods in Enzymology.* Wu, R. (ed.), Academic Press, New York, Vol. 155, p. 537.
24. Sakonju, S. and Brown, D. D. (1982) *Cell,* **31**, 395.
25. Singer, B. (1975) *Prog. Nucleic Acids Res. Mol. Biol.,* **15**, 219.
26. Suck, D., Lalm, A. and Oefner, C. (1988) *Nature,* **332**, 465.
27. Fairall, L., Rhodes, D. and Klug, A. (1986) *J. Mol. Biol.,* **192**, 577.
28. Vrana, K. E., Churchill, M. E., Tullius, T. D. and Brown, D. D. (1988) *Mol. Cell Biol.,* **8**, 1684.
29. Drew, H. R. and Travers, A. A. (1985) *J. Mol. Biol.* **186**, 773.

CHAPTER 8

Identification and purification of sequence-specific DNA-binding proteins

PETER K. SORGER, GUSTAV AMMERER and DAVID SHORE

1. INTRODUCTION

In this chapter we will describe methods for the identification of sequence-specific DNA-binding proteins in cell-free extracts and for the purification of these proteins by DNA affinity chromatography. The use of these methods will be illustrated with a number of yeast proteins that have been studied in this laboratory. These proteins include the heat shock transcription factor (HSF; 1, 20), repressor/activator protein 1 (RAP1; 2), pheromone receptor transcription factor (PRTF; 3), silencer binding factor-B/ARS binding factor 1 (abbreviated as SBF-B; 4) centromere-binding protein 1 (CP1; 5) and a yeast homologue of the mammalian activator protein 1 (AP1; 6). Although this chapter deals exclusively with yeast proteins, the techniques described should be useful in the analysis of a broad range of sequence-specific DNA-binding proteins from a variety of sources, including animal cells.

DNA-binding proteins can be detected and analysed most conveniently using either of two electrophoretic assays: footprinting gels and bandshift gels (also known as probe retardation or mobility shift gels; 7,8); the application of these techniques to the study of purified proteins is described in the preceding chapter. Unfractionated cell-free extracts contain, in addition to the protein of interest, however, a large number of other sequence-specific DNA-binding proteins, plus abundant proteins that bind DNA with little sequence specificity. It is necessary to adjust the conditions in which extract and binding-site DNA (sometimes called probe DNA) are reacted so that complexes containing the correct protein are preferentially stabilized. Below we describe the composition of binding buffers and methods for discriminating between the various types of DNA–protein complexes.

A further complication of studies with unfractionated extracts is that many DNA sequences for which there exist functional assays (such as the ability of the sequence to promote transcription *in vivo* or *in vitro*) interact with several proteins. Effective resolution of the various protein binding sites in such sequences usually involves a combination of footprinting and bandshift analysis, as well as functional studies. The latter is beyond the scope of this chapter, so the discussion that follows pre-supposes the availability of a short functional sequence capable of binding a protein of interest. As discussed below, however, it is often helpful to construct mutant binding sites that have no activity *in vivo* or *in vitro* and differ from the optimal site in as few positions as possible. The formation of protein–DNA complexes with functional and closely-related non-functional sequences can then be compared.

2. PREPARATION OF WHOLE-CELL EXTRACTS FROM S.CEREVISIAE

Although in this chapter we will describe only the preparation of extracts from *S.cerevisiae*, extracts have been successfully prepared from animal cells grown in culture, from solid tissues such as thymus, from *Drosophila* larvae and from a large number of other sources. In most cases, extracts are prepared from isolated nuclei. Nuclei can be isolated from yeast cells following enzymatic degradation of the cell wall (9), but a major problem associated with the use of cell wall-degrading enzymes is their contamination with proteases; a second limitation is the expense involved in making spheroplasts of large numbers of cells. Mechanical disruption has neither of these drawbacks and we have had excellent success in purifying proteins from whole-cell yeast extracts.

To maximize the yield of cells during large-scale purification, we typically grow cells to late log phase (5–10 $\times$ 10^7cells/ml). In some cases however, it may be necessary to harvest cells at quite different densities (either early log phase or stationary phase) or after a specific treatment (heat shock or α-factor arrest, for example). The mechanical disruption technique described here has the advantage that specially-treated cells can be harvested and frozen rapidly and that the efficiency of breakage is independent of cell density.

We routinely prepare extracts from as little as 0.1 g of cells (10 ml mid-log culture) and as much as 5 kg (400 litres) of culture as described in *Tables 1* and *2*. Yeast vacuoles, which are disrupted during cell breakage, contain a variety of powerful proteolytic enzymes. For this reason, the use of strains deficient in vacuolar proteases has been essential, in our hands, for the reliable preparation of whole-cell extracts. In particular, we have detected little or no degradation of HSF, RAP1 or PRTF in extracts made from the strain BJ2168 (*leu*2, *trp*1, *ura*3-52, *prb*1-1122, *pep*4-3, *prc*1-407, *gal*2, reference 4). This strain contains a mutation in protease A (encoded by the PEP4 locus, reference 10), a major protease that is required for the activation of several other vacuolar proteases. Additionally, BJ2168 is deficient in proteases B and C (loci PRC1 and PRB1, respectively), but has only a slightly increased generation time relative to wild-type yeast. We have found that extracts made from this strain are stable at room temperature even in the absence of protease inhibitors. If the DNA-binding protein of interest must be isolated from a strain other than BJ2168 it may be advisable to introduce a mutation into the PEP4 locus by gene disruption.

Table 1 describes a procedure suitable for the preparation of analytic-scale cultures using glass beads. This method permits routine isolation of 2–50 mg of total soluble protein from 5–100 ml of culture. *Table 2* describes a protocol for the preparation of large-scale extracts and is adapted from an earlier procedure by Klekamp and Weil (11). It involves the disruption of frozen cells in a blender

Table 1. Preparation of small-scale yeast extracts by agitation with glass beads.

1. Grow cells (5–100 ml) in YEPD medium [1% (w/v) yeast extract, 2% (w/v) peptone, 2% (w/v) glucose] to mid to late log phase (A_{600} = 2–10)[a].
2. Harvest cells by centrifugation and re-suspend in breakage buffer [200 mM Tris–HCl, pH 8.0, 10% (w/v) glycerol, 10 mM $MgCl_2$, 1 mM DTT (or 10 mM 2-mercaptoethanol), 1 mM PMSF, 0.5 mM TPCK and 0.02 mM TLCK, 2 μg/ml pepstatin A][b] at 2 ml of buffer per gram of cells.
3. Add an equal volume of ice-cold acid-washed glass beads (0.4–0.5 mm diameter), and lyse the cells by vortexing at maximum speed. Some experimentation is necessary to maximize the amplitude of the oscillations, but between 1 and 3 mins of vortexing is usually sufficient to achieve greater than 70% breakage[c]. For small cultures (up to 25 ml of cells) breakage can be done in 1.5-ml Eppendorf tubes. In this case, the best results will be obtained with 0.5 ml of beads and 0.7 ml of cells. Larger amounts of cells are broken in 15-ml Corex glass tubes[d].
4. Centrifuge for 15 min in a standard bench-top microfuge. The supernatant can be used directly in DNA binding assays[e].

[a] The yield of any given binding protein, and its extent of degradation, may depend critically on the density of cells at harvesting. Whenever possible, a strain deficient in one or more proteases should be used (see text).
[b] For some proteins, $MgCl_2$ may be omitted and substituted with 1 mM EDTA, which may help to reduce proteolysis.
[c] The efficiency of breakage is monitored by examining cells by phase contrast microscopy in a 1% SDS solution at 100–200× magnification. Broken cells will appear as empty shells.
[d] If large numbers of samples are to be broken, the IKA Vibrax VXR2 appears to be the shaker of choice. It allows efficient breakage of up to 36 samples in parallel.
[e] Although this simple procedure has proven adequate for a number of different sequence-specific DNA-binding proteins, improved yields may be achieved by further extraction in 0.4 M ammonium sulphate, followed by preparation of an S-100 supernant (see *Table 2*).

cooled with liquid nitrogen. Much higher yields are obtained with this procedure than with glass bead breakage, and many kilograms of cells can be processed rapidly. Two hundred litres of culture typically yield 2.5 kg of packed cells and 75 g of total protein. An indefinite period can elapse between harvesting and disrupting the cells if they are stored frozen at −70°C.

3. PROTEIN–DNA BINDING CONDITIONS

Proteins bind to DNA in both a sequence-specific and in a sequence-independent manner (12). Histones, for example, bind with moderate affinity to DNA, forming relatively salt-stable complexes, but exhibit little sequence specificity. Regulatory proteins, in contrast, usually bind much more tightly to specific DNA sequences than to bulk DNA. When binding-site DNA and unfractionated cell-free extracts are mixed in a suitable buffer, complexes containing both types of protein are formed. We will refer to complexes containing the sequence-specific DNA-binding proteins as 'specific complexes' and those containing proteins with little sequence specificity as 'non-specific complexes'. Because the non-specific proteins are often abundant and compete with specific proteins for interaction with radiolabelled binding-site DNA, conditions are chosen to maximize the yield of specific complexes. Not all specific complexes contain the protein of interest,

Table 2. Preparation of large-scale yeast extracts: disruption by blending in liquid nitrogen.

A. *Harvesting*

1. Grow cells in 100–400 litres of YEPD to mid to late log phase (*Table 1*) and harvest in a Sharples continuous-flow centrifuge[a].
2. Convert the compacted cell pellet (typically 1 kg wet weight per 100 litres of medium) into a thick paste by the addition of a minimum volume of ice-cold water and load into 100-ml syringes.
3. Extrude the paste directly into liquid nitrogen with the aid of a caulking gun (the paste should be too thick to allow manual operation of the syringe). The frozen cell spaghetti can be stored indefinitely at −70°C before disruption.

B. *Disruption*

1. Disrupt cells mechanically by blending batches in a 3.8-litre stainless steel vessel cooled with liquid nitrogen. We employ a Waring blender with a 1725-W motor at a blade speed of 22000 r.p.m.
2. Add approximately 500 g of frozen cells to the pre-cooled blender vessel and sufficient liquid nitrogen just to cover the cells. Allow the vessel to equilibrate for 1 min; breakage is achieved by three successive bursts of 2 min each. Between bursts, mix the frozen powder and add additional liquid nitrogen until the powder has a paste-like consistency.
3. If insufficient liquid nitrogen is present, the powder will begin to thaw. On the other hand, an excess of liquid nitrogen results in copious venting of gas and powder from the beginning of the run and should be avoided.
4. This regime of blending typically results in greater than 95% breakage as judged by microscopy of the thawed powder, (*Table 1*, footnote c).
5. Suitable precautions should be taken during operation of the blender, and due to the inevitability of venting, we have found it best to perform the operation outdoors.

C. *Preparation of the extract*

1. Thaw the blended powder in the presence of 1–1.5 litres of room-temperature breakage buffer (*Table 1*) per kg of cells.
2. Make the lysate 0.4 M in ammonium sulphate by the addition of a 4 M stock solution (pH 8.0) and stir for 1 h. This and subsequent steps are performed in a coldroom at 4°C or on ice.
3. Prepare a high-speed supernatant (S-100) from the solubilized cell extract by spinning the lysate for 45 mins at 42 000 r.p.m. in Beckman-type 45Ti rotors. Pour the supernatant through cheesecloth into a large measuring cylinder.
4. Precipitate the S-100 by the addition of 0.35 g of ammonium sulphate per ml of extract. Allow the precipitate to form by stirring for at least 1 h at 4°C, and harvest by centrifugation at 10 000 *g* for 45 min.
5. Re-suspend the pellet in approximately 0.4 litres of A50 per kg of cells (see text) and dialyse exhaustively against A50 containing protease inhibitor stock solutions (Section 5.3) at a dilution of 1:1000.
6. Clarify the dialysed extract by centrifugation and load directly onto a 1.5-litre heparin–agarose column. Alternatively, freeze 200-ml batches of extract in liquid nitrogen and store at −70°C for future use.
7. We typically obtain about 1.8–2 litres of extract containing about 100 g of protein from 5 kg of cells (400 litres of initial culture).

[a] With three people working together, we are able to process 4–5 kg of previously frozen cells (from a 400-litre fermenter run) in 1 day. This procedure could be scaled down by the suitable choice of blender and vessel and is probably the method of choice with as little as 10–50 g of cells.

however, and sequence-specific interactions between binding-site DNA and contaminating proteins are often observed. These interfering sequence-specific binding proteins either may be closely related to the correct protein or may fortuitously find a close match to their binding sites in the labelled DNA. In this section we will discuss various parameters of the binding conditions than can be altered to minimize interference from non-specific DNA-binding proteins, and in Section 4 we discuss methods for identifying, amongst the set of specific complexes, those that contain the protein of interest.

Both the concentration of salt and the pH of the binding buffer can be adjusted to alter the stability of various DNA–protein complexes. In addition, unlabelled 'carrier DNA' is usually added in substantial excess to radiolabelled binding-site DNA. Proteins with little sequence specificity will bind to this carrier, thereby freeing the binding-site DNA for interaction with sequence-specific proteins. The yield of specific complexes can also be increased by the addition of carrier protein and by the inclusion of specific divalent and trivalent cations. Finally, there may be instances in the analysis of complex formation on bandshift gels in which the resolution of protein–DNA complexes from each other and from unbound DNA may depend critically on the pH of the binding buffer and of the gel electrophoresis buffer.

3.1 **Carrier DNA**

The ratio of the sequence-specific to sequence-independent DNA-binding affinities of regulatory proteins varies over several orders of magnitude. In cases where it is low, carrier DNA will compete with binding-site DNA for binding to the protein of interest. For this reason, the correct choice of both the amount and composition of the carrier can be critical. Typical carrier DNAs include mixed-sequence DNA, such as sonicated salmon sperm, calf thymus or *Escherichia coli* DNA, plasmids, and synthetic co-polymers such as poly[d(I-C)] and poly[d(A-T)]. Synthetic co-polymers are used most widely, presumably because they do not interfere appreciably with sequence-specific binding. The inclusion of plasmid DNA is usually essential if the binding-site DNA contains plasmid-derived sequences (see *Figure 1*); in this case, the plasmid carrier eliminates interference from proteins that bind to the plasmid rather than to the cloned binding-site sequences. The sensitivity of various proteins to the composition and amount of carrier DNA is variable: HSF and RAP1 are very tolerant, whereas hormone receptors, such as the glucocorticoid receptor and its relatives (13), can be very sensitive. The amount of carrier DNA must be determined empirically, but it is typically used at a 100- to 5000-fold (w/w) excess over binding-site DNA.

The optimal binding conditions for different proteins vary considerably and depend both upon the properties of the protein itself and on the characteristics of various contaminating proteins. Because of this, the conditions must usually be adjusted when preparations of different purities are analysed (*Figure 1* and *Table 3*). In general, the amount of carrier DNA should be lower in binding reactions using purified protein than in those containing crude extract, and it may be helpful to change the ratio of plasmid DNA to poly[d(I-C)] if both are being used.

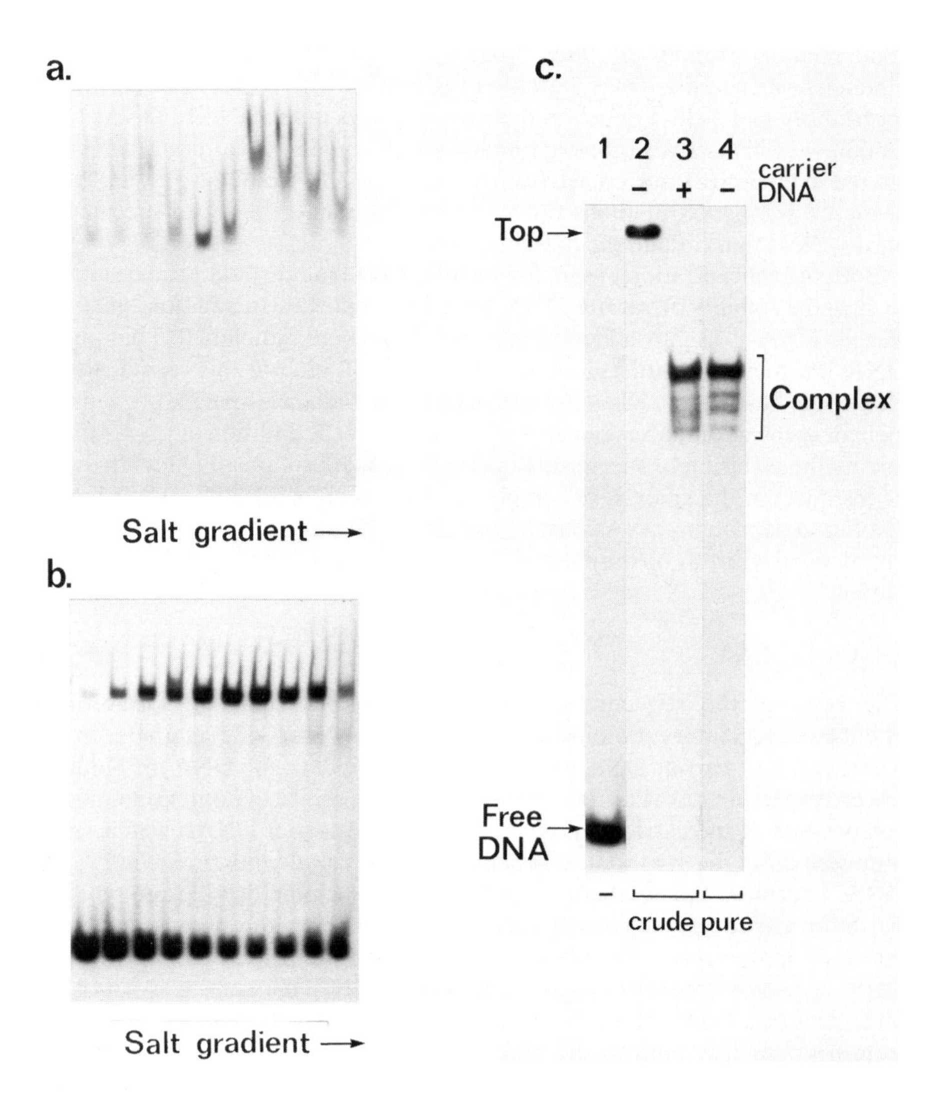

Figure 1. Effect of carrier DNA on resolution of complexes in bandshift gels. (**a**) Heparin–agarose column fractions reacted with radiolabelled DNA containing an SBF-B binding site and sequences derived from the pUC19 polylinker. The binding reactions also contained 1.0 μg of poly[d(I-C)] and 0.1 μg of uncut pUC19 DNA. The lanes contain equal volumes of successive fractions from a heparin–agarose column. Free DNA is not visible on this gel. (**b**) Identical samples assayed in the presence of 2 μg of poly [d(I-C)] and 1 μg of pUC19 DNA, 10 times the amount used in '**a**' (**c**) Pure and impure samples containing PRTF reacted with a 45-bp radiolabelled oligonucleotide. **Lane 1**, binding buffer alone; **lane 2**, 0.5 μg of protein from a calf thymus DNA–Sepharose column fraction reacted without carrier DNA; **lane 3**, the same fraction reacted in the presence of 1.0 μg of poly[d(I-C]; **lane 4**, purified PRTF reacted without carrier DNA.

Table 3. Typical binding conditions.

A. *Heat shock factor* (1)

1. Add 25–125 μg of unfractionated whole-cell extract containing HSF to a 25-μl reaction containing 0.5 ng of radiolabelled binding-site DNA, 2 μg of pUCl9 DNA, 2 μg of poly[d(I-C)], 20 mM Hepes, pH 8.0, 1 mM EDTA, 60 mM KCl, 12% glycerol, 0.1% Nonidet P-40 and 1 mM DTT.
2. DNA affinity column fractions containing up to 2 M NaCl are assayed by adding 4 μl to a 50 μl reaction containing 0.5 ng of radiolabelled binding-site DNA, 250 ng of poly[d(I-C)], 50 μg of BSA, 1 mM spermidine, 20 mM Hepes, pH 8.0, 20% glycerol, 0.1% Nonidet P-40 and 1 mM DTT. The final NaCl concentration should be 160 mM or less.

B. *RAP1* (2)

1. Both whole-cell extracts and purified fractions are analysed in 20-μl binding reactions containing 0.5 ng of radiolabelled binding-site DNA, 1 μg of pUC19 DNA, 1 μg of poly[d(I-C)], 20 mM Tris–HCl, pH 8.0, 5 mM $MgCl_2$, 100 mM NaCl, 5% glycerol and up to 20 μg of BSA.

3.2 Binding buffers

A wide variety of binding buffers based on Tris, Hepes, Pipes, etc. have been employed with success. Buffers of pH 8.0 and 50–150 mM NaCl or KCl are typical, but a variety of different conditions should be checked. The failure to detect a specific complex may be due either to de-stabilization of the specific complex or to excessive interference from non-specific binding proteins. In the latter case, different combinations of carrier DNA should be tested for their effects on the various complexes. Alternatively, the resolution of protein–DNA complexes from each other and from free DNA may depend critically on the pH of the electrophoresis buffer (for example, see reference 14).

Carrier protein is typically added to binding reactions at a concentration of 0.25–1 mg/ml, although its inclusion is not essential with crude extracts. Bovine serum albumin (BSA) is used most frequently, but other proteins are also effective provided that they do not bind appreciably to DNA. The carrier protein presumably prevents complexes from adhering to reaction vessels (the use of siliconized tubes also helps) and stabilizes DNA-binding proteins by lowering the chemical activity of the binding buffer. The addition of glycerol to between 5 and 20% (v/v) appears to have a similar effect. Variability in complex formation can often be attributed to the failure to add sufficient carrier protein.

Other common components of binding buffers include detergent and polycations such as spermidine. We routinely add either Nonidet NP-40 at 0.1% (v/v) or octyl-β-D-glucopyranoside at 0.025% (w/v) to binding reactions containing either crude extracts or purified proteins. The addition of spermidine to between 0.25 and 5 mM appears to stabilize some DNA–protein complexes. Moreover, some proteins require the presence of divalent cations such as Mg^{2+} in order to bind DNA, and proteins containing the 'zinc finger' motif require Zn^{2+}. It is customary to incubate binding reactions for about 30 min before analysis (or nuclease digestion). Some protocols call for initial reaction with carrier DNA on ice, followed by the addition of binding-site DNA and further incubation at room temperature. With HSF, PRTF, RAP1, AP1, SBF-B and CP1, however, we simply add extract to a mixture containing buffer, carrier DNA, BSA and binding-site DNA and incubate for 5–30 min at room temperature. There may, nevertheless, be cases in which DNA–protein complexes are stable only at low temperatures.

4. IDENTIFICATION OF SEQUENCE-SPECIFIC DNA-BINDING PROTEINS

4.1 **Cross-competition analysis**

As mentioned above, not all of the complexes detected in a binding reaction with crude extract will contain the protein of interest. In Section 3 we considered interference from abundant proteins that exhibit little sequence specificity. A second source of interference arises from the presence of sequence-specific binding proteins that, although not the protein of interest, show a significant affinity for the radiolabelled binding-site DNA. In footprinting it may be sufficient to determine that the expected sequences are protected; when bandshift gels are used, however, complexes containing the correct protein must be indentified by cross-competition analysis. This is done by adding to the binding reaction a 10-fold or greater excess of unlabelled binding-site DNA prior to the addition of extract, so that the unlabelled and labelled DNAs compete for binding to proteins. It is good practice, moreover, to include binding-site DNA in molar excess over the protein of interest; if competitor DNA is identical to the labelled DNA (and in excess), all of the label should, in principle, be eliminated from the complexes. In practice, the addition of competitor DNA often removes the label only from a subset of the complexes. The complexes unaffected by the addition of competitor are thought to arise from low-affinity binding by abundant proteins that are present in excess to binding-site DNA.

Complexes containing contaminating proteins that find a fortuitously good match to their own binding sites in the labelled DNA, or which are similar to the protein of interest, can be identified by performing a series of competitions with functional and non-functional binding sites. Those complexes in which the labelled DNA is competed out by the addition of DNA containing functional binding sites, but are unaffected by the addition of DNA containing non-functional mutant sites, are likely to contain the correct protein. This type of analysis can be used with a set of related binding sites to compare their activities in a functional assay with their binding affinities. In the example shown in *Figure 2*, the ability of a series of five variant HSF binding sites, or heat shock elements (HSEs), to activate transcription *in vivo* in a synthetic promoter is seen to correlate well with the ability of these sites to bind HSF *in vitro* as measured by cross-competition (see also references 2 and 15). This experiment provided good initial evidence that the putative HSF polypeptide being studied was in fact involved in activating the transcription of heat shock genes.

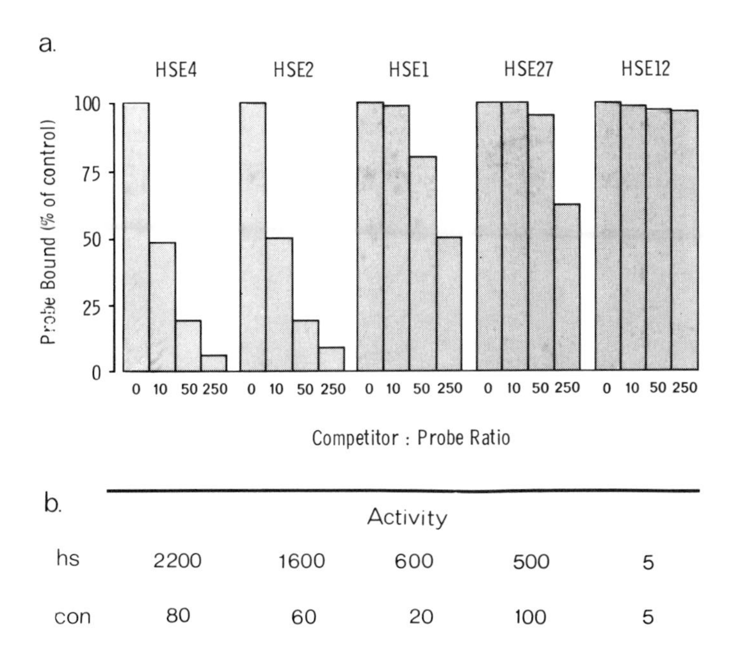

Figure 2. Comparison of *in vitro* binding affinity of variant HSF binding sites with *in vivo* activity. (**a**) Competitive DNA binding assay with variant HSF binding sites (HSEs). Parallel reactions with varying amounts of each competitor DNA, 1 μg of protein from the calf thymus DNA–Sepharose column and 0.5 ng of radiolabelled HSE2 binding-site DNA were analysed on bandshift gels, and the bound and free probe were excised and counted. The total amount of DNA in each reaction was adjusted to 5.5 μg by the addition of pUC19 carrier. The amount of probe bound is expressed as a per cent of that bound in the absence of competitor (50% of input), and represents the average of two determinations. See reference 1 for more details. (**b**)Transcriptional activity of variant HSE sequences *in vivo* in a heterologous promoter linked to the *Lac-Z* gene (3). β-Galactosidase activity (expressed in arbitrary units) was determined at normal growth temperatures (con) and following a heat shock to 39°C for 20 min (hs).

4.2 **Estimating dissociation rates**

A comparison of the affinities of variant binding sites can also be made by estimating the dissociation rates of pre-formed protein–DNA complexes. In this case binding-site DNA is first incubated with extract and a 100-fold or greater excess of unlabelled competitor DNA subsequently added. The dissociation rate of the pre-formed complexes is monitored on bandshift gels by determining the rate of disappearance of protein-bound label. As shown in *Figure 3*, the rate of dissociation of HSF complexes containing a DNA sequence with a single HSE is much greater than that of complexes containing DNA with two overlapping HSEs. The difference in the binding affinities of the two types of sequences is more apparent in this assay than in cross-competition experiments (compare HSE1 and HSE2 in *Figures 2* and *3*).

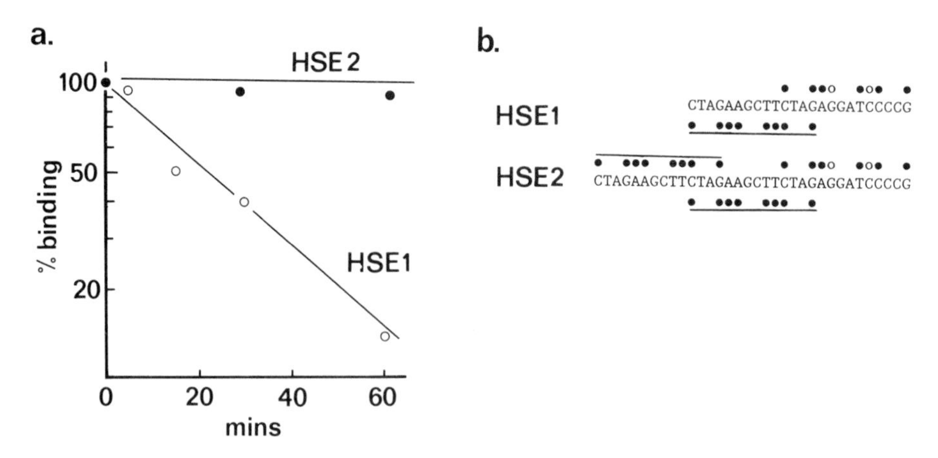

Figure 3. Dissociation rates of pre-formed HSF–DNA complexes. (**a**) Partially-purified HSF was reacted with radiolabelled HSE1 or HSE2 binding-site DNA in standard conditions (*Table 3*) for 30 min, and then incubated for the indicated times with a 50- to 10-fold molar excess of plasmid containing cloned HSE2 sequences. Bound radiolabelled DNA was separated from free DNA on bandshift gels and is expressed as a per cent of that bound in the absence of added competitor. (**b**) Sequence of HSE1 and HSE2 binding-site DNAs with matches to the HSF consensus binding site indicated by filled circles (HSE2 contains two overlapping sites).

4.3 **Photo-activated cross-linking**

Once a protein has been identified on the basis of its ability to bind a DNA sequence of interest, it is often useful to estimate its molecular weight. This can be done most easily by cross-linking complexes containing labelled DNA with UV light and then resolving the proteins on an SDS–polyacrylamide gel (16; *Table 4* and *Figure 4*). The label is incorporated continuously in the binding-site DNA, and bromodeoxyuridine is substituted for thymidine. The choice between [^{32}P]dCTP or [^{32}P]dATP and coding or non-coding strand is made empirically. Labelled DNA is mixed with extracts in standard binding buffer, and protein and DNA are cross-linked by exposure to 300-nm UV light. Following cross-linking, the complexed DNA can be isolated on non-denaturing gels to reduce interference from free DNA; alternatively, the reaction may be extracted with phenol, and the proteins recovered from the organic phase. The presence of covalently-bound nucleic acid increases the apparent size of binding proteins, so it is customary to treat cross-linked complexes with DNase I to trim off excess DNA but, at least in some cases, this can substantially reduce the strength of the signal.

As shown in *Figure 4*, several bands are visible following cross-linking of extracts to an HSF binding site; the correct protein is identified by competition analysis as previously described. In this case the band corresponding to the HSF polypeptide is seen to diminish in intensity with the addition of an excess of a good competitor (HSE2) but not a weak competitor (HSE12).

There exist two additional methods for determining the molecular weight of an unpurified DNA-binding protein. The first involves separating an extract on

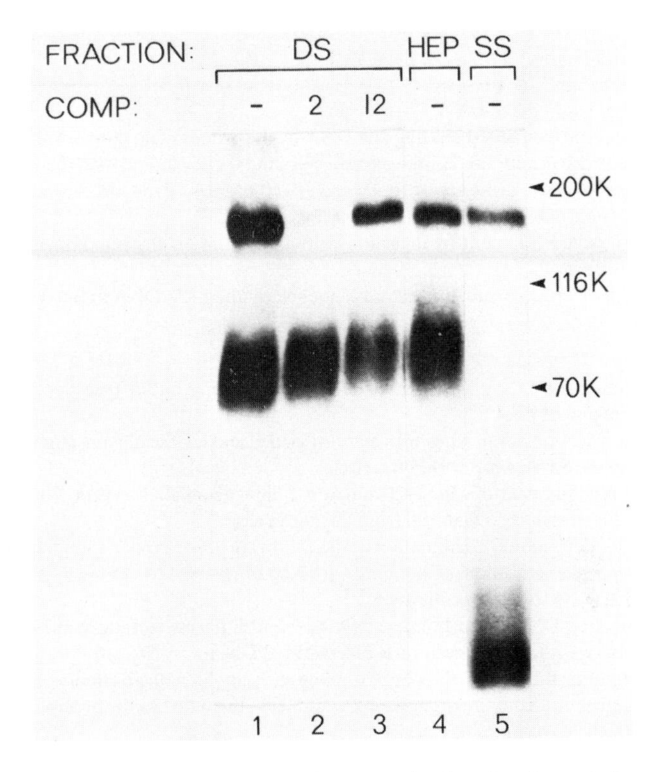

Figure 4. UV cross-linking of HSF binding-site DNA to protein. Samples of protein from a calf thymus DNA–Sepharose fraction (DS; 2 μg per lane), heparin–agarose column (HEP; 8 μg protein) and the HSF affinity column (SS; 6 μl) were cross-linked to 0.6 ng of labelled bromodeoxyuridine-substituted binding-site DNA as described in *Table 4*. 'Comp' indicates the presence of a 40-fold molar excess over radiolabelled DNA, of either a strong ('2') or a weak ('12') competitor DNA. Reactions with the HEP and DS fractions were phenol-extracted and ethanol-precipitated following cross-linking; there was no nuclease digestion. Samples containing the SS fraction were precipitated with 10% trichloroacetic acid without extraction: unbound probe is therefore seen only in this lane. The positions and molecular weights of the markers in kd are indicated (1).

SDS-containing gels, eluting the proteins from a series of gel slices, renaturing the eluted protein pools and then testing them for binding activity (see refs 16 and 17 for more details). A second method involves electrophoretic transfer of proteins from an SDS-containing gel to nitrocellulose (Western blotting), denaturation and subsequent renaturation of the blot, and finally incubation with radiolabelled binding-site DNA (Chapter 3). This 'southwestern' blotting method is described in reference 18.

5. DNA AFFINITY CHROMATOGRAPHY

5.1 **Introduction**

The development in the last few years of DNA affinity chromatography has made it possible to purify rapidly DNA-binding proteins of low abundance from either

Table 4. UV cross-linking[a].

A. *Preparation of binding-site DNA*

1. Prepare continuously-labelled binding-site DNA by the prime cut method[b]. Anneal single-stranded M13 templates with primer[b] and initiate synthesis of the second strand with the Klenow fragment of DNA polymerase in the presence of 50 μM dGTP, 50 μM dATP, 50 μM 5-bromo-2′-deoxyuridine triphosphate (Pharmacia) and 5 μM (α-^{32}P]dCTP[c].
2. Release the labelled fragment by digestion with appropriate restriction enzymes and purify on a polyacrylamide gel.
3. Store bromodeoxyuridine-substituted radiolabelled binding-site DNA in the dark and use within a few days.

B. *Cross-linking*

1. Incubate binding-site DNA with extract in 50 μl of binding buffer (see Section 3.2 for details) containing 1 mg/ml of BSA.
2. Draw 1.5-cm circles on Saran wrap with a grease pencil and lay the film on a light box. Spot binding reactions onto the Saran wrap within the circles.
3. Irradiate the reaction mixtures for 1–20 min at 302 nm with a light box (UV Photoproducts TM20) and recover the cross-linked material from the Saran wrap.
4. An empirically-determined combination of the following procedures is then followed:[d]
   (a) Adjust samples to 5 mM $CaCl_2$, digest with 2 μg of DNase I for 30 min at 30°C, and precipitate with 10% (w/v) trichloroacetic acid.
   (b) Separate free DNA from bound DNA on bandshift gels to reduce the background smear caused by irregular electrophoresis of irradiated DNA.
   (c) Separate free and bound DNA by extracting the samples with an equal volume of phenol, and recover proteins from the organic phase and from the interface by precipitation on dry ice for 15 min with 2.5 volumes of ethanol.
5. Dissolve the proteins in SDS-sample buffer and boil briefly before electrophoresis.

[a] Adapted from the method of Chodosh *et al.* (16)
[b] Binding-site DNA can also be prepared from collapsed super-coiled templates.
[c] The choice of [^{32}P]dCTP and dATP as well as between labelling coding versus non-coding strand is made empirically.
[d] The classical method involves digestion of complexes with DNase I to remove excess DNA. In some cases this can significantly reduce the strength of the labelling and it may be helpful to omit the nuclease digestion.

nuclear or whole-cell extracts. Although a variety of different approaches have been successfully employed, we advocate the use of a simple three-column procedure. In the first step cell-free extracts are fractionated on heparin–agarose, a polyanionic resin. Fractions containing the protein of interest are then further separated on calf thymus DNA–Sepharose and final purification is accomplished on a DNA affinity resin consisting of DNA containing a tandem array of high-affinity binding sites coupled to a Sepharose support.

All of the DNA-binding proteins that we have examined bind to heparin–agarose; because approximately 99% of the protein in whole-cell extracts flows through this column, a substantial purification (up to 250-fold) can be achieved by eluting the bound proteins with a salt gradient. Furthermore, careful elution of the heparin–agarose column can separate many different DNA-binding proteins, so that several proteins can be purified simultaneously from the same extract. The critical part of the purification, however, involves successive chromatographic steps, first on calf thymus DNA–Sepharose and then on the specific DNA affinity resin. In practice we have found that most sequence-specific DNA-binding proteins elute from calf-thymus DNA columns at low salt concentrations and that passage through this resin therefore removes proteins that bind very tightly but non-specifically to DNA (these tight-binding proteins comprise up to 0.5% of the total protein in the pooled fractions from the first column). Chromatography on calf thymus DNA produces a fraction of proteins with a very similar over-all electrostatic affinity for non-specific (or more correctly, multi-specific) DNA. Proteins that elute from the non-specific column at low salt concentrations, but from the subsequent affinity column at high salt, necessarily show sequence specificity, in that their binding to the affinity column is more salt-resistant than their non-specific binding to calf thymus DNA. A careful salt elution of the affinity column should also resolve the correct protein from possibly contaminating sequence-specific binding proteins with related, but different, binding specificities. This procedure can eliminate the need for multiple passes over the affinity column and reduces the problem of identifying the correct protein.

5.2 **Assays**

Detection of the protein of interest during purification is most easily accomplished either on footprinting or bandshift gels. We have found DNA binding assays to be non-linear and quantification of the activity present at each step inexact. One possibility is to construct a DNA fragment with three or four non-interacting binding sites. The number of proteins bound per DNA molecule can be determined on non-denaturing or footprinting gels and is a sensitive measure of the concentration of DNA-binding protein (at least within the same set of samples, *Figure 5*). It is occasionally noted that the electrophoretic mobility of a protein–DNA complex is different in purified fractions and in crude extracts, perhaps due to the high protein concentrations of the latter (15). Finally, as described above, the optimal binding conditions can change considerably as the purification proceeds.

5.3 **Buffers**

The buffers described below have been used in all of the purifications described in this section and use ammonium sulphate as the principle salt. Ethylenediamine tetraacetic acid (EDTA) is used throughout, but in cases where metals such as zinc are thought to be essential for binding, these buffers should be suitably supplemented (e.g. with 1 mM $ZnCl_2$).

(i) *A50.* 50 mM Tris–HCl, 1 mM EDTA, 50 mM $(NH_4)_2SO_4$, 10% (v/v) glycerol, 1 mM dithiothreitol (DTT), pH 8.0, protease inhibitor stock solutions (see below) each diluted 1:1000.
(ii) *A600.* Identical to A50 but containing 600 mM $(NH_4)_2SO_4$.
(iii) *A50N2000.* Identical to A50 but supplemented with 2.0 M NaCl.
(iv) *Protease inhibitor stock solutions.* 1, 100 mM phenylmethylsulphonyl fluoride (PMSF) in ethanol; 2, 50 mM *N*-tosyl-L-phenylalanine chloromethyl ketone

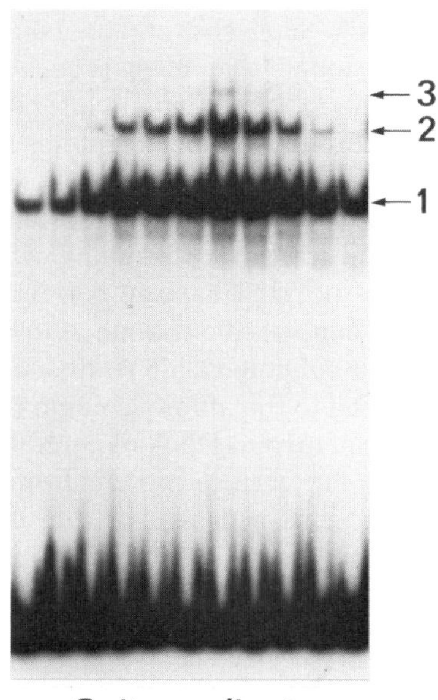

Figure 5. Quantification of sequence-specific DNA-binding activity by the use of radiolabelled DNA containing multiple non-interacting binding sites. The DNA consists of four copies of a synthetic CP1 binding site cloned into the polylinker of pUC13. Samples of 1 μl from successive 50 ml fractions of a heparin–agarose column elution were assayed as described in *Table 3* for RAP1. Complexes containing three CP1 molecules bound per molecule of DNA can be clearly detected in the peak fractions.

(TPCK), 5 mM N_{α}-*p*-tosyl-L-lysine chloromethyl ketone (TLCK) in methanol; 3, 2 mg/ml pepstatin A in 1 : 1 (v/v) methanol : water.

5.4 Initial fractionation

Initial fractionation by cation-exchange chromatography results in a significant purification of the crude extract and separates different DNA-binding proteins from each other. Ideally, the great majority of DNA-binding proteins should bind to the resin, but the bulk of the cellular protein (>99%) should be left in solution. When used to fractionate yeast whole-cell extracts, heparin–agarose meets those criteria (*Table 5*).

The high-speed supernatant from a whole-cell extract is precipitated with $(NH_4)_2SO_4$ and dialysed exhaustively against A50 as described in *Table 2*. The dialysed preparation is applied to heparin–agarose, eluted with a linear gradient of A50–A600 (*Figure 6*) and possibly also with a further gradient of A50–A50N2000; the column is always regenerated with an A50N2000 wash. Experience will suggest appropriate modifications to the elution procedure. In practice we have found that the majority of DNA-binding proteins are eluted between 50 and 600 mM $(NH_4)_2SO_4$; PRTF however, is eluted only at much higher salt concentrations. The capacity of heparin–agarose for different proteins varies significantly

Table 5. Preparation of heparin–agarose[a].

1. Heparin–agarose can be conveniently prepared in 500-ml batches. The complete procedure is performed in a suitable fume cupboard and all CNBr waste is discarded into 5 M KOH.
2. Wash 500 ml of agarose A 15-m (100–200 mesh) with several volumes of distilled water. For this and subsequent washing steps, the resin is collected under vacuum on a large Buchner funnel covered with a porous sheet of polyethylene.
3. Combine the washed agarose beads (500 ml) with 1.5 litres of 2 M sodium carbonate and mix by gently stirring in a plastic beaker at room temperature.
4. Add 50 ml of an acetonitrile solution of CNBr (2 g CNBr per ml of dry acetonitrile, best prepared by adding acetonitrile directly to bottles of CNBr) to the agarose beads and stir the mixture vigorously for 2 min.
5. Collect the CNBr-activated beads and wash them consecutively with two litres each of 0.1 M sodium bicarbonate (pH 9.5), water and 0.2 M sodium bicarbonate (pH 8.5).
6. Quickly transfer the activated agarose to a plastic bottle containing 6.0 g of heparin (500 000 units Sigma grade I at 170 units/mg) dissolved in 500 ml of 0.2 M sodium bicarbonate, pH 8.5.
7. Seal the bottle carefully and allow the coupling to proceed with end-over-end mixing at 4°C for 20 h. This and subsequent steps are performed outside the fume cupboard.
8. Add glycine to 1 M and incubate at room temperature for 4 more hours. This step blocks unreacted groups on the gel.
9. Collect the agarose beads and wash with 4 litres each of (a) 0.1 M sodium acetate (pH 4), (b) 2 M urea, (c) 0.1 M sodium bicarbonate (pH10) and (d) water. The wash solutions also contain 0.5 M NaCl.
10. Store the washed resin in 20 mM Tris–HCl (pH 8.0) and 0.02% sodium azide at 4°C. After repeated use the resin should be regenerated by stirring with 0.1 M NaOH for 10 min at 4°C, and washed extensively with cold water and several volumes of 0.1 M Tris base containing 2 M ammonium acetate.

[a] Method of Davison *et al.* (19)

and is not related in a direct way to their affinities for the resin. A 1.5-litre column, for example, is capable of binding more than 90% of the HSF present in 100 g of whole-cell yeast extract (from 2.5 kg of cells), but binds much less than 10% of the AP1 present in the same extracts, despite the fact that AP1 and HSF elute at similar salt concentrations. Although the reasons for this behaviour are poorly understood, we have observed it with other proteins, including CP1 and SBF-B. In practice we have found that AP1, SBF-B and CP1 can be quantitatively recovered by up to three successive passages over the same heparin–agarose column, suggesting that this may apply in general to proteins that fail to bind efficiently in the initial run.

The salt concentration at which DNA-binding proteins elute from heparin–agarose depends upon characteristics such as pI and net charge and is sensitive to their state of covalent modification. For example, the highly-phosphorylated state of HSF (present in heat-shocked cells) elutes at a lower salt concentration than the non-phosphorylated protein present in control cells (*Figure 7*). Finally, in cases where the species capable of binding DNA is composed of multiple subunits or polypeptide chains, it is possible that activity will be lost as a result of their separation during chromatography.

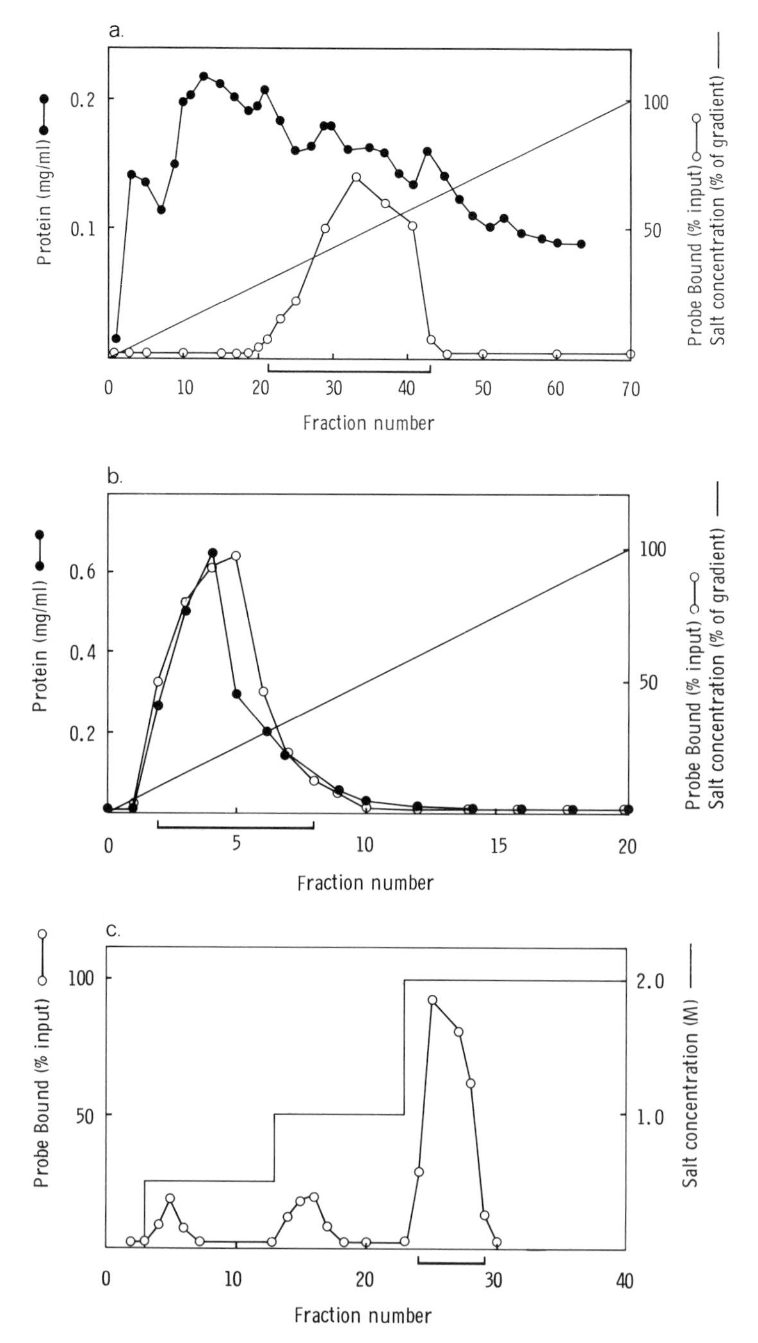

Figure 6. Purification of HSF. (**a**) Elution profile of activity eluted from heparin–agarose column between 50 and 400 mM ammonium sulphate (indicated as 0–100% of gradient). Ten microlitres of each 10-ml fraction were assayed as described in *Table 3*. The peak HEP pool is indicated by the brackets. (**b**) Elution profile of the calf-thymus DNA–Sepharose column between 50 (0%) and 240 mM (100%) ammonium sulphate. Aliquots of 4 μl of each 1-ml fraction were assayed as in **a**. (**c**) Profile of the HSE DNA–Sepharose affinity column eluted with steps of 0.5 M KCl, 1.0 M KCl and 2.0 M NaCl and assayed as described in *Table 3* (1).

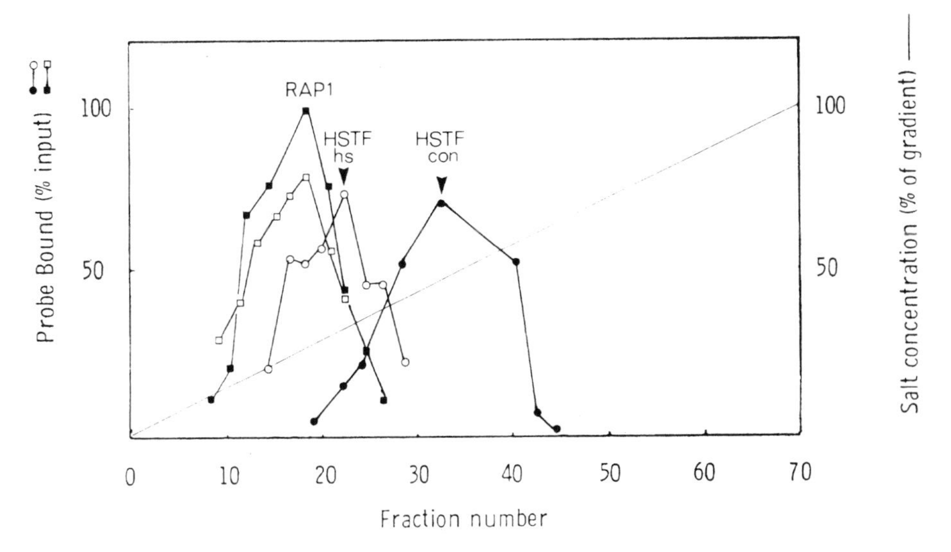

Figure 7. Effect of HSF phosphorylation on elution from heparin–agarose (20). Elution profile on heparin–agarose of HSF and RAP1 from heat shocked (hs) and control (con) cells: profile of activity in extracts from cells heat shocked 20 min at 39°C for RAP1 (□———□) and HSF (○———○) and from control cells for RAP1 (■———■) and HSF (●———●). A salt gradient of 50 mM to 400 mM ammonium sulphate is shown (indicated as 0–100%); fractions were assayed as in *Figure 6*. The profiles have been aligned so that the RAP1 peaks are coincident

5.5 Chromatography on calf thymus DNA–Sepharose

As mentioned above, the principle objective of chromatography on calf thymus DNA–Sepharose ('non-specific' DNA column) is the separation of sequence-specific DNA-binding proteins from proteins than bind to DNA tightly but non-specifically. In practice, the former class of proteins elutes from the DNA resin at relatively low salt concentrations and the latter only at much higher salt concentrations. A second benefit of this step is that it typically achieves a 10- to 20-fold purification, since most of the proteins in the applied heparin fractions do not bind to the DNA column. Preparation of DNA–Sepharose is detailed in *Table 6*.

Fractions from the heparin–agarose column containing the protein of interest are pooled, precipitated by addition of 0.35 g/ml ammonium sulphate, re-suspended in 10–20% of the original volume of A50 and dialysed exhaustively against the same buffer. A precipitate often forms at this stage and should be removed by centrifugation; although comprising up to one-third of the total protein in pooled heparin–agarose column fractions containing HSF and PRTF, the precipitate does not contain significant specific DNA-binding activity. When the dialysed protein is applied to a DNA–Sepharose column in A50, about 90% of the total protein flows through. The resin is then washed and eluted with a linear gradient of A50–A600. The capacity of the column must be determined empirically, but we have found 100 ml of resin to be adequate for use with 0.5–1 g of pooled protein

Table 6. Preparation of calf thymus DNA–Sepharose[a].

1. Dissolve 1 g of calf thymus DNA (Sigma grade I) in 50 ml of water with vigorous shaking[b]. Sonicate the solution until it is no longer viscous. Add one-tenth volume of 100 mM potassium phosphate (pH 8.0). Tris buffers will interfere with the coupling reaction and must be avoided.
2. Suspend 30 g of CNBr-activated Sepharose 4B (Pharmacia) in 500 ml of 1 mM HCl, following the manufacturer's recommendations[c]. Transfer the resin to a sintered glass funnel and wash successively with 2 litres of 1 mM HCl, 2 litres of water and 1 litre of 10 mM potassium phosphate (pH 8.0).
3. Immediately transfer the activated Sepharose to a 250 ml polypropylene screw-cap tube, mix with enough phosphate buffer to give a thick slurry (final volume 150–200 ml) and add the DNA solution.
4. Couple for 16 h at room temperature with gentle end-over-end agitation.
5. Collect the resin on a sintered glass funnel and wash successively with 1 litre of water and 500 ml of 1 M ethanolamine–HCl (pH 8.0) resuspend in 150 ml of ethanolamine. Mix 4–6 h at room temperature to block unreacted groups on the resin.
6. Collect the blocked resin and wash successively with 500 ml each of (a) 10 mM potassium phosphate (pH 8.0), (b) 1 M KCl, (c) water, and finally (d) with storage buffer (10 mM Tris–HCl, pH 7.6, 0.3 M NaCl, 1 mM EDTA and 0.02% sodium azide).
7. Store the resin at 4°C in storage buffer or in A50 (Section 5.3) supplemented with 0.02% sodium azide.

[a] Method according to Kadonaga and Tjian (21), Arndt-Jovin *et al.* (22), and Pharmacia Fine Chemicals.
[b] The quantities given here are suitable for the preparation of 100 ml of resin.
[c] Alternatively, Sepharose CL-2B can be activated with CNBr (21).

from heparin–agarose chromatography (i.e. from 75 g of extract). Almost all of the proteins we have studied (RAP1, HSF, ATF, CP1, SBF-B) elute from calf thymus DNA–Sepharose between 50 mM and 150 mM ammonium sulphate (*Figure 6*). The single exception we observed is PRTF, which elutes at 400 mM sodium chloride. Tight-binding proteins, that is, those that elute from the non-specific column at salt concentrations greater than 600 mM ammonium sulphate, comprise up to 0.5% of the total protein. Oligonucleotide DNA affinity resins typically have low capacities, and it appears that the failure to remove non-specific tight-binding proteins prior to affinity chromatography results in rapid saturation of the affinity columns and failure of the purification. As mentioned below, a possible alternative (or supplement) to chromatography on a non-specific DNA resin is the addition of carrier DNA to the sample during loading on the affinity column (21).

5.6 Purification by DNA affinity chromatography

The final and most powerful purification step involves passing fractions from the non-specific DNA column over a DNA affinity resin containing high-affinity DNA binding sites coupled to Sepharose (specific column). This is the most troublesome stage in the procedure, and success depends critically on the correct choice of DNA binding site, as well as on the conditions in which the column is loaded and eluted. In the following sections we discuss a number of variables that affect the success of DNA affinity chromatography. At the same time we must point out that the methods presented here may not be applicable to all sequence-specific DNA-binding proteins for reasons that may be unclear. Although we attempt to present a theoretical overview of the technique, the majority of our hypotheses have not been tested rigorously and may be incorrect. Thus, the process of designing an effective purification strategy remains critically dependent on empirical experience.

5.6.1 *Selecting a DNA binding site*

To maximize the chances of success, two parameters of the DNA binding sequence must be optimized: selectivity and affinity. As discussed in Section 4, one major source of contamination will be sequence-specific binding protein(s) that bind most tightly to a related sequence but also show a significant affinity for the binding site being used.

The minimal binding site must be determined by a combination of functional assays and binding studies; an optimal sequence will contain a binding site for only a single protein species and will interact, if at all, with other proteins much more weakly. In most cases, the use of a high-affinity site will ensure that sequence-specific binding to the column (by the correct protein) will be more salt-stable than non-specific binding (by contaminating proteins) and elution with increasing concentrations of salt will therefore achieve an effective separation. There may be cases, though, in which the highest affinity site is not the most selective. In this case, a binding site should be chosen for which the difference between the salt-stability of complexes containing the correct versus the contaminating protein(s) is the highest, thereby permitting greater separation of the two species during elution of the affinity column.

The relative affinities of known binding sites can be estimated by cross-competition analysis, using either bandshift gels or footprinting (see Section 4.1 and Chapter 7). It may also be useful to test synthetic sites; in cases where 2-fold symmetry is apparent, for example, creation of a perfect palindrome often increases the binding affinity. There will inevitably be a trade-off between the time spent optimizing the binding sequence and the potential gain in ease of purification.

As discussed below, affinity resins are usually constructed with DNA molecules consisting of multiple binding sites arranged in a tandem array. If possible, it is the binding affinity of the tandem array that should be optimized and not just that of a single site. It seems likely that the spacing between the sites is important. Binding sites for HSF, for example, often occur *in vivo* in overlapping pairs (so called double heat shock elements of double HSEs; 23). The time required *in vitro* for dissociation of complexes between HSF and a single HSE is approximately 15 min whereas that for complexes containing double HSEs is greater than 24 h (*Figure 3*). This indicates that double sites have a much higher binding affinity than single sites, presumably as a result of cooperative interaction between proteins bound to adjacent sequences. A practical consequence of this cooperativity is that HSF pre-bound to an affinity column containing overlapping HSEs is eluted only at 2 M NaCl (or higher concentrations) even though the protein will not bind

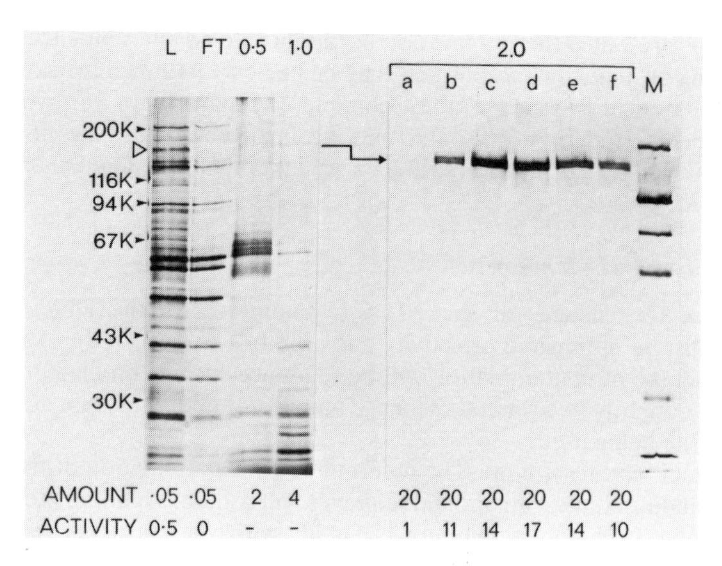

Figure 8. Purification of HSF: silver-stained SDS–polyacrylamide gel of fractions from an HSF DNA affinity column. Samples are as follows: affinity column load (L) and flowthrough (FT), 0.5 M KCl eluate (0.5), 1.0 M KCl eluate (1.0), six successive fractions of the 2.0 M NaCl eluate (2.0, a–f) and markers (M). Amount refers to the per cent of each sample loaded onto the gel: 4 μl of 7 ml for L and FT, 100 μl of 5 ml for 0.5 M eluate, 200 μl of 5 ml for 1.0 M eluate and 200 μl of six 1-ml fractions from the 2.0-M eluate. 'Activity' indicates the amount of DNA-binding activity as assayed on bandshift gels (arbitrary units) that was loaded onto the SDS gel. Molecular weight markers are labelled in kilodaltons and the position of HSF is indicated by the open triangle (3).

initially at a salt concentration above 300 mM (*Figures 6* and *8*). Because naturally occurring heat shock promoters contain binding sites with seemingly optimal overlap, the correct design of the affinity resin was obvious. In general, however, the relative importance of the binding site affinity itself and of the spacing between sites is unclear. Nevertheless, it seems prudent to determine the relative affinities of several tandem arrays in which the spacing between successive sites is varied. We have observed that the relative dissociation rates of pre-formed DNA–protein complexes is a much more sensitive measure of this affinity than analysis by cross-competition (see Section 4.2).

5.6.2 *Preparing the affinity resin*

As mentioned above, DNA affinity columns are typically constructed by coupling tandem arrays of the binding site sequence to an appropriate resin, such as cyanogen bromide-activated Sepharose. The DNA consists either of plasmid DNA with cloned multiple copies of the binding site (2, 24) or of ligated concatamers of synthetic oligonucleotides (21). Because they contain only binding site sequences, resins made with the latter type of DNA are expected to be more specific than those made with the former, and affinity columns containing plasmid DNA are unlikely to achieve as clean a separation between the protein of interest and contaminants. In at least some cases (e.g. the purification of SP1 from HeLa cells, 21), plasmid columns have proven ineffective. A vexing problem with oligonucleotide columns, however, is their low capacity; typically only 20–50 μg of DNA is coupled per ml of resin. A protocol for attaching DNA to Sepharose is given in *Table 7* but there is clearly room for improvement in the coupling efficiency, so it may soon be superseded by new chemistries.

An alternative approach to constructing a DNA affinity column involves reacting biotinylated binding-site DNA with a partially-purified extract in solution and then passing the mixture over a streptavidin–agarose column (16). The biotinylated DNA binds tightly to the streptavidin–agarose, and then the DNA-binding proteins are released by eluting with increased salt concentration. This

Table 7. Preparation of oligonucleotide DNA–Sepharose[a].

1. If two complementary oligonucleotides are used, they must be annealed. Add about 25–50 nmol of each to 100 μl of water, heat to 90°C and slowly cool the solution to room temperature over a period of 30–45 min.
2. Add 20 μl of 10 × kinase buffer (10 × kinase buffer is 700 mM Tris–HCl, pH 7.6, 100 mM $MgCl_2$, 1 mM spermidine, 1 mM EDTA, 100 mM DTT), 5–10 μCi of [γ-^{32}P]-ATP[b]; adjust to a final volume of 200 μl and add 100 units of polynucleotide kinase. Incubate at 37°C for 10 min.
3. Add ATP to 0.5 mM and continue the incubation for 2 more hours.
4. Ligate the oligonucleotides by the addition of sufficient T4 DNA ligase (as determined from the manufacturers recommendation; we use 4000 units of New England Biolabs ligase) and incubate overnight (12 h) at 15°C.
5. Analyse 1 μl of the ligation mixture by electrophoresis on a 2% agarose or 8% polyacrylamide gel (DNA may be detected either by autoradiography or with ethidium bromide). The DNA should be mainly in concatamers with an average length of 100–400 bp. The ligation may be continued if necessary[c,d].
6. Extract the ligation mixture with an equal volume of phenol, re-extract with choloroform, add 1/10 volume of 4 M NaCl, and precipitate with 2.5 volumes of ethanol. After centrifugation re-suspend the DNA in 200 μl of water (it is very important that primary amines such as Tris be avoided).
7. Prepare CNBr-activated Sepharose (Pharmacia) as described in *Table 6*, but use 1.5 g (dry weight) of resin and about one-tenth to one-twentieth the volume of buffers[e].
8. Add the ligated DNA to the slurry and couple for 16 h at room temperature with gentle end-over-end agitation.
9. Collect the resin and save the coupling solution.
10. Proceed as described in *Table 6*, steps 5–7, but with one-tenth the volume of buffers.
11. Calculate the coupling efficiency by determing the ratio of radioactive label in the resin and in the coupling buffer. Coupling of up to 60% of the ligated DNA has been observed but the efficiency is often lower.

[a] Adapted from Kadonaga and Tjian (21).
[b] ^{32}P is incorporated into the DNA in order to permit the coupling efficiency to be estimated.
[c] If ligation is too extensive, circles may be formed. These will not react efficiently with the CNBr-activated Sepharose. If the presence of circles in the ligation mixture is suspected and the ligated DNA contains a restriction site, it can be partially digested to linearize the molecules.
[d] If the oligonucleotide DNA is ligated to an average of about 10 monomer units, a high concentration of ends (which react with the CNBr-activated resin) is maintained.
[e] The quantities given here are suitable for the preparation of about 5 ml of resin.

technique has the theoretical advantage that reaction conditions optimized in solution can be used without significant modification during purification. A possible drawback, however, is that streptavidin–agarose adsorbs non-specifically considerably more contaminating proteins than Sepharose. Nevertheless, this may be a useful technique in some cases.

5.6.3 *Running the column*

Purification on oligonucleotide affinity columns appears to involve discrimination in favour of the correct protein at two levels. First, the great majority of the proteins recovered from the calf thymus DNA column (>95%) do not bind to this column, presumably because they prefer a DNA sequence or geometry that can be found amongst the genomic sequences in the non-specific resin, but which is different from that of synthetic oligonucleotide. Second, the protein of interest has a higher affinity for the binding-site DNA coupled to the column than do other proteins in the preparation. In practice, it is the increased salt stability of specific interactions relative to non-specific interactions that can be exploited. Because affinity resins have limited capacities, however, it is not sufficient to allow the bulk of the proteins to bind to the column and then to elute with successively higher salt concentrations. Instead, we have found it essential to favour the high-affinity interactions initially by loading the affinity column at the highest possible salt concentration. In principle, this salt concentration should be higher than that present in the highest salt fraction (containing the protein of interest) from the non-specific column. Thus, if the protein of interest elutes from the calf thymus DNA column at 100–200 mM salt, the affinity column should (if possible) be loaded with greater than 200 mM salt concentration.

The salt-stability of DNA–protein complexes is most easily determined using a modification of the procedure described in Section 4.2. Binding-site DNA and extract are mixed in a buffer adjusted to an appropriate concentration of salt and incubated for 30 min. A vast molar excess of unlabelled binding-site DNA is then added, and the reactions are analysed immediately on a bandshift gel. We have observed that if the unlabelled binding-site DNA is omitted, proteins bind the radiolabelled DNA even at high salt concentrations, presumably because the salt is diluted out during electrophoresis.

The optimization of loading conditions is demonstrated by the purification of HSF. At concentrations less than 100 mM ammonium sulphate, very little HSF binds to an HSE affinity column, presumably because the resin is saturated by the weak binding of other proteins. At 140 mM ammonium sulphate, however, HSF is quantitatively retained; at yet higher salt concentrations, no proteins, including HSF, are able to bind. Thus, there is a window of salt concentrations between about 140 mM and 200 mM in which the effective discrimination between high-affinity and low-affinity binding can be achieved. In practice, the optimal conditions may be difficult to determine, and the addition of carrier DNA (such as sheared salmon sperm DNA) may make the purification less dependent on the concentration of salt (21). This would be expected because proteins will partition between the carrier DNA and the resin, which will reduce non-specific binding to the synthetic sequences.

HSF was purified to homogeneity in a single pass over an HSE affinity resin (*Figures 6* and *8*). In most cases, however, it is not possible to achieve a single-step purification, and multiple passes are necessary (25). During the purification of PRTF, for example, it was found that the factor elutes from the affinity and calf thymus DNA columns at essentially the same concentration of salt. Purification of PRTF to near homogeneity was nevertheless achieved in three successive passages over the affinity resin (*Figure 9*).

A possible modification of the protocol described here is to replace (or supplement) the calf thymus DNA–Sepharose resin with a column containing synthetic mutant sites coupled to Sepharose. As described in Section 4.1, such a mutant site should be as similar to the optimal binding site as possible, but should have no activity in a functional assay. In principle, successive chromatography on DNA–Sepharose resins differing in sequence at only a few positions should permit discrimination between the correct protein and closely-related contaminants.

Finally, we have noted that many DNA-binding proteins become almost insoluble when pure. The solubilities of both yeast HSF and mammalian serum response factor (15) are increased by the addition of either 0.1% (w/v) Nonidet P-40 or 0.05% (w/v) octyl-[β]-D-glucopyranoside to the elution buffers. The latter detergent has the advantage that it is not precipitated by 10% trichloroacetic acid. Despite our precautions, we found that up to one-quarter of the purified HSF adhered to the siliconized plastic tubes in which fractions were collected. This

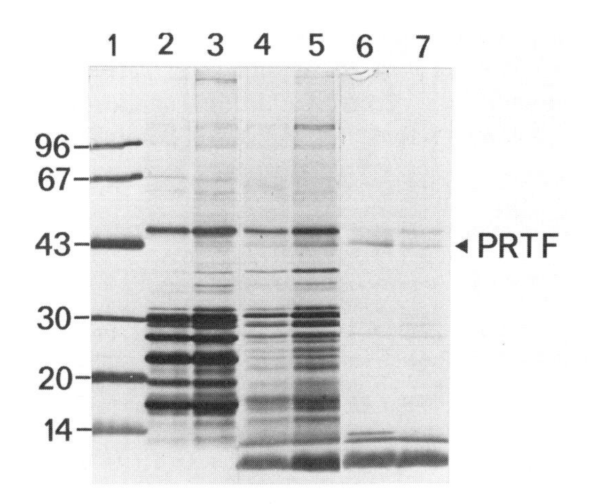

Figure 9. Purification of PRTF: silver-stained SDS–polyacrylamide gel of fractions from a PRTF DNA affinity column. The purification of PRTF involved successive runs over the affinity column in the presence of poly[d(I-C)] as a carrier. The column was loaded in A50 supplemented with 150 mM NaCl and eluted with a single step of A50 plus 400 mM NaCl. The position of the PRTF polypeptide is indicated by the triangle; **lane 1**, markers with molecular weights in kd; **lane 2**, affinity column flow-through; **lane 3**, affinity column load; **lanes 4** and **5**, fractions containing the peak of sequence-specific DNA-binding activity during the first pass over the affinity column; **lanes 6** and **7**, peak activity from the second pass over the affinity column. The strong band at the bottom of the gel in **lanes 4–7** is human insulin, which was used as a carrier. The PRTF in **lanes 6** and **7** is roughly 50% pure.

protein could be efficiently recovered by washing the tubes in 0.1% sodium dodecyl sulphate (SDS) and was used to prepare antibodies. Although a description of the methods used to clone purified proteins is beyond the scope of this chapter, it is worth noting that following the precipitation of HSF and serum response factor from very dilute solutions (~10 μg/ml) with 10% (w/v) trichloroacetic acid, both good polyclonal antisera and clear protein sequence data were obtained (20, 26).

5.7 Conclusion

Once a protein has been purified, it is important to demonstrate that it is in fact responsible for the activity initially characterized in crude extracts. Depending on the system, a combination of functional, genetic and biochemical assays can be used. UV cross-linking, for example, can be profitably employed to demonstrate which of the polypeptides visible on a gel is capable of binding a particular sequence (Section 4.3). Similarly, detailed footprinting studies and competition analysis will reveal whether the purified protein has the expected DNA-binding properties.

6. ACKNOWLEDGEMENTS

Above all we would like to thank the people who have been responsible for the development of the methods described here and, in particular, those whose work has not been fully cited. We would also like to thank Kim Nasmyth, Hugh Pelham, Richard Treisman, Nick Jones, Rob Jones, Collin Young, Daniela Rhodes and Gotthold Schaffner. Finally, we would like to thank Mark Bretscher for having graciously sacrificed his coldroom. G.A. and D.S. were supported by EMBO long-term fellowships and D.S. received additional support from the Jane Coffin Childs Foundation. P.K.S. is a Research Fellow of Trinity College, Cambridge.

7. REFERENCES

1. Sorger, P. K. and Pelham, H. R. B. (1987), *EMBO J.*, **6**, 3035.
2. Shore, D. M. and Nasmyth, K. A. (1987), *Cell*, **51**, 721.
3. Bender, A. and Sprague, J. F. Jr., (1987) *Cell*, **50**, 681.
4. Shore, D. M., Stillman, D. J., Brand, A. H. and Nasmyth, K. A. (1987) *EMBO J.*, **6**, 461.
5. Bram, R. J. and Kornberg, R. D. (1987) *Mol. Cell. Biol.*, **7**, 403.
6. Harshman, K. D., Moye-Rowley, W. S. and Parker, C. S. (1988), *Cell*, **53**, 321.
7. Garner, M. M. and Revzin, A. (1981), *Nucleic Acids Res.*, **9**, 3047.
8. Fried, M. and Crothers, D. M. (1981), *Nucleic Acids Res.*, **9**, 6505.
9. Wiederrecht, G., Shuey, D. J., Kibbe, W. A. and Parker, D. S. (1987) *Cell*, **48**, 507.
10. Ammerer, G., Hunter, C., Rothman, J., Saari, G., Valls, L. and Stevens, T. H. (1986), *Mol. Cell. Biol.*, **6**, 2490.
11. Klekamp, M. S. and Weil, P. A. (1982) *J. Biol. Chem.*, **257**, 8432.
12. von Hippel, P. H. (1979) In *Biochemical Regulation and Development.* Goldberger, R. F. (ed.), Plenum, New York, Vol. 1, p. 279.
13. Yamamoto, K. R. (1985), *Annu. Rev. Genet.*, **19**, 209.
14. Carey, J., (1988) *Proc. Natl. Acad. Sci. USA*, **85**, 975.
15. Treisman, R. (1987) *EMBO J.*, **6**, 2711.
16. Chodosh, L. A., Carthew, R. W. and Sharp, P. A. (1986), *Mol. Cell. Biol.*, **6**, 4723.
17. Hager, D. A. and Burgess, R. R. (1980), *Anal. Biochem.*, **109**, 76.

18. Singh, H., LeBowitz, J. H. Baldwin, A. S. Jr. and Sharp, P. A. (1988) *Cell,* **52**, 415.
19. Davison, B. L., Leighton, T. and Rabinowitz, J. C. (1979), *J. Biol. Chem.,* **254,** 9220.
20. Sorger, P. K. and Pelham, H. R. B. (1988) *Cell,* **54**, 855.
21. Kadonaga, J. T. and Tjian, R. (1986) *Proc. Natl. Acad. Sci. USA,* **83**, 5889.
22. Arndt-Jovin, D. J., Jovin, T. M., Bahr, W., Frischauf, A. M. and Marquardt, M. (1975) *Eur. J. Biochem.,* **54**, 411.
23. Bienz, M. and Pelham, H. R. B. (1989) *Adv. Genet.,* in press.
24. Rosenfeld, P. J. and Kelly, T. J. (1986) *J. Biol. Chem.,* **261**, 1398.
25. Briggs, M. R., Kadonaga, J. T., Bell, S. P. and Tjian, R. (1986) *Science,* **234**, 47.
26. Norman, C., Runswick, M., Pollock, R. and Treismon, R. (1989) *Cell*, in press.

CHAPTER 9

Chemical characterization of functional groups in proteins by competitive labelling

N. MARTIN YOUNG and HARVEY KAPLAN

1. INTRODUCTION

Competitive labelling is a technique for determining the chemical properties of individual residues in a protein (1). Since the pK and reactivity of each functional group reflect its micro-environment, the technique can be used to study structural and functional aspects of proteins. For example, if a group is involved in hydrogen bonding or ionic interactions, or is buried in the interior of the protein, its chemical properties will differ from those of the same group interacting freely with solvent. Therefore, measurement of the chemical properties of functional groups provides a basis for making deductions regarding the local structure and topography of a protein. Changes in these properties can indicate the nature of its association with other macromolecules or ligands.

In competitive labelling, or differential chemical modification (2), far less than stoichiometric ratios of reagent to protein are used. This 'trace labelling' overcomes a general problem with most chemical modification procedures, which use relatively high levels of derivatization. Consequently, several of the modifying groups will be introduced into each protein molecule, which may result in alteration of the native structure. In order to establish the validity of any conclusions that may be drawn, it is often necessary to obtain additional evidence that no major structural alterations have occurred. This difficulty is eliminated in the competitive labelling approach, since reaction conditions are such that only a fraction of the protein molecules, and hence an even smaller fraction of each functional group, reacts with the reagent. The reagent is therefore always reacting with protein in its unaltered state, and the properties determined apply to the native state of the protein under the particular experimental conditions employed. The name given to this technique refers to the competition between the various residues in the protein for the available reagent, which is governed by their individual reactivities.

Since most conformational transitions are much more rapid than the rates of the chemical modification reactions, an instantaneous average will be obtained of the properties of a residue in the equilibrium mixture of the various conformational states available to the protein. For example, a functional group that is exposed to solvent only 2% of the time in the various equilibrium conformational states will have only 2% of its 'normal reactivity', and we can conclude that in the predominant conformational state it is buried. Whether the protein is in the free state or associated with other macromolecules, the chemical properties of any functional group will always reflect its micro-environment and the dynamic conformational equilibria that exist. The technique of competitive labelling, with appropriate modifications for each application, provides a relatively simple experimental approach for quantifying chemical properties of functional groups, and hence for elucidating structure–function relationships. In instances where a crystallographic structure is available, the details of fine structure, such as hydrogen bonding, and the dynamic properties of the protein can be determined.

There are several other advantages of the competitive labelling approach. The labelled functional groups of interest are isolated as derivatized peptides or amino acids that can be characterized further and their location in the primary structure determined. Hence there is no ambiguity in the assignment of parameters to specific groups. Relatively modest amounts of protein are required, in most cases much less than 500 nmol for the complete analysis. Very dilute protein concentrations can be employed, for example, 10^{-8} M in a study of insulin (3), which approaches the physiological concentration of the circulating hormone. Also, all the equipment required is normally present in a protein laboratory so that minimal additional resources are required to employ this technique.

2. DESIGNING A COMPETITIVE LABELLING EXPERIMENT

The procedures detailed below follow the same over-all format, with variations to suit the particular requirements of each type of experiment. Exact amounts will not be given, since they will depend upon the aim of the experiment, the supply of the protein of interest, its molecular weight and composition and the sensitivity of the amino acid analyser available. The literature cited should be consulted to help in designing procedures based upon those described in this chapter. We describe first the general organization of an experiment followed by comments on each of the main aspects.

2.1 **General scheme**

The steps followed are listed below.

(i) Select the modifying reagent and the internal standard.
(ii) Establish a method for separating the expected amino acid or peptide derivatives.
(iii) Choose the approach for obtaining uniformly-labelled, ^{14}C-modified protein and ^{14}C-modified internal standard.
(iv) Trace-label aliquots of the protein and internal standard with [^{3}H]reagent, varying the required parameters such as pH, concentration, presence of associating protein or ligand, etc.
(v) Convert sample and internal standard to completely and uniformly ^{14}C-labelled form. A second type of chemical modification such as maleylation may be necessary here. Alternatively, add separately prepared ^{14}C-labelled protein and ^{14}C-labelled internal standard.

(vi) Hydrolyse the modified protein or prepare peptides by chemical or proteolytic digestion. The internal standard derivative may be removed before or after this step, depending upon the procedure.

(vii) Separate and purify the amino acid or peptide derivatives and the internal standard derivatives.

(viii) Obtain the $^{3}H/^{14}C$ ratios of each by liquid scintillation counting and calculate the results by means of the equations in Section 2.5.

If the reagent chosen is not available in both ^{3}H and ^{14}C forms, the design will be more straightforward, with radiolabelling only at step (iv) and non-radioactive reagent in steps (iii) and (v). A second type of assay at step (viii) will be needed to measure the specific radioactivity.

2.2 Choice of labelling reagent

Any reagent that reacts with the type of functional group of interest can be used. However, there are some general criteria to be considered. The reagent should be uncharged to minimize electrostatic interactions and be as small as possible to minimize steric interference with the reaction. It may not always be possible to eliminate these factors completely, especially the latter. A second labelling reagent can sometimes be used as a check to avoid mis-interpretation of the results (4). The reagent chosen should be available in radioactive form, preferably in both ^{3}H and ^{14}C forms. If only one isotopic form is available, a spectroscopic signal or amino acid analysis can be used to establish the specific radioactivity.

The most useful reagents that have been employed are acetic anhydride and 1-fluoro-2,4-dinitrobenzene (DNP-F). Acetic anhydride reacts rapidly with protein amino groups and forms a sufficiently stable derivative for peptide isolation, but not for acid hydrolysis. It may also react with histidine, cysteine, tyrosine, serine and threonine residues, but the acetyl group is usually lost rapidly by hydrolysis of these derivatives. This may not, however, be the case for threonine residues (5). The other reagent, DNP-F, forms stable derivatives with all the above functional groups, except for serine and threonine, and has the advantage that the dinitrophenyl (DNP) derivatives are stable in the 6 M HCl used to hydrolyse proteins. Its disadvantages are that some DNP-derivatives are sensitive to light, particularly imidazolyl-DNP-histidine, and fully dinitrophenylated proteins are not readily digested by enzymes due to their extreme insolubility. However, with appropriate modifications to the basic procedures, it is relatively easy to overcome these difficulties.

Both acetic anhydride and DNP-F are available in ^{3}H and ^{14}C forms. To our knowledge, the only other reagent employed so far in a double-radioisotope system (6) is formaldehyde, which can be incorporated into amino groups by reductive methylation. Formaldehyde is available in ^{14}C-form and $NaB[^{3}H]_4$ is used for the reductive step. Iodoacetic acid and iodine are also available in two isotopic forms. Other reagents are available in only one isotopic form and therefore must be used in conjunction with another form of assay. Ghélis (7) has used ^{14}C-labelled forms of several reagents, iodoacetamide, dimethylsulphate and ethoxyformic anhydride and examined methionine and cystine residues in addition to tyrosine, lysine and histidine residues.

2.3 **Choice of internal standard**

The primary reason for including an internal standard is to provide a reference when comparing the relative reactivities of groups from separate labelling experiments. In addition, by choosing a standard whose absolute reactivity is known, the absolute value of the second-order rate constant for the reaction of groups in the protein with the reagent can be obtained. However, for most purposes, it is sufficient to determine the reactivity relative to the internal standard.

The internal standard should, if possible, contain the same class of reactive functional group with a similar pK value to the protein functional groups under investigation. While this is not an absolute requirement, the determination of its ^{3}H/^{14}C ratio provides a preliminary indication of how well the labelling of the protein has proceeded before the more extensive and laborious work on the modified protein is undertaken. It is beneficial to include two or more standards with different pK values in the reaction mixture when several types of functional groups are being studied simultaneously.

There are two other factors to consider when choosing an internal standard. It should not interact strongly with any component of the system under investigation. Also, it is very helpful if the internal standard is readily separated from the protein and, in particular, from the low molecular weight by-products of the labelling reaction.

It is always possible to use a functional group in the protein as an internal reference. In some instances only the changes in the relative reactivities of groups within the protein are of interest, and any group can be chosen as the internal standard. It may be selected on the basis of its being distant from the interaction site being studied or because it is believed to have 'normal' properties on the basis of other evidence, such as the crystallographic structure.

2.4 **Quantification of reaction**

The isolation procedures are unlikely to give quantitative recoveries of the derivatized internal standard or derivatized amino acids and peptides. For precise quantification of the extent of reaction, it is necessary to employ an isotope dilution procedure that yields a specific radioactivity that is independent of the recovery.

In practice there are three suitable methods. In the first method, after the trace labelling with the [^{3}H]reagent, the sample is completely modified with an excess of the same reagent in ^{14}C-form. The ratio of the two isotopes in any isolated derivative is an exact measure of the extent of reaction with the ^{3}H-labelled reagent and will be independent of the recovery. In the second method, the [^{3}H]sample is completely modified with non-radioactive reagent. A separate portion from the stock solution of protein and internal standard is completely modified with [^{14}C]reagent, and aliquots of this preparation are added to each

^{3}H-labelled sample. In the third method, only one radioactive reagent is used for the trace labelling, and the protein and internal standard are completely modified with non-radioactive reagent. The quantity of sample is finally determined by other techniques such as amino acid analysis or by UV-visible spectroscopy when the derivative has suitable spectral properties.

Other approaches to quantification are possible. Amino acid sequencing can be used, and it is particularly valuable when two modifiable residues are side-by-side or close to each other in the sequence. In this case, the phenylthiohydantoin (PTH)-derivative generated by the Edman procedure at a given cycle can be quantified and its specific radioactivity determined, or its ^{3}H/^{14}C ratio determined. Another approach is to prepare the individual [^{14}C]derivatives of the amino acids of interest and to add a suitable mixture of them to the ^{3}H-labelled sample (8).

2.5 Theory and calculation of results

Since only trace amounts of radiolabelled reagent are added, the reactive functional groups in the protein will always be in excess, and only a small fraction of each group can react. The reaction of each individual functional group will be pseudo-first-order with respect to the reagent so that the fraction of radiolabelled reagent reacting with each functional group per unit time is constant. Therefore, for each functional group, the amount of radioactive label incorporated will be proportional to the second-order rate constant for its reaction with the reagent. After complete derivatization with excess [^{14}C]reagent, it follows that

$$k_A/k_S = (^3H/^{14}C)_A/(^3H/^{14}C)_S \qquad (1)$$

where k_A is the second-order rate constant for reaction of the functional group being studied; k_S is the second-order rate constant for reaction of the internal standard; $(^3H/^{14}C)_A$ and $(^3H/^{14}C)_S$ are the experimentally determined ^{3}H/^{14}C ratios for the derivatized functional group and derivatized internal standard.

If A and S have ionizable groups with acid dissociation constants K_A and K_S, Equation 1 becomes:

$$\alpha_A k_A/\alpha_S k_S = (^3H/^{14}C/(^3H/^{14}C)_S \qquad (2)$$

where α_A is degree of ionization of the functional group A being studied and $\alpha_A = K_A/([H^+] + K_A)$; α_S is the degree of ionization of the internal standard S and $\alpha_S = K_S([H^+] + K_S)$.

It is convenient to define the following: r $= k_A/k_S$; $R = (^3H/^{14}C)_A/(^3H/^{14}C)_S$.

Equation 2 becomes:

$$\alpha_A r = \alpha_S R \qquad (3)$$

The reactivities, given by the $\alpha_A r$ values, are calculated from Equation 3 using the experimentally determined ^{3}H/^{14}C ratios and α_S, which is calculated from the pK of internal standard. A plot of the $\alpha_A r$ values versus pH gives a sigmoidal curve with an inflexion point at a pH equal to the pK value and a limiting value of $\alpha_A r$ equal to r. It is desirable to fit the data by means of a non-linear least-squares regression procedure applied to the following equation:

$$y = \frac{C_1}{1 + 10^{-x}/C_2} \tag{4}$$

to determine C_1 and C_2 where y is the experimentally determined $\alpha_A r$ value; x is the pH; $C_1 = r = k_A/k_S$; $C_2 = K_A$.

In experiments where interactions with ligands are being studied, measurements are made of two different reaction mixtures, with and without the ligand, at the same pH and with the same internal standard. It follows from Equation 3 that:

$$R_1/R_2 = \frac{(\alpha_A k_A)_1}{(\alpha_A k_A)_2} \tag{5}$$

Therefore the relative reactivity of a functional group under two sets of conditions can be determined readily using either another group in the protein as the internal standard or an external compound as the standard.

2.5.1 *Precision of experimental data*

Histidine and cysteine residues and N-terminal amino groups have pK values within the range where most proteins retain their native structures, so their reactivity and pK values will usually be obtained with greater accuracy than the corresponding values for lysine and tyrosine residues. The high pK values of the latter usually result in titration curves that are not completely delineated within an acceptable pH range. The errors that result will be much greater for the r values than for the pK values. For the first group, r values can be obtained within 5% accuracy and pK values to better than 0.1 pH unit. For tyrosine and lysine residues, the pK values will be within 0.2 unit and the r values within 25%. Since the r values are normally compared on a logarithmic basis (see Section 2.6), this error is acceptable.

For experiments comparing free and complexed forms of proteins, values of the reactivity ratio (Equation 5) are usually in error by 10–15% (9). This is far less than the reactivity changes produced for residues involved in binding of proteins of ligands, which can be greater than an order of magnitude.

2.6 **Interpretation of the reactivity data**

Competitive labelling is usually applied to proteins of known sequence, and often of known three-dimensional structure. The results are interpreted in the light of this structural information, to obtain insight into the dynamic properties and to supplement the static structural picture.

When data are obtained for the reactivity of a group at various pH values, there is no *a priori* reason for them to conform to a titration curve. Proteins are complex entities and the response to pH change is not a strictly local phenomenon. Quite possibly the protein may undergo a major conformational change in the pH range used. However, such a change will be seen as a discontinuity in the reactivity–pH plot, which can provide information on the nature and extent of any structural change. An example is shown in *Figure 1* where two of the N-terminal residues of α-chymotrypsin follow a regular titration curve until pH 10 and then undergo a

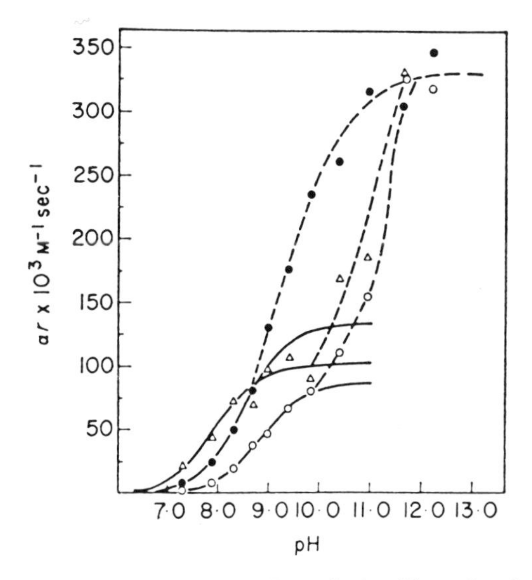

Figure 1. The pH–reactivity profiles for the N-terminal residues of α-chymotrypsin (10). The solid lines are theoretical titration curves with the following parameters: half-cystine 1 (△) pK = 7.9, r = 0.10; Ile16 (○) pK = 8.9, r = 0.087; ala149 (●) pK = 8.5, r = 0.13. Reproduced with permission from Academic Press Ltd.

marked change in reactivity. Frequently the data do fit a simple titration curve, and a pK value and absolute reactivity can be deduced. In this case, it is useful to compare these numbers to the behaviour of standard compounds with the same modifiable group. Brønsted plots have been constructed for primary amines (1, 4) and for imidazole groups (4, 11) reacting with DNP-F. These reactivity–pK data fit relationships of the form:

$$\log k = a + b \times \mathrm{p}K \tag{9}$$

where k is the rate constant for the reaction and a and b are known as the Brønsted coefficients. Deviations from such linear relationships for groups in proteins indicate special environmental effects. For example, it is often found that the imidazole groups of histidine residues have higher reactivities than free imidazole compounds with comparable pK values (12). This may be ascribed to their interaction with neighbouring residues through hydrogen bonds. In the case of α-chymotrypsin, when the data from the solid lines of *Figure 1* are compared with those of standard compounds on a Brønsted plot (*Figure 2*), the unusually low reactivity of the amino group of Ile16 is apparent.

Interactions between neighbouring ionizable groups can give rise to unique chemical behaviour, in which the reactivity of a group is dominated by the state of the neighbour, and reflects the latter's pK value rather than its own. The N-terminal histidine residue of glucagon (8, 13) is a case in point, the unusual reactivity of its imidazole group being directly related to the ionization of its amino group. Effects such as these are of extra interest, since they may help in understanding the unique properties of active sites in proteins.

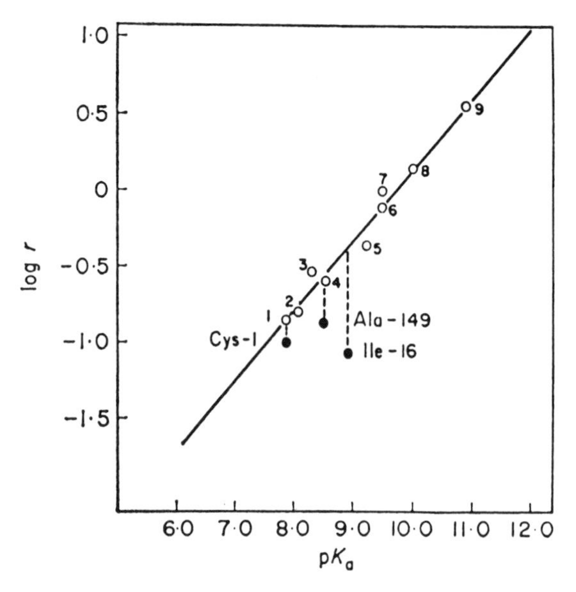

Figure 2. Comparison of the pK values and reactivities of the three N-terminal groups of α-chymotrypsin with those of standard compounds using a Brønsted plot (10). The compounds are (**1**) Glu-Gly (**2**) Phe-Gly (**3**) Leu-Ala-Gly (**4**) Ala-Gly (**5**) Asn (**6**) Gln (**7**) Phe (**8**) Ala (**9**) Bz-Gly-Lys. Reproduced with permission from Academic Press Ltd.

2.7 **Hazards**

In the procedures described below, the reagents used and the reaction products are radioactive, volatile, and may be carcinogenic. Appropriate precautions must therefore be taken to ensure safe laboratory practice. Volatile reagents, such as acetic anhydride, should be used in an appropriate fume-hood. Pipettes, test-tubes and other glassware should be treated with 1 M sodium hydroxide to destroy any remaining traces of radioactive reagent prior to washing or disposal. Some procedures generate radiolabelled acetic acid, for example the acid hydrolysis of acetylated peptides, which should be trapped with sodium hydroxide pellets in any drying step.

3. CHARACTERIZATION OF SPECIFIC GROUPS IN PROTEINS

3.1 **Determination of average chemical properties**

The following procedure uses DNP-F and the modified protein is hydrolysed completely to yield the side-chain DNP derivatives and the ε-amino DNP derivative of the N-terminal residue(s). Thus the average properties (pK and reactivity) of the various lysine, tyrosine, histidine and cysteine residues will be obtained, unless only one residue of any one of these is present. This is more likely to occur with smaller proteins or peptides, though a single cysteine is not uncommon. The N-terminal residue will usually be unique, but some multi-chain proteins such as immunoglobulins may have the same type of residue at the N-termini of more than one constituent chain. Another complication that may arise is the presence at the N-terminus of one of the four residues whose side-chains are modifiable. Information about both of the modifiable groups on such a residue can still be obtained, by a modification of the procedure that is described after the main method.

3.1.1 *Separation procedure for DNP-amino acid derivatives*

The first requirement for the method is a means of separating the expected products from hydrolysates of dinitrophenylated proteins. High-pressure liquid chromatography (HPLC) is very suitable (12), though earlier work used paper electrophoresis, which is entirely satisfactory but is now less common in protein laboratories. The species to be separated are ε-DNP-lysine, *O*-DNP-tyrosine, imidazolyl-DNP-histidine and *S*-DNP-cysteine from side-chain modification, together with the α-DNP-amino derivative of the N-terminal amino acid, the DNP-derivative of the internal standard and the by-product dinitrophenol (DNP-OH). Since C18 columns from different manufacturers vary in properties, the buffer will need adjustment for the particular column to be used. Relatively small samples are used, so column sizes need not be the 25 cm length.

(i) Prepare stock solutions of the required DNP-derivatives above. These are commercially available, except for imidazolyl-DNP-histidine and possibly *S*-DNP-cysteine, which may be prepared by dinitrophenylation of the *N*-acetyl compounds, followed by acid hydrolysis. Since they are only used to establish the elution positions, high purity is not necessary.

(ii) Equilibrate the column at 60°C with a 27:73 (v/v) mixture of acetonitrile–35 mM ammonium formate (pH 3.0). Inject mixtures of the derivatives to establish their elution positions, using isocratic conditions. There is a considerable difference between the hydrophobicities of *O*-DNP-tyrosine and ε-DNP-lysine, compared with imidazolyl-DNP-histidine and *S*-DNP-cysteine; the latter two elute early, and are not as clearly separated from by-products. A second system in which these fractions can be re-cycled isocratically must therefore be established; an 8:92 (v/v) acetonitrile–buffer mixture is suitable.

3.1.2 *Modification procedure with DNP-F*

In the variation described here, uniformly ^{14}C-labelled protein is prepared separately, rather than the ^{3}H-labelled protein being modified completely with [^{14}C]DNP-F. This is much more economical on [^{14}C]reagent.

(i) Prepare a stock solution of the protein and the internal standard (imidazolyl-lactic acid or alanylalanine) using 0.1–1 mol of each in a buffer that will be unreactive with DNP-F (e.g. 5 mM *N*-ethylmorpholine, 5 mM borate). Divide into two portions.

(ii) To one portion add $NaHCO_3$ (0.5 g/ml) and urea (1 g/ml) followed by a 20-fold excess of [^{14}C]DNP-F. (200 mCi/mol) in 50% acetonitrile. React in the dark for 18 h with stirring or gentle shaking. Add HCl to bring the pH to 2.

(iii) Divide the second part of the solution into aliquots and adjust each to a different pH in the range 5.5–10.5. Intervals of 0.35–0.5 pH units are suitable.

(iv) Add [^{3}H]DNP-F in acetonitrile (1 nmol, sp. act. 5 Ci/mmol) to each sample and leave stirring in the dark at the selected temperature for 18 h.

(v) React each ^{3}H-labelled sample with excess non-radioactive DNP-F according to the procedure described in (ii) above. Add HCl to bring the pH to 2.

(vi) Add to each sample an equal aliquot of the ^{14}C-reaction from (ii) above.

(vii) Dialyse the samples against a small volume of water, to extract an aliquot of the internal standard derivative. Continue the dialysis against a large volume of water to remove the urea completely. Freeze-dry the protein samples.

(viii) Purify the internal standard derivatives by the HPLC procedure and measure their ^{3}H/^{14}C ratios.

(ix) Hydrolyse the DNP-protein in 6 M HCl at 110°C for 18 h. Purify the DNP-amino acids by HPLC and measure their ^{3}H/^{14}C ratios.

(x) Analyse the data using Equation 3. An example of the calculation is given in *Table 1*, using data from a study of glucagon (8).

3.1.3 *Alternative forms of the procedure*

As mentioned above, instead of adding separately-prepared [^{14}C]DNP-protein and internal standard, the [^{3}H]samples can instead be treated with saturating amounts of [^{14}C]DNP-F, diluted with non-radioactive DNP-F. In principle the ^{14}C-labelled protein can be replaced by an appropriate mixture of the [^{14}C]DNP-amino acids, although this is likely to be less accurate.

A somewhat different approach is necessary if the N-terminal amino acid carries a reactive side-chain. This was encountered in studies of glucagon (8, 13) which has an N-terminal histidine. To obtain data for both functional groups, full reaction of the protein cannot be used since this would give only the di-DNP-derivative, scrambling the information. The approach used was to label partially when preparing the [^{14}C]DNP-protein, then measure the ratios of the *O*-DNP-tyrosine, imidazolyl-DNP-histidine, ε-DNP-lysine and α-DNP-histidine in a hydrolysed sample. The relative proportions of ^{14}C counts can later be used to correct the experimental data for the true molar ratios in the calculations of reactivity. Alternatively the separately prepared [^{14}C]DNP-α-amino and [^{14}C]side-chain derivatives can be added prior to the hydrolysis step (8). The data obtained in this way for the N-terminal histidine of glucagon (*Table 1*) are shown in *Figure 3*.

The lower limits of sensitivity of the procedure have not yet been reached. It is certainly possible to do a relatively complete study on 1–2 mg of protein. Since more than half the sample is used to form the [^{14}C]DNP-protein, the mixed [^{14}C]DNP-derivative approach could require less than 0.5 mg of protein.

Table 1. Calculation of reactivity data for the N-terminal histidine residue of glucagon[a].

pH	$^3H/^{14}C$			α_s	$\alpha_A r$	
	im-lac	*im-His*	*NH_2-His*	*im-lac*	*im-His*	*NH_2-His*
5.00	14.7	25.0	155	0.0053	0.009	0.058
5.25	16.2	25.9	137	0.0095	0.021	0.081
5.50	21.2	66.9	88.9	0.0167	0.053	0.074
5.75	16.9	48.5	92.1	0.0293	0.084	0.160
6.00	13.6	32.7	39.8	0.0597	0.144	0.171
6.25	7.7	9.5	24.9	0.0872	0.108	0.284
6.50	14.9	23.0	56.9	0.145	0.224	0.547
6.75	15.6	22.8	58.7	0.232	0.339	0.873
7.00	15.2	21.8	56.2	0.349	0.501	1.31
7.25	13.0	14.5	54.8	0.488	0.542	2.02
7.50	13.7	25.9	48.8	0.629	1.19	2.21
7.75	17.9	39.5	54.7	0.751	1.66	2.37
8.00	14.3	29.1	59.9	0.843	1.72	3.58
8.25	12.2	19.6	56.2	0.905	1.42	4.38
8.50	13.2	23.3	60.2	0.944	1.66	4.48
8.75	10.9	24.5	51.8	0.968	2.15	4.60
9.00	8.7	17.7	44.5	0.982	1.84	5.00
9.25	12.8	23.5	65.3	0.990	1.81	5.09
9.50	17.1	43.3	86.0	0.994	2.41	5.00
9.75	14.5	28.1	77.6	0.997	1.93	5.41
10.00	50.6	150	110	0.998	2.96	2.31
10.25	26.6	70.2	57.3	0.999	2.64	2.27
10.50	19.8	36.8	44.5	0.999	1.86	2.18
10.75	25.0	40.7	80.7	1.00	1.63	3.14
11.00	19.9	28.4	96.3	1.00	1.42	5.17

[a] Columns 2–4 are the ratio data obtained from the $^3H/^{14}C$ contents of the labelled internal standard β-imidazolyllactic acid (im-lac), the imidazole group of the N-terminal histidine (im-His) and its amino group (NH_2-His). Column 5 is the calculated values for the ionization of the internal standard from its pK of 7.27. Columns 6 and 7 were calculated using Equation 3.

3.2 Determination of the chemical properties of individual tyrosine and histidine residues

This procedure is very similar to the previous one, but with peptide mapping replacing acid hydrolysis, to permit analysis of the individual residues when more than one residue of a given type is present. It is useful to combine both procedures into one, since the N-terminal amino acid is better investigated by the hydrolysis method and other single residues may often be present in smaller proteins. The other major change arises from the insolubility of a fully dinitrophenylated proteins. This is not a problem when acid hydrolysis is used, but greatly affects digestion with proteolytic enzymes. A convenient way to avoid the problem is to treat the protein sample with maleic anhydride, modifying the lysines and terminal amino group, prior to the complete dinitrophenylation step (14).

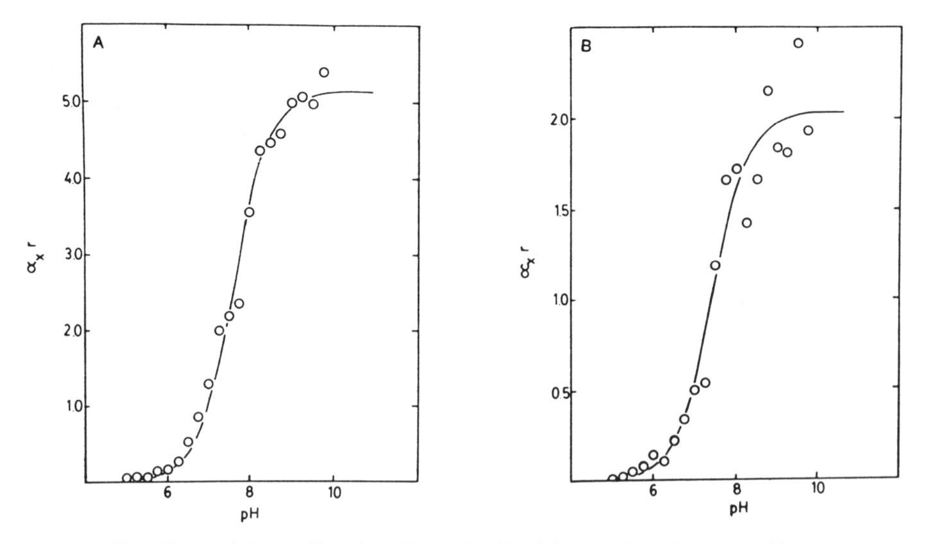

Figure 3. The pH–reactivity profiles of the N-terminal histidine residue of glucagon (8). The lines are theoretical titration curves with the following p*K* values and *r* values relative to L-β-imidazolyllactic acid: (**a**) the α-amino group, pK = 7.60 and r = 5.13; (**b**) the imidazole group, pK = 7.43 and r = 2.02. The raw data and calculation are shown in *Table 1*. Reproduced with permission from the American Chemical Society, copyright 1985.

For the proteolytic digestion we favour thermolysin, which yields relatively small peptides, hence reducing the likelihood of more than one modifiable residue occurring in a peptide.

3.2.1 *Peptide mapping of doubly-modified protein*

The general procedures for peptide mapping by HPLC are used. It is useful, if the equipment is available, to measure the absorbance of the column effluent at two wavelengths. The various side-chain derivatives have distinct $A_{270}:A_{320}$ ratios that may be used to identify the modified residue in the eluted peptide (14, 15).

(i) Treat a sample of the protein (2.5 mg) in pH 8.0 buffer (veronal or $NaHCO_3$ are suitable) with maleic anhydride (1–2.5 mg). Maintaining the pH with a pH-stat system is desirable but not essential.
(ii) Add urea (1 g/ml), $NaHCO_3$ (0.5 g/ml) and DNP-F (1:1 in acetonitrile) and leave at room temperature overnight in the dark, with stirring.
(iii) Dialyse the reaction mixture against 0.1 M NH_4HCO_3 to remove urea and DNP-OH etc.
(iv) Digest the DNP-protein with thermolysin [1:25 (w/w), 37°C, 4 h] then freeze-dry.
(v) Using an HLPC column and buffer of choice, separate the modified peptides. Since the DNP-peptides, particularly those containing *O*-DNP-tyrosine are more hydrophobic than many normal peptides, C3 or C4 columns may be advantageous. Re-cycle the peptides if necessary.

(vi) Determine the amino acid compositions of the recovered DNP-peptides. If two different modified residues, such as histidine and tyrosine, are thought to be present, an aliquot of the hydrolysate can be checked using the isocratic HPLC system described in Section 3.1.1

(vii) Assign the peptides on the basis of their amino acid compositions to the appropriate segment of the amino acid sequence. In cases of ambiguous compositions, sequencing may be necessary.

3.2.2 *Competitive labelling with 2,4-dinitrofluorobenzene*

The procedures followed resemble those in Section 3.1.2 closely, so only a brief outline and the major changes are given.

(i) In the manner used to prepare the sample for the peptide map, prepare maleyl-[^{14}C]DNP-protein and add [^{14}C]DNP-internal standard (alanylalanine).

(ii) Label aliquots of the native protein and the internal standard with [^{3}H]DNP-F, as in (iii) and (iv) of Section 3.1.2.

(iii) Add aliquots of the [^{14}C]mixture to the samples.

(iv) Remove the internal standard derivatives by dialysis against a small volume of water, purify them and measure their ^{3}H/^{14}C ratios.

(v) After further dialysis, fully maleylate and dinitrophenylate the [^{3}H]DNP-proteins, and dialyse into 0.1 M NH_4HCO_3.

(vi) Digest the modified proteins with thermolysin, then freeze-dry.

(vii) Separate and purify the resulting peptides by the peptide mapping procedure, and determine their ^{3}H/^{14}C ratios.

(viii) Analyse the data using Equation 3.

3.2.3 *Comments on the over-all method*

This procedure was developed for characterization of histidine and tyrosine residues, and data from a study of concanavalin A (14) are shown in *Figure 4*. However, it should be suitable for cysteine residues also, in view of the recovery of an *S*-DNP-peptide reported from sainfoin lectin (15). Although the lysine and N-terminal amino groups are also labelled, in most cases they can only be characterized by the acid hydrolysis procedure. An exception would be the case of a protein that retained solubility when fully dinitrophenylated. Information on lysine residues can be obtained by a procedure with acetic anhydride to be described below, or by reductive methylation (6).

The presence of a lysine and a tyrosine residue in the same peptide does not compromise the method, since the procedure described will give two different final peptide products, the desired *N*-maleyl-*O*-[^{3}H]DNP-product from tyrosine modification and the *N*-[^{3}H]DNP-*O*-DNP-product from lysine modification. These two forms would have dramatically different HPLC properties, and only the former will elute in the position determined in Section 3.2.1. It is worth noting that the DNP group can be removed from cysteine, tyrosine and histidine

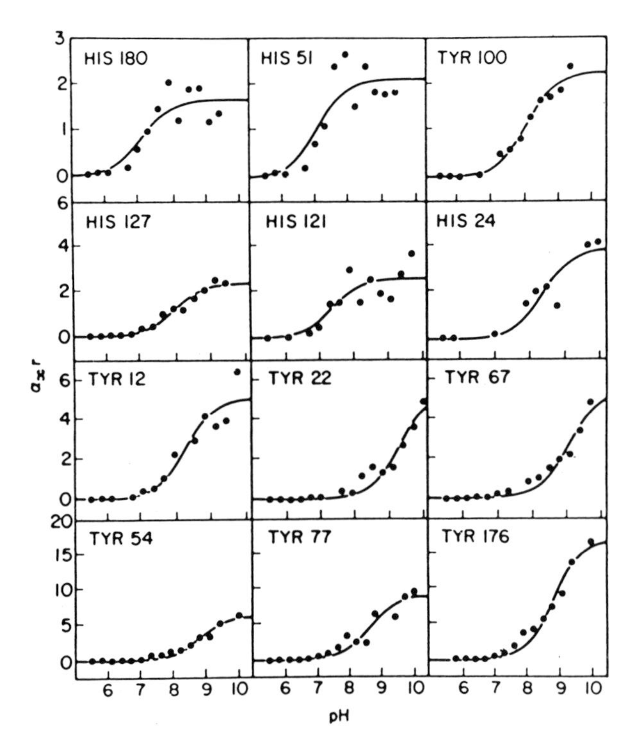

Figure 4. The pH–reactivity profiles of individual histidine and tyrosine residues in concanavalin A (14), with the theoretical curves obtained by least squares fitting to the data points. Reproduced with permission from the American Chemical Society, copyright 1986.

side-chains by treatment with thiol compounds such as 2-mercaptoethanol (16). This can be used in some cases to obtain information on amino groups (8).

3.3 Determination of the chemical properties of individual amino groups

As discussed above, dinitrophenylation is not well suited to the study of amino groups because of the low solubility of DNP-proteins. An added complication in studying lysine residues is that they more frequently occur close together in the amino acid sequence than histidine or tyrosine residues. Despite the disadvantages of ε-*N*-acetyl-lysine being unstable to acid hydrolysis and lacking a unique spectroscopic signature, acetic anhydride is nevertheless the reagent of choice. Among its advantages are: the rapidity with which it reacts; the clean nature of the reaction giving fewer by-products to be removed; the lower hydrophobicity and hence greater solubility of acetylated peptides compared with DNP-peptides, which leads to better recovery from HPLC or paper electrophoresis. Acetic anhydride is normally considered specific for amino groups, but stable derivatives can be formed with tyrosine and threonine residues. These esters, however, can be selectively destroyed by treatment with hydroxylamine or dilute base.

3.3.1 *Peptide mapping of acetylated protein*

Due to the greater number of lysine residues compared with tyrosine or histidine residues, and their frequent proximity to each other, a two-stage fragmentation is desirable. The first cleavage can be carried out with an enzyme such as pepsin or with CNBr. These procedures give relatively large peptides with multiple lysine residues that are purified either by gel filtration (17) or paper electrophoresis (4) and are assigned within the sequence from their amino acid compositions. These peptides are then further digested with an enzyme such as Pronase to obtain smaller fragments with predominantly single lysine residues. This process gives a much more reliable assignment of the positions of the individual lysine residues within the sequence than total digestion of the protein. Peptides from a competitive labelling experiment can be sequentially cleaved by the Edman procedure to quantify the $^{3}H/^{14}C$ ratios at each lysine residue from the PTH-derivatives. This is the only approach by which the properties of two or more residues occurring together can be measured individually.

Unlike the DNP-F experiments, this preliminary mapping should be done with ^{14}C-labelled protein to facilitate identification of the ε-*N*-acetyl-lysine peptides and the acetylated amino-terminal peptide. The procedure is as follows.

(i) Dissolve a sample of protein (~350 nmol) in 8 M urea at pH 2–3 (to avoid reaction of amino groups with cyanate).
(ii) Adjust the pH to 9.0 and add 5 μCi of [^{14}C]acetic anhydride (100 Ci/mol) in acetonitrile with stirring at room temperature. The [^{14}C]acetic anhydride is used undiluted with [^{12}C]reagent to obtain the most efficient incorporation of radiolabel.
(iii) Add excess non-radioactive acetic anhydride and maintain the pH at 9.0 with a pH-stat or by manual addition of 5 M NaOH.
(iv) Dialyse the reaction mixture against distilled water, then freeze-dry.
(v) Digest the labelled protein with the chosen proteolytic enzyme and purify the peptides by HPLC, using liquid scintillation counting to identify the [^{14}C]acetyl-peptides. Paper electrophoresis with autoradiography can also be used.
(vi) Determine their amino acid compositions to locate the peptides within the sequence.
(vii) Re-digest peptides that have more than one lysine residue and repeat steps (v) and (vi).

3.3.2 *Competitive labelling with acetic anhydride*

A typical procedure is given in Kaplan *et al.* (4).

(i) Prepare a stock solution of the protein (35 nmol/ml) with phenylalanine as the internal standard (35 nmol/ml) and divide into two portions.
(ii) Make one portion 8 M in urea, adjust the pH to 9 and react with 2.5 mmol of [^{14}C]acetic anhydride (100–300 mCi/mol), using a pH-stat to maintain the reaction pH.

(iii) Label aliquots of the stock solution at selected pH values from 6 to 11 with [^{3}H]acetic anhydride (200 nmol, 6.6 Ci/mmol, in 50 μl of acetonitrile).
(iv) Add urea (1 g/ml) and adjust the pH to 8. React with excess non-radioactive acetic anydride using a pH-stat.
(v) Add aliquots of the ^{14}C-labelled acetylated protein and ^{14}C-labelled acetylated internal standard mixture. Add HCl to bring the pH to 2.
(vi) Extract the internal standard derivative from each sample with ethyl acetate. Rotary evaporate the extracts using a dry ice trap to collect the radioactive acetic acid. Remove remaining radioactive acetic acid by keeping the samples in an evacuated desiccator containing NaOH pellets. Purify the internal standards by HPLC and determine their ^{3}H/^{14}C ratios.
(vii) Digest enzymatically the acetylated proteins. Purify the peptides by the procedure developed above and determine their ^{3}H/^{14}C ratios.
(viii) Analyse the data using Equation 3.

3.3.3 *Variations in the procedure*

As mentioned above, amino acid analysis can be used to obtain specific radioactivity data for labelled peptides, instead of using a second isotopically labelled reagent. The data shown in *Figures 1* and *2* were obtained in this manner.

In the above examples the main variable was pH. Similar protocols can readily be devised for the study of other variables such as temperature, ionic strength, etc. Two experiments of this type that illustrate these uses of competitive labelling are the investigation of protein folding by Ghélis (7) and the studies of insulin oligomerization by Kaplan *et al.* (3). In the first case, elastase denatured in guanidinium chloride (GdmCl) was mixed with [^{14}C]reagent in more dilute GdmCl, in a pulsed quenched-flow apparatus. The refolding of the protein in different segments of its sequence could thus be investigated. In the insulin case, by studying the variation in reactivity of specific groups as a function of protein concentration, the sites involved in the association of insulin monomers to form dimers could be investigated. The lack of any structural differences between the free and associated monomeric units was also demonstrated.

In the case of a protein with a very large number of lysine residues or a multi-protein complex such as the ribosome, it is usually preferable to examine the properties of only a few selected amino groups. This can be accomplished by preparing the [^{14}C]acetyl peptides containing the residue of interest and using them to track the corresponding [^{3}H]acetyl peptide from the competitive labelling studies. This approach is given in a study of the ribosomal L12 protein in intact ribosomes (18). In principle, rather than adding [^{14}C]protein and internal standard at step (v), equimolar amounts of selected [^{14}C]acetyl peptides and internal standard could be used.

4. INVESTIGATION OF PROTEIN INTERACTION SITES

The main use of competitive labelling has been the investigation of the interactions of proteins with other proteins or other classes of macromolecules, or with small ligands such as substrates or inhibitors. A full study of the chemical properties of all the functional groups in a protein in the presence and absence of the interacting ligand demands considerable laboratory work. It is usual, therefore, to use one of two simpler approaches. In the first, the reactivity and pK value of only certain key groups within the protein are determined. Examples are studies of the effect of oxygen on the N-terminal groups of haemoglobin (4) and of the inhibitor indole on the active-site histidine of chymotrypsin (11). The pK data that were obtained served to test theories of the roles of these residues in each protein's function. In the second approach, reactivity data for each individual functional group are obtained in the presence and absence of the interacting species, but no pH studies are undertaken. The data show which residues are changing in properties when the interaction occurs, and this is an excellent means of locating the actual site of interaction.

4.1 **Protein–protein interactions**

Acetic anhydride has been used in the majority of experiments of this type. The main additional complication in studying protein interaction by competitive labelling is the need to separate the proteins that form the complex of interest after they have been trace labelled so that they can be individually analysed. In experiments by Bosshard and co-workers on the interaction of cytochrome *c* with various proteins such as cytochrome *c* oxidase (19), flavocytochrome c_{552} (20) and specific antibodies (5), the internal standard used was [^{14}C]phenylalanine. This allowed correction to be made for variations in the trace labelling conditions between cytochrome *c* alone and the cytochrome–protein complex. These chiefly arise from the presence of the extra protein in the sample containing the protein complex. However, the ^{3}H-labelled proteins were separated before addition of the ^{14}C-labelled protein, and variations in recovery of the ^{3}H-labelled protein could not be corrected. The protocol below is based upon these experiments with some adjustments. Other laboratories have chosen to use one protein residue as the reference, omitting internal standards altogether (20), or to compare each residue's reactivity to the average reactivity of all the residues (21). The ratios of the reactivities of residues in the free and complexed protein have been termed protection or shielding factors (Equation 5), though increased reactivity in the bound state is possible (19). Data obtained by Bosshard *et al.* (9) for three cytochrome *c* complexes are shown in *Figure 5*.

4.1.1 *Separation procedure for labelled proteins*

Since trace-labelled proteins are minimally changed from their native state, any suitable procedure based upon protein charge or size can be used, for example, ion-exchange chromatography (9) or gel filtration (5), and denaturants can be present. Affinity chromatography should not be used, since it may lead to distortion of the data for active site residues. In cases where the average properties of functional groups of a constituent protein in a multi-protein complex have been examined, rather than individual residues, micro-techniques such as polyacrylamide gel electrophoresis have been used (22).

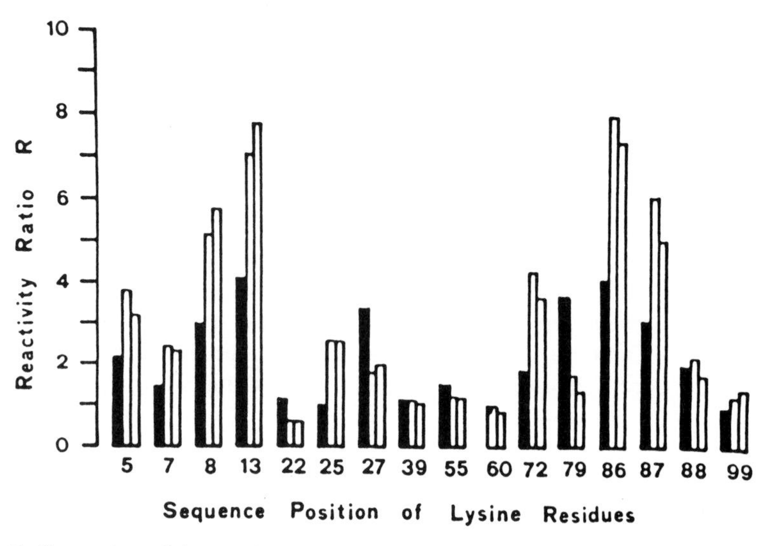

Figure 5. Comparison of the reactivity ratios of the lysine groups of cytochrome *c* (9) in the presence and absence of flavocytochrome c_{552} (solid bars), mitochondrial cytochrome bc_1 complex (left open bars) and mitochondrial cytochrome *c* oxidase (right open bars). Reproduced with permission of the authors and the American Society for Biological Chemistry.

4.1.2 *Modification procedure*

(i) Prepare a stock solution of protein (25–100 μmol) and internal standard (e.g. phenylalanine) preferably in a pH 8.0 buffer such as triethanolamine. Divide into two portions.

(ii) React one portion with [^{14}C]acetic anhydride, in 8 M urea as in Section 3.2.2.

(iii) Divide the second portion into a set of four or six equal aliquots and add an excess of the interacting protein to half of them. Equilibrate at 0°C to ensure as complete formation of complex as possible.

(iv) React both sets of aliquots with [^{3}H]acetic anhydride (5–14 Ci/mmol), using a 2- to 4-fold molar excess over protein.

(v) Add urea (1 g/ml) and adjust the pH to 8; add excess non-radioactive acetic anhydride, maintaining the pH at 8.

(vi) Add equal amounts of ^{14}C-labelled protein and internal standard to the two sets of reactions.

(vii) Separate the protein species by gel filtration in 8 M urea. Recover the double-labelled protein fraction and dialyse. Retain the internal standard fraction for step (ix) below.

(viii) Fragment the protein samples by proteolytic or chemical means and purify the peptides by HPLC. Determine the ^{3}H/^{14}C ratios.

(ix) Acidify the internal standard fractions and extract the [^{3}H]/[^{14}C]acetyl phenylalanine with ethyl acetate. Purify by HPLC and determine the ^{3}H/^{14}C ratios.

(x) Analyse the data using Equation 5.

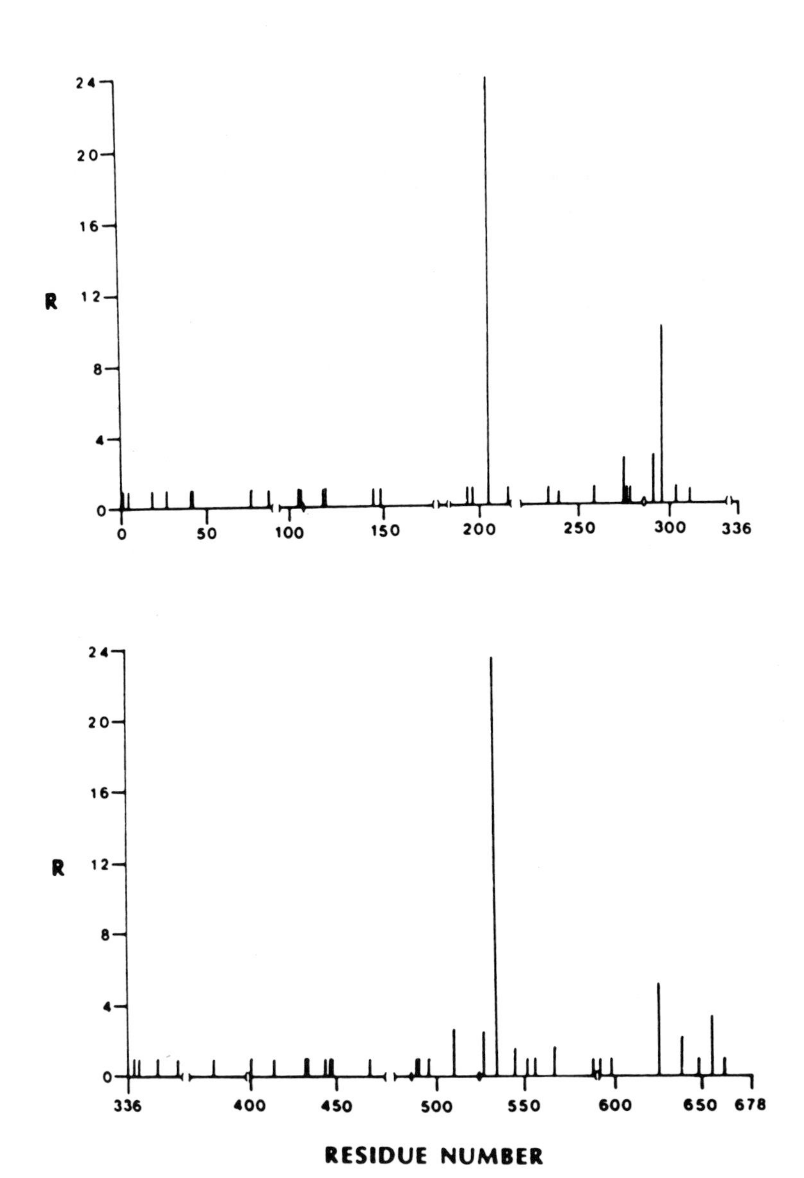

Figure 6. The effects of iron binding on the reactivities of the amino groups of human serum transferrin (17). The protection factor, R, is the ratio of the reactivity of the group in apo-transferrin to that in holo-transferrin. The upper and lower parts of the diagrams represent the two homologous domains of the proteins, with some gaps to align corresponding residues. Reproduced with permission of the authors and the American Society for Biological Chemistry.

4.1.3 *Variations in the procedure*

The roles of carboxyl groups have been studied in the cytochrome *c*–cytochrome *c* peroxidase complex using a mixture of carbodiimide and taurine for the trace labelling (23). There have been studies of proteins interacting with other classes of macromolecules, such as the DNA–histone interaction in nucleosomes (24) and protein–lipid interaction (25).

The separation of the interacting proteins in step (vii) (Section 4.1.2) can be eliminated by using tracking peptides, as described in Section 3.3.3, to isolate peptides containing the funtional groups of interest. This variation is most suited to systems that contain a large number of interacting proteins, such as the ribosome (18).

4.2 **Protein–ligand interactions**

This type of experiment is simpler than the previous one, since there is usually no need to remove the interacting inhibitor, substrate or other ligand from the labelled protein before peptide analysis. Step-by-step protocols will not be given since those described above are readily adapted to compare reactivities in the presence and absence of ligand.

The site of interaction with a ligand will be smaller than that for interaction with a macromolecular species, so relatively few residues will undergo major changes in reactivity when the ligand is bound. While over-all conformational changes in the bound state may affect many residues to some degree, it should not be difficult to distinguish residues actually at the binding site. A good example of this procedure is the identification in transferrin of amino groups affected by iron binding. Shewale and Brew (17) found two amino groups were highly affected and two moderately affected, out of a total of 59 amino groups in the protein (*Figure 6*). These workers also introduced an interesting variant to the usual competitive labelling design by trace labelling the apo-transferrin and di-ferric transferrin with [^{3}H]acetic anhydride, then adding ^{14}C-trace-labelled apo-transferrin to each, rather than uniformly ^{14}C-labelled protein.

Other examples helpful in designing this type of experiment include measurements of the effect of an inhibitor, indole, on the active-site histidine residue of chymotrypsin (11), using DNP-F; the effect of oxygenation on N-terminal residues of haemoglobin, using both DNP-F and acetic anhydride (4), and the effect of cadmium loading in the 1–7 mol/mol range on the 20 cysteine residues of metallothionein, using iodoacetamide (26). One caution must be added, however, and that is that the occurrence of a residue at an interaction site does not guarantee that it will change in properties upon ligand binding. It was found with concanavalin A that two tyrosine residues known to be in the carbohydrate binding-site from X-ray crystallographic studies did not alter their chemical properties upon ligand binding (27).

5. REFERENCES

1. Kaplan, H., Stevenson, K. J. and Hartley, B. S. (1971) *Biochem. J.*, **124**, 289.
2. Bosshard, H. R. (1979) *Methods Biochem. Anal.*, **25**, 273.
3. Kaplan, H., Hefford, M. A., Chan, A. M.-L. and Oda, G. (1984) *Biochem, J.*, **217**, 135.
4. Kaplan, H., Hamel, P. A., Chan, A. M.-L. and Oda, G. (1982) *Biochem. J.*, **203**, 435.
5. Burnens, A., Demotz, S., Corradin, G., Binz, H. and Bosshard, H. R. (1987) *Science*, **235**, 780.
6. Kraal, B. and Hartley, B. S. (1978) *J. Mol. Biol.*, **124**, 551.
7. Ghélis, C. (1980) *Biophys. J.*, **32**, 503.
8. Hefford, M. A., Evans, R. M., Oda, G. and Kaplan, H. (1985) *Biochemistry*, **24**, 867.
9. Bosshard, H. R., Davidson, M. W., Knaff, D. B. and Millett, F. (1986) *J. Biol. Chem.*, **261**, 190.
10. Kaplan, H. (1972) *J. Mol. Biol.*, **72**, 153.
11. Cruickshank, W. H. and Kaplan, H. (1975) *Biochem. J.*, **147**, 411.
12. Kaplan, H., Long, B. G. and Young, N. M. (1980) *Biochemistry*, **19**, 2821.
13. Cockle, S. A., Kaplan, H., Hefford, M. A. and Young, N. M. (1982) *Anal. Biochem.*, **125**, 210.
14. Jackson, G. E. D. and Young, N. M. (1986) *Biochemistry*, **25**, 1657.
15. Jackson, G. E. D. and Young, N. M. (1987) *Anal. Biochem.*, **162**, 251.
16. Shaltiel, S. (1967) *Biochem. Biophys. Res. Commun.*, **29**, 178.
17. Shewale, J. G. and Brew, K. (1982) *J. Biol. Chem.*, **257**, 9406.
18. Hasnain, S., Visentin, L. P. and Kaplan, H. (1977) *Eur. J. Biochem.*, **80**, 35.
19. Rieder, R. and Bosshard, H. R. (1980) *J. Biol. Chem.*, **255**, 4732.
20. Hitchcock-De Gregori, S. E. (1982) *J. Biol. Chem.*, **257**, 7372.
21. Giedroc, D., Sinha, S. K., Brew, K. and Puett, D. (1985) *J. Biol. Chem.*, **260**, 13406.
22. Visentin, L. P., Yaguchi, M. and Kaplan, H. (1973) *Can. J. Biochem.*, **51**, 1487.
23. Bechtold, R. and Bosshard, H. R. (1985) *J. Biol. Chem.*, **260**, 5191.
24. Malchy, B. L. and Kaplan, H. (1976) *Biochem, J.*, **159**, 173.
25. Oomen, R. P. and Kaplan, H. (1987) *Biochemistry*, **26**, 303.
26. Bernhard, W. R., Vasak, M. and Kagi, J. H. R. (1986) *Biochemistry*, **25**, 1975.
27. Young, N. M. and Jackson, G. E. D. (1986) In *Lectins*. Bog-Hansen, T. C. and van Driessche, E. (eds), Walter de Gruyter and Co., Berlin, Vol. V, p. 177.

CHAPTER 10

Chemical modification

TAIJI IMOTO and HIDENORI YAMADA

1. INTRODUCTION

Chemical modification is one of the most useful methods of identifying the functional groups of a protein. Whether or not the various types of amino acid side-chains are involved in a protein's function can be determined readily by whether or not chemical modification of that type of amino acid affects the function. This requires only chemical reactions that are reasonably specific for each type of amino acid side-chain. Suitably specific reactions are available for aspartic acid, glutamic acid, histidine, lysine, arginine, methionine, tryptophan, tyrosine and cysteine residues. A few, less satisfactory, reactions are available for serine, threonine, asparagine and glutamine residues, but none are available for glycine, alanine, valine, leucine, isoleucine and proline, which have no reactive groups on their side-chains. Fortunately, reactive groups are usually required for important functional roles, so chemical modification is able to test all the likely functional amino acids.

Having determined which type of amino acid side-chains are involved, it is also desirable to identify which specific residues of that type in a protein are responsible. This requires that the different residues be distinguished by their reactivities. Fortunately, the residues involved in protein function often have unusual reactivities or environments that make them distinguishable in this way.

The recently developed technology of gene engineering using site-directed mutagenesis (Chapter 11) has the advantage of altering protein structure much more specifically than does chemical modification normally, because only certain amino acid residues are altered, using the specificity of the genetic apparatus. This approach is limited, however, by the need for information about what residues of the protein are likely to be involved in its function. Otherwise, there are too many possible genetic alterations to be tested. Chemical modification can provide just this information, so the two techniques are complementary. Chemical modification can identify the residues likely to be involved, and gene engineering can be used to elucidate in detail their roles in the biological function.

2. GENERAL APPROACH TO CHEMICAL MODIFICATION

2.1 **Modification of activity**

Residues that participate in activity are usually accessible to the solvent and can be modified only by accessibility selection. Sometimes we can design specific reagents by considering the characteristic environment of the active site of the protein. Affinity labelling is a sophisticated way of modifying active site residues selectively, which utilizes the original capacity of protein to form the complex to exert its biological function (see Chapter 4). Modification in the presence and absence of ligand is a good method to identify residues in the active site (Chapter 9).

In the case of enzymes, catalytic or vicinal residues can be modified selectively by using a suicide substrate that is converted to a reactive species to modify residue(s) during the catalytic turnover [reviewed by Walsh (1)].

2.2 **Modification of stability**

We can stabilize a protein by stabilizing the native state, for example by introducing new interactions (such as hydrophobic or electrostatic interactions or hydrogen bondings), or by releasing any strain in the native state. Another way to stabilize a protein is to destabilize the denatured state by cross-linking to decrease the entropy of the denatured state. Intra-molecular cross-links can be effectively introduced by two-stage reactions with bifunctional reagents when the two functional groups of the reagent react with different residues or under different conditions (e.g. different pH, etc.).

Proteins tend to be stabilized by immobilization. Applications of the method to elucidate the structure and function of protein were recently reviewed by Martinek and Mozhaev (2) and in Chapter 1. Proteins can be stabilized by oligomerization (Chapter 6) and by attaching polymers on their surfaces (Chapter 1).

2.3 **Purification of the modified protein**

Gel filtration is a mild method but is effective only in separating the modified protein from oligomerized proteins or from the reagents employed. When the modification causes changes in the net charge of the protein, ion-exchange chromatography is a very powerful method to fractionate the modified proteins. Ion-exchange high-performance liquid chromatography (HPLC) is recommended for purifying small quantities of proteins (i.e. micrograms to milligrams, as is reversed-phase HPLC. Representative examples of ion-exchange and reversed-phase HPLCs are shown in *Figure 1*.

Affinity column chromatography can be effectively applied to the separation of modified proteins whose binding site residues have been modified (*Figure 2*).

2.4 **Identification of the modified residue**

Peptide maps of the unmodified and modified proteins on reversed-phase HPLC can usually identify the modified residue (*Figure 3a* and *b*). The peptide that contains the modified residue is eluted at a different position from the unmodified one. If the peptide contains only one residue that reacted with the reagent employed, we can identify the modified residue from the displacement of the position of the peptide on the chromatogram (*Figure 3b*). Co-chromatography

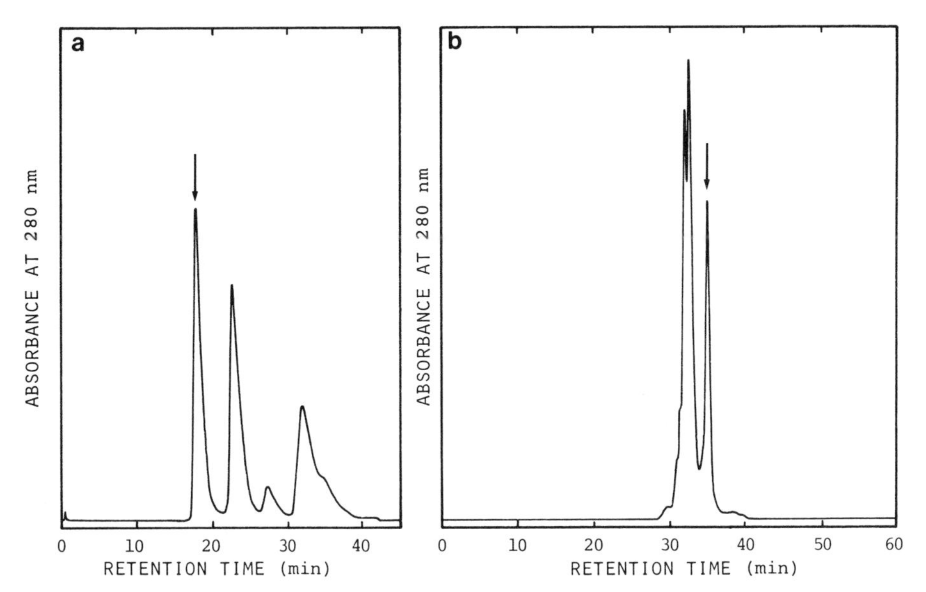

Figure 1. HPLC patterns of the reaction mixture obtained in the reaction of hen egg lysozyme with ethylenediamine catalysed by EDC. The carboxyl groups of lysozyme were only partially amidated with ethylenediamine. (**a**) Ion-exchange HPLC on a carboxylic cation exchanger column of TSKgel CM-2SW (4 × 50 mm). The column was eluted at a flow-rate of 0.8 ml/min and with a gradient of 20 ml of 0.05 M phosphate buffer and 20 ml of the same buffer containing 0.5 M NaCl at pH 7.0. (**b**) Reversed-phase HPLC on a column of YMC-Pak A-212 C8 (6 × 150 mm). The column was eluted with a gradient of 25 ml of 20% acetonitrile and 25 ml of 45% acetonitrile, both containing 0.1% concentrated HCl, at a flow-rate of 0.8 ml/min. Arrows indicate the peaks of unmodified lysozyme.

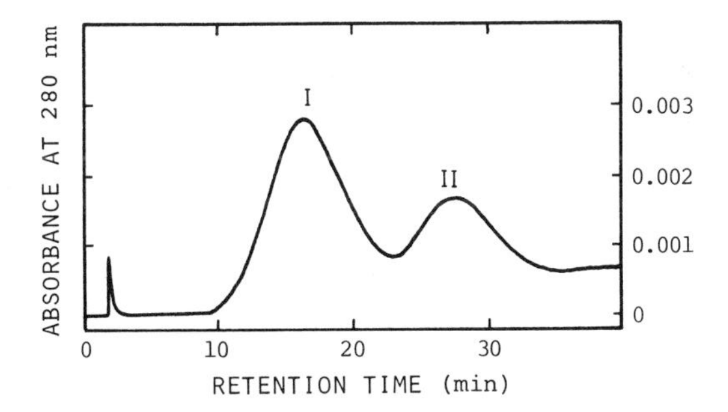

Figure 2. Affinity HPLC of hen egg lysozyme (II) and its derivative (I) that possesses the β-aspartyl sequence at Asp101 (3, 4). About 10 μg of protein was injected onto a chitin-coated Celite column (4 × 50 mm), and the column was eluted at 0°C with a gradient of 10 ml of 0.1 M acetate buffer containing 0.5 M NaCl and 10 ml of 1 M acetic acid at a flow-rate of 0.5 ml/min.

with unmodified peptides is a sensitive method to identify the modified peptides with altered properties (*Figure 3c*).

The peptides can be identified from their amino acid compositions. If the modified peptide contains more than one reactive residue, it should be analysed by a peptide sequencer to identify the residue modified.

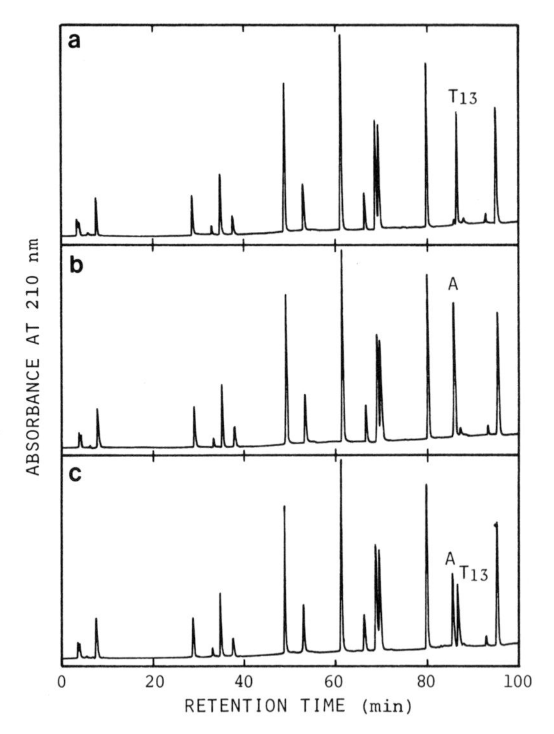

Figure 3. Reversed-phase HPLC of tryptic peptides derived from reduced and *S*-carboxymethylated hen egg lysozyme and its derivative on a column (4.6 × 250 mm) of TSKgel ODS-120A (5 μm, Toyo Soda, Japan). The column was eluted with a gradient of 40 ml of 1% acetonitrile and 40 ml of 40% acetonitrile, both containing 0.1% concentrated HCl at a flow-rate of 0.8 ml/min. (**a**) From native lysozyme (5). (**b**) From the derivative where Asp101 is amidated with ethanolamine. The Asp101-containing peptide (T13) is replaced by peptide A. (**c**) Co-chromatography of a mixture of **a** and **b**.

3. CONSEQUENCES OF MODIFICATION

Several comments on the application of chemical modifications to elucidate the function of a protein must be made.

(i) Chemical modifications are apt to cause side-reactions that can be critical, and sometimes it is very hard to discover them. For example, the side-reaction of conversion of an α- to β-peptide bond took place between Asp101 and Gly102 in lysozyme when Asp52 was esterified with triethyloxonium tetrafluoroborate or when Lys13 (ε-amino) and Leu129 (α-carboxyl) were cross-linked by 1-ethyl-3-[3-(dimethylamino)propyl]carbodiimide hydrochloride (EDC) in the presence of 1 M imidazole. The derivatives formed by the side-reaction were separated from normal derivatives only by affinity chromatography (4).

(ii) When chemical modification causes little change in biological function, we can safely eliminate the participation of the modified residues in the function. However, when the modification does affect the biological function, it is not certain that the modified residue is involved in the function, because chemical modifications are apt to cause conformational changes in proteins. It is necessary to check that no substantial conformational change occurred in the protein as a result of the modification.

(iii) To analyse the results of modification of an enzyme, employ as simple a substrate as possible. Complex substrates sometimes give misleading results. For example, acetylation of lysozyme caused a dramatic decrease in lytic activity but did not affect the activity against a simple substrate like glycol chitin (6). Thus, analysis of the results of the modification should be performed under conditions where the effect on the function of interest is sensitively reflected.

3.1 **Binding groups**

Modifications of residues that are responsible for substrate binding by an enzyme cause changes in the Michaelis constant, K_m, but sometimes also a decrease in the catalytic rate constant, k_{cat}. The modification of Trp62 in lysozyme by *N*-bromosuccinimide, which is one of the binding residues, caused not only a dramatic decrease in substrate binding ability, but also a decrease in k_{cat} of the enzyme (7).

3.2 **Catalytic groups**

Modification of catalytic groups of an enzyme should cause a decrease in k_{cat}. However, as mentioned above, the modification of binding residues sometimes also causes a decrease in k_{cat}, and we cannot directly relate the decrease in k_{cat} to the modification of a catalytic group. In this sense, the strict proof of a catalytic group can be obtained only by blocking the group with a reagent that causes little or no change in size. As a consequence, there should be little change in the binding ability of the substrate or substrate analogues after modification of catalytic groups.

Good examples for reasonable modifications of catalytic groups are conversions of –SH to –OH in the active site cysteine residue in papain (8), –OH to –SH in subtilisin (9, 10) and trypsin (11) and –COOH to $–CONH_2$ in lysozyme (12).

3.3 **Stability**

Generally, we can improve the activities of proteins by stabilizing them. Catalytic efficiency increases with increase of temperature, and stabilized proteins work for longer periods under drastic conditions. If we can use proteins in organic solvents, their usefulness will expand enormously. Stabilized proteins last longer when they are used clinically. One example of the effectiveness of the stabilization of a protein in enhancement of its activity is shown in *Figure 4* (5), where lysozyme was intra-molecularly cross-linked between Lys1 (ε-amino) and His15 with *bis*-

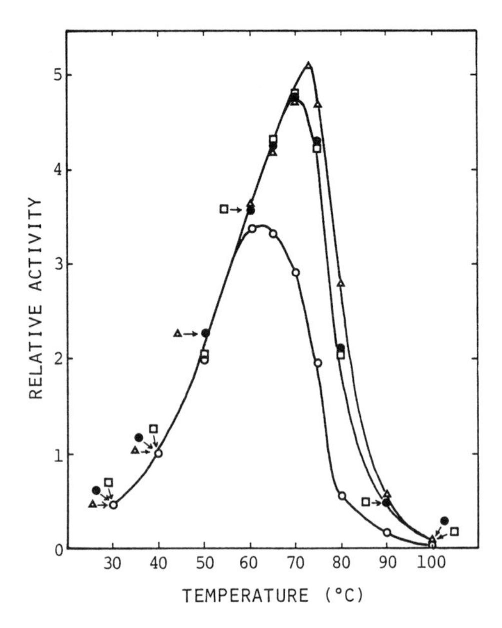

Figure 4. Temperature dependence of the activities of hen egg lysozyme (○) and its derivatives cross-linked with $BrCH_2CONH(CH_2)_nNHCH_2Br$ [n = 2 (●), 4 (△) and 6 (□)] between Lys1 and His15 (5). Activities were measured against glycol chitin at pH 5.5.

bromoacetamide derivatives [$BrCH_2CONH(CH_2)_nNHCOCH_2Br$, $n = 2$, 4 and 6]. The order in n of stabilization was 4 (9.6 kJ/mol more stable than native) $>2 = 6$ (6.7 kJ/mol). The more stable derivative is active at the higher temperature and showed the more efficient activity.

Amino groups of some proteins were modified with polyethylene glycol; some of the derivatives were de-sensitized from immune reactions (13, 14) and others were active in organic solvents (15–17).

4. SOME USEFUL SPECIFIC CHEMICAL MODIFICATIONS

In the use of chemical modification to investigate the role of a particular residue in the function of a protein, it is important to modify the residue specifically. Although many specific chemical modifications of proteins have been developed, in most cases specificity results either from enhanced reactivity of the residue due to the local environment in the native protein, or from specific binding of a particular reagent to a site in the protein or perhaps from a combination of both effects. Therefore, reagents and reaction conditions for the modification of proteins must be chosen depending upon the properties of the proteins and the purpose of the investigation.

Since it is impossible to cover all chemical modifications in the limited space available, this section describes some selected chemical modifications focused on carboxyl groups, amino groups, thiol groups, tyrosine residues, tryptophan residues and intra-molecular cross-linking of particular proteins. Readers who need more information should refer to refs 18–21.

4.1 **Carboxyl groups**

Amidation and esterification are known as useful chemical modifications of carboxyl groups of proteins.

4.1.1 *Amidation*

(i) *Carbodiimide method*. In 1966, Hoare and Koshland reported the amidation of carboxyl groups of protein with various amines by water-soluble carbodiimides under mild conditions (22). EDC is now commercially available and the most widely used water-soluble carbodiimide. The reaction mechanism is illustrated in *Scheme I*. First, the carboxyl groups are activated by a carbodiimide, and the activated carboxyl groups are then reacted with an added amine (a nucleophile) to give amidated carboxyl groups and the urea derivative of the carbodiimide. In the absence of an amine, the activated carboxyl groups are usually hydrolysed to regenerate free carboxyl groups or react with amino groups of the same or other protein molecules to give either an intra-molecular cross-link or inter-molecularly cross-linked polymer. The reaction is usually carried out at slightly acidic pH (4.75–5.0). Depending on the conditions, both complete and partial amidations are possible.

Scheme I

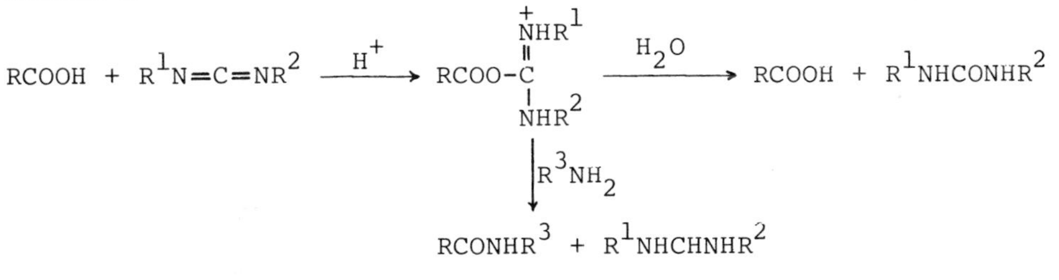

$$RCOOH + R^1N{=}C{=}NR^2 \xrightarrow{H^+} RCOO\text{-}\underset{\displaystyle NHR^2}{\overset{\displaystyle \overset{+}{N}HR^1}{\overset{\|}{C}}} \xrightarrow{H_2O} RCOOH + R^1NHCONHR^2$$

$$\downarrow R^3NH_2$$

$$RCONHR^3 + R^1NHCHNHR^2$$

(a) *Complete amidation of carboxyl groups of a protein* (23, 24).

(1) Dissolve an amine (1 M) in a 7.5 M urea or 5 M guanidinium chloride (GdmCl) solution in a vial or small beaker. Adjust the pH to 4.75 with HCl.
(2) Add the protein to a concentration of 1–10 mg/ml.
(3) Add EDC as solid or concentrated solution to bring its concentration to 0.1 M.
(4) Stir the solution at room temperature and maintain the pH at 4.75 by the addition of 1 M HCl manually or by use of a pH-stat for 1–2 h.
(5) Remove reagents by exhaustive dialysis or gel filtration.

In this manner, all of the carboxyl groups of the protein can be amidated with the amine employed. When an amine that is detectable after hydrolysis by amino acid analysis, such as glycine methyl ester, glycinamide, methylamine or ethanolamine etc., is employed, the number of carboxyl groups of the protein can be determined directly. When some of the carboxyl groups have been already modified, this method is used for the determination of the extent of the modification by measuring the number of the unmodified carboxyl groups. The extent of the reaction can also be determined by use of a radioactive amine. The reaction is terminated by dialysis or gel filtration of the reaction mixture. The addition of acetic acid or acetate buffer to terminate the reaction is not recommended because it may lead to the acetylation of amino groups of the protein.

Carraway and Triplett (25) and Perfetti *et al.* (26) noticed the formation of adducts of hydroxyl groups of tyrosine residues or thiol groups of cysteine residues with the carbodiimide as a side-reaction. Tyrosine reacts with EDC slowly, but free tyrosine can be regenerated by a 5-h treatment with 0.5 M hydroxylamine at pH 7.0 and 25°C (27). The thiol of Cys25 in papain was reported to react with EDC irreversibly (26). The formation of inter-molecularly cross-linked polymers is another side-reaction (28). Polymer formation becomes predominant in the absence of an amine nucleophile or with high concentrations of the protein (29).

(b) *Limited amidation.* In the absence of the denaturant and with a reduced amount of EDC, more limited amidation of carboxyl groups can be achieved. The selective modification of Asp101 of hen egg lysozyme with ethanolamine (30) is given as an example of limited amidation.

(1) Dissolve 100 mg of lysozyme (final concentration 0.7 mM) and ethanolamine (to 0.1 M) in about 5 ml of water in a 20-ml vial containing a magnetic stirring bar.
(2) Adjust the pH of the solution to 5.0 with 1 M HCl. If the temperature is increased by the addition of HCl, cool the mixture in an ice-water bath or dissolve the lysozyme after the adjustment of the pH in order to prevent the denaturation of the enzyme.
(3) Adjust the volume to 10 ml with water, then add 5 mg of EDC (to 2.6 mM) as a solid with stirring at room temperature.
(4) Maintain the pH at 5.0 by the addition of 1 M HCl through a capillary for 3 h.
(5) Dialyse the reaction mixture against distilled water.

After reaction of lysozyme, the dialysate is chromatographed on Bio-Rex 70, a carboxylic cation-exchanger, and the fractions of the major product collected. The protein is precipitated with 85% saturation of $(NH_4)_2SO_4$, collected by centrifugation, re-dissolved in a small amount of water, dialysed against distilled water exhaustively, and then lyophilized to give 40–60 mg of the modified lysozyme. The presence of one ethanolamine per molecule was shown by amino acid analysis of the modified lysozyme, indicating that one carboxyl group out of 10 is amidated with ethanolamine in this derivative.

For determination of which carboxyl group was amidated, reduced and *S*-carboxymethylated lysozyme (31) was digested with TPCK-treated trypsin at pH 8.0 and 40°C for 2 h, and then the digestion was stopped by lowering the pH to 2.0 (4). The resulting tryptic peptides were analysed with reversed-phase HPLC. The chromatographic patterns from the unmodified lysozyme, modified lysozyme and a mixture of the two are shown in *Figure 3a–c*, respectively. Clearly, only peptide T13 was changed to peptide A by the modification. T refers to the Canfield's nomenclature of tryptic peptides of lysozyme (32). The amino acid composition of peptide A was that of peptide T13 plus one ethanolamine. Since peptide T13 in the original lysozyme contains only one carboxyl group, Asp101, the site of the amidation was concluded to be the β-carboxyl group of Asp101. Similarly, various amines were found to be introduced into Asp101 of lysozyme by the carbodiimide reaction (30, 33).

Under carefully controlled conditions, Leu129 (C-terminal carboxyl group) and Asp119 were also amidated by the carbodiimide reaction (34).

(ii) *Liquid ammonia method* (12, 35). When carboxyl groups are already esterified, ammonolysis of the esters in liquid ammonia can be used for the conversion of aspartic acid and glutamic acid residues to the corresponding amides (asparagine and glutamine), respectively. Since these conversions are those to other amino acids, gene engineering using site-directed mutagenesis could make similar changes.

The following procedure was employed for the ammonolysis of Glu35- or Asp52-esterified hen egg lysozyme (12).

Caution: carry out this reaction only in a well-ventilated fume hood to avoid exposure to ammonia

(1) Put 0.5 g of NH_4Cl and a magnetic stirring bar into a 100-ml two-necked flask equipped with drying tubes, both containing anhydrous $CaSO_4$ (Drierite) and NaOH (pellets).
(2) Cool the flask to below −45°C in a thermostatted cooling bath, and introduce through one of the drying tubes commercial anhydrous ammonia gas into the flask with stirring, until about 20 ml of liquid ammonia is trapped in the flask.
(3) Add the esterified protein (10–50 mg) as lyophilized powder into the flask with vigorous stirring to dissolve it. Replace the drying tube through which ammonia has been introduced with a stopper. When the protein floats on the surface of liquid ammonia, the ammonolysis does not proceed.
(4) Keep the temperature below −45°C for several days with gentle stirring.
(5) Evaporate the liquid ammonia to dryness by aspiration, keeping the temperature below −45°C.
(6) Add about 30 ml of 10% acetic acid in the flask with stirring and then remove the cooling bath to allow the temperature to rise.
(7) Dialyse the resulting solution against 10% acetic acid and then against distilled water exhaustively.

In this manner, Glu35-esterified lysozyme and Asp52-esterified lysozyme were converted to Gln35 lysozyme and Asn52 lysozyme, respectively (12). At −55°C the reaction was complete after 3 days, and 10–15 mg of the original 50 mg of lysozyme became insoluble. At temperatures above −45°C, the reaction went faster but formation of the insoluble protein was greatly increased. The formation of insoluble protein was also increased by contamination by water or in the absence of NH_4Cl. The peptide analysis on reversed-phase HPLC after reduction, *S*-carboxymethylation and tryptic digestion indicated that the insoluble protein was identical to the soluble protein. Furthermore, the insoluble protein could be solubilized by the reduction of its disulphide bonds with mercaptoethanol followed by the re-oxidation in a mixture of reduced and oxidized glutathione at pH 8 (unpublished observation; for the refolding of reduced protein, see ref. 36). From these observations, the side-reaction of the ammonolysis in liquid ammonia appears to be the interchange of protein disulphide bonds.

Ammonolysis of esterified protein in concentrated aqueous ammonium acetate and ammonia (both 7 M) at 25°C has also been reported (37).

(iii) *Isoxazolium salt method.* The carboxyl groups of proteins can be activated by isoxazolium salts including Woodward's Reagent K (2-ethyl-5-phenylisoxazolium 3′-sulphonate) at pH 3–5. The activated carboxyl, an active ester, could be isolated but is not stable. In the presence of an amine it is converted to the corresponding amide (38, 39). In this way, the isoxazolium salt method is similar to the carbodiimide method.

4.1.2 *Esterification*

Carboxyl groups of proteins were first modified by esterification of dry protein with methanol–HCl (40). The conditions used are so drastic, however, that this method is not suitable for investigating the functions of carboxyl groups. Carboxyl groups can be esterified with trialkyloxonium salts, and diazomethyl compounds under mild conditions (pH 4–7).

(i) *Trialkyloxonium salts.* Triethyloxonium tetrafluoroborate is commonly used for ethyl esterification of carboxyl groups of proteins. A solution of 1 M triethyloxonium tetrafluoroborate in dichloromethane is commercially available (Aldrich Chemical Co.). To obtain the solid salt, excess hexane is added to the solution. The separated oil is transferred to a bottle by use of a dry syringe and dried to the solid by evacuation under nitrogen. Solid triethyloxonium tetrafluoroborate can also be prepared from diethyl ether, epichlorohydrine and boron trifluoride etherate (41). The salt is very hygroscopic, so it should be stored in a tightly closed screw-cap bottle at 0–5°C and should be used within a few days. It should be handled in a dry box. Triethyloxonium hexachloroantimonate and hexafluorophosphate are more stable and are commercially available as solids (Aldrich Chemical Co.). If methyl esterification is required, trimethyloxonium tetrafluoroborate (42) or the hexachloroantimonate derivative (Aldrich Chemical Co.) may be used (43).

The following procedure was used for ethyl esterification of hen egg lysozyme (44).

(a) Prepare a solution of 188 mg of the protein in 15 ml of distilled water in a titration vessel. Adjust the pH to the desired value (4.0–4.5) by the addition of dilute perchloric acid. Place the vessel on a pH-stat.
(b) Dry approximately 0.7 g of triethyloxonium tetrafluoroborate under a stream of nitrogen in a dry box and weigh it.
(c) Add approximately 0.2 g of dry acetonitrile to the salt and weigh the solution.
(d) Take up the solution of the oxonium salt into a dry calibrated syringe and inject the weight that contains 0.57 g of the salt (final concentration of the oxonium salt is to be 0.2 M) into the vigorously stirred lysozyme solution.
(e) Maintain the pH by the addition of about 1 ml of 4 M NaOH with the aid of the pH-stat. The reaction is over in about 20 min and a slightly cloudy solution results.
(f) Dialyse the reaction mixture against several changes of distilled water and recover the modified protein by lyophilization.

Under these conditions, two carboxyl groups of lysozyme were esterified sequentially. The initially esterified carboxyl group was identified as that of Asp101 (4) and the second was that of Asp52 (45). The mono-esterified and di-esterified lysozymes were isolated by chromatography on a cation-exchange resin. (45).

Esters are usually hydrolysed at alkaline pH and the determination of the esterified residue may be very difficult. In the case of lysozyme, the ethyl ester of Asp101 is very labile and easily hydrolysed at pH 7.2 and room temperature. However, in this case, the peptide bond between Asp101 and Gly102 was partially inverted from α to β simultaneously (discussed in Section 4.6.1) to reveal the esterified residue to be Asp101 (4).

(ii) *Diazomethyl compounds.* Diazomethyl compounds react with carboxyl groups of proteins at acidic pH to form esters. The reagents often used are the derivatives of diazoacetic acid (for example diazoacetamide, methyl diazoacetate, *N*-diazoacetylglycinamide, etc.) (46, 47). The procedures for the preparations of the reagents are described elsewhere (46–48). Diphenyldiazomethane (49) and 4-diazomethyl-*N,N*-benzenesulphonamide (12) are also used for esterification of the carboxyl groups of proteins. The latter reagent is commercially available (Wako Pure Chemical Ind. Ltd) and was used for the esterification of Glu35 of hen egg lysozyme (12).

Since diazomethyl compounds react with water and various inorganic anions, as well as carboxylic acids (*Scheme II*), excess reagent should be used for the

Scheme II

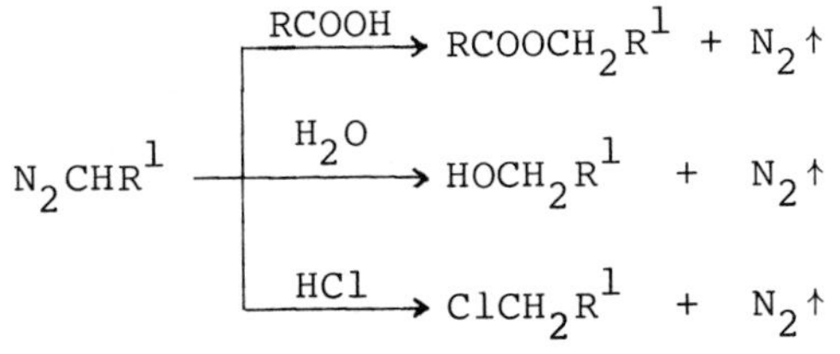

$$N_2CHR^1 \xrightarrow{RCOOH} RCOOCH_2R^1 + N_2\uparrow$$
$$N_2CHR^1 \xrightarrow{H_2O} HOCH_2R^1 + N_2\uparrow$$
$$N_2CHR^1 \xrightarrow{HCl} ClCH_2R^1 + N_2\uparrow$$

esterification of carboxyl groups. Perchloric acid is known not to react with diazomethyl compounds and can be used to adjust the pH of the reaction mixture. Using a diazoacetylated amino acid, the extent of modification can be determined by amino acid analysis of the modified protein. Diazomethyl compounds exhibit a degree of specificity for carboxyl groups, except for the alkylation of free sulphydryl groups (46).

The following procedure was employed for the esterification of RNase A with *N*-diazoacetylglycinamide (48).

(a) Dissolve 250 mg of the protein in 5 ml of water in a jacketed titration vessel, in which the temperature is kept constant by circulating water.
(b) Place the vessel on a pH-stat, adjust the pH of the solution to 4.5 with 1 M $HClO_4$ and maintain the temperature at 10°C.
(c) Add with stirring a solution of 500 mg of *N*-diazoacetylglycinamide in 5 ml of water and maintain the pH at 4.5 by adding 1 M $HClO_4$.
(d) Add additional quantities of diazoacetoglycinamide (500 mg) after 6 and 12 h.
(e) After 24 h, dialyse the mixture against six changes of HCl at pH 3.0, and then recover the modified protein by lyophilization.

The Asp53-esterified RNase A was isolated by chromatography on a cation-exchange resin as a major product in this reaction (48).

(iii) *Other reagents for esterification.* Active halogen compounds (halomethyl ketones, including haloacetyl derivatives) can esterify glutamic or aspartic acid residues, although they are normally reactive toward the side-chains of cysteine, histidine and lysine residues (50). For example, iodoacetic acid (or bromoacetic acid) was used for the esterification of Glu58 of RNase T1 (51).

Ethylenimine has also been used for the esterification of Glu35 and/or Asp52 of hen egg lysozyme (52).

4.2 **Amino groups**

There are two kinds of amino groups in a protein. One is the terminal α-amino group with a pK of 6–8 and the other is the ε-amino group of lysine residues with pK values of 8–10.5. An amino group, being positively charged under physiological conditions, tends to be located on the surface of the protein molecule and its unprotonated form is a strong nucleophile, so it can be modified with various reagents at slightly acidic to alkaline pH values. Because of the difference in pK values between α- and ε-amino groups, the former may be selectively modified by controlling the pH of the reaction medium. However, a protein contains a relatively large number of ε-amino groups and their pK values are usually spread widely depending upon their micro-environments. As a result, even the selective modification of the α-amino group is not easy. On the other hand, since many amino groups are exposed and not involved in the function of the protein, the modification of amino groups is suitable to introduce reporter groups, such as chromophores and radioisotopes, into proteins.

In *Table 1*, some representative modifications of amino groups are shown.

Table 1. Chemical modification of amino groups of proteins.

Reaction	*Reagent*	*pH*	*Reference*
Guanidination	*O*-Methylisourea	10–11	53
	1-Guanyl-3,5-dimethylpyrazole nitrate	9.5	54
Amidination	Methyl acetimidate	7–10.5	55–57
Acylation	Acetic anhydride	5.5–8	58
	N-Acetylsuccinimide	>4	59
	N-Hydroxysuccinimide acetate	6.9–8.5	60
	Acetylimidazole	>5	59, 61
	Succinic anhydride	7–10	62
	Maleic anhydride	6–10	63
	Citraconic anhydride	8.2	64
Carbamylation	Potassium cyanate	>7	65
Alkylation and arylation	Iodoacetic acid	7.5–9	50
	1-Fluoro-2,4-dinitrobenzene	7–11	66, 67
	2,4,6-Trinitrobenzenesulphonic acid	9.5	68
Reductive alkylation	Formaldehyde + sodium borohydride	8–10	69

Among them, trinitrophenylation with 2,4,6-trinitrobenzenesulphonic acid (TNBS) is used for the determination of the extent of the modification of amino groups with other reagents (66). Since most of the modifications listed in *Table 1* have been well described elsewhere, only acetylation with acetic anhydride, and then a rather special subject, attachment of polyethylene glycol (PEG) to amino groups, are described here.

4.2.1 *Acetylation*

Acetylation of amino groups with acetic anhydride is one of the most common reactions employed in the chemical modification of proteins (see also Chapter 9). It reduces the positive charges of proteins and changes the interactions between the proteins and other components of the environment.

The following procedure is recommended as a first trial for acetylation of any protein.

(i) Prepare a solution or suspension of the protein at a concentration of 2–10% (w/v) in half-saturated sodium acetate in a vial containing a magnetic stirring bar.
(ii) Cool the mixture in an ice-water bath, stirring with a magnetic stirrer.
(iii) Add over 1 h a weight of acetic anhydride equal to that of the protein in five equal portions at 0°C and continue the stirring for an additional hour.
(iv) Remove the salt and acetic acid by gel filtration or dialysis.

Although this procedure has general applicability, it may lead to protein denaturation, probably because of the high concentrations of acetic anhydride employed. This amount of the anhydride is required, however, because the reagent is unstable and undergoes spontaneous hydrolysis. In addition, the pH falls from about 8 to 5.5 soon after addition of the anhydride and the rate of acetylation decreases during the course of the reaction because only non-protonated amino groups are acetylated. An excess of the anhydride is therefore required to achieve an appreciable rate of acetylation.

Under these conditions, hydroxyl groups of tyrosine residues are also acetylated. However, such high concentrations of acetate ion catalyse the hydrolysis of the phenolic esters, thereby increasing the specificity of the reaction for amino groups.

When the reaction is carried out at constant pH (7.5–8.2) in a lightly or non-buffered solution using a pH-stat, a 30- to 60-fold molar excess of the anhydride is sufficient for the acetylation of amino groups and the reaction is completed within 20–30 min. Of course, the optimum conditions must be determined in each instance (70). It is important to note that these conditions tend to acetylate the side-chains of tyrosine, serine and threonine residues. These esters can be hydrolysed by incubation at pH 10 (70) or by treatment with 1 M hydroxylamine at pH 10.5 (71). It must be also noted that the peptide bond of Asn-Gly residues may be cleaved by 1–2 M hydroxylamine at pH 9–10.5 (72). Thiol groups of cysteine residues and imidazole groups of histidine residues are also acetylated by acetic anhydride, but the products are not stable and free thiol and imidazole groups are regenerated under the reaction conditions.

4.2.2 *Attachment of polyethylene glycol*

A protein modified with polyethylene glycol (a PEG-protein) is stable and active in organic solvents (16), and also is stable against clearance from the serum when it is injected (13). Furthermore, the PEG-protein does not provoke an immune response (13) and its injection suppresses immunization with the unmodified protein (14).

Scheme III

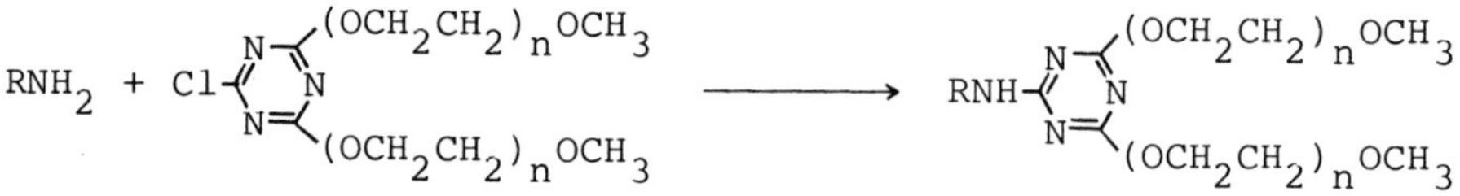

$$RNH_2 + Cl\text{-}C_3N_3[(OCH_2CH_2)_nOCH_3]_2 \longrightarrow RNH\text{-}C_3N_3[(OCH_2CH_2)_nOCH_3]_2$$

Modification of amino groups of a protein with PEG can be achieved either by mixing the protein with cyanuric chloride and PEG (14) or by treatment of the protein with the activated PEG_2 (*Scheme III*), a conjugate of cyanuric chloride and PEG (13). The following describes the latter method.

(i) *Synthesis of 2,4-bis(O-methoxypolyethylene glycol)-6-chloro-s-triazine (activated PEG_2)* (13).

(a) Dissolve 20 g of monomethoxypolyethylene glycol (M_r = 5000) in 100 ml of dry benzene containing 5 g of Molecular Sieve 3A and 10 g of anhydrous sodium carbonate in a 500-ml round-bottomed flask equipped with a magnetic stirrer, an oil bath and a reflux condenser attached to a drying tube.

(b) Add 365 mg of cyanuric chloride to the mixture and reflux it at 80°C for 44 h with vigorous stirring.
(c) Stop the refluxing and stirring, and decant the benzene solution into a 500-ml beaker.
(d) Precipitate crude activated PEG_2 by the addition of 200 ml of petroleum ether to the solution and then collect the precipitate by filtration.
(e) Dissolve the precipitate in 100 ml of dry benzene and repeat step (4) six times to remove unreacted cyanuric chloride and its hydrolysates.

Determination of the chloride content of the activated PEG_2 prepared in this way demonstrated that two of the three chloride atoms in the cyanuric chloride molecule were substituted with monomethoxypolyethylene glycol (13).

(ii) *Modification of a protein with activated PEG_2.*

(a) Dissolve 20 mg of the protein in 3.2 ml of 0.4 M borate buffer (pH 10.0). Add activated PEG_2 (a 20 fold molar excess over the total amino groups of the protein).
(b) Incubate the mixture at 37°C for 1 h with gentle stirring.
(c) Dilute the mixture with 100 ml of 0.1 M borate buffer (pH 8.0).
(d) Remove the unreacted PEG by ultrafiltration 10 times with Amicon Diflo PM-30 membrane against the borate buffer.
(e) Dialyse the sample solution against distilled water and recover the modified protein by lyophilization.

By this procedure, 55% of the total seven amino groups in lipoprotein lipase from *Pseudomonas fluorescens* were modified with activated PEG_2 (16).

4.3 Thiol groups

Thiol groups of cysteine residues of proteins can be modified by almost as many procedures as can amino groups. The thiol group is the strongest nucleophile among the functional groups of amino acids, so the specificities of many reagents to thiol groups are generally high. In *Table 2*, representative reactions of thiol groups are listed. Various reagents may be designed based on these reactions (79). The number of thiol groups of a protein is often determined spectrophotometrically after the reaction with 5,5′-dithiobis(2-nitrobenzoic acid) (74), 2,2′-dithiodipyridine (or 4,4′-dithiodipyridine) (75) or with *N*-ethylmaleimide (73).

While some proteins contain only thiol groups of cysteine residues, many proteins contain disulphide bonds between them (known as cystine residues). Disulphide bonds can be reduced to thiol groups by excess thiol reagents, although the native secondary and tertiary structures as well as biological activities of proteins are often lost by the reduction. Reduction of disulphide bonds and blocking of the resultant thiol groups are often used for determination of the amino acid sequences of disulphide-containing proteins. Artificial thiol groups can also be introduced into proteins by chemical modification.

This section describes some chemical modifications of thiol groups, reduction of disulphide bonds and introduction of artificial thiol groups.

Table 2. Reactions of thiol groups of proteins.

Reaction	*Reagent (Example)*	*Product*	*Reference*
Metal binding	HOCO–C_6H_4–HgCl (PCMB)	R–S–Hg–C_6H_4–COOH	73
Mixed disulphide formation	O_2N–C_6H_3(COOH)–S–S–C_6H_3(COOH)–NO_2 (DTNB)	R–S–S–C_6H_3(COOH)–NO_2	74
	2-pyridyl–S–S–2-pyridyl, 4-pyridyl–S–S–4-pyridyl (Dithiodipyridine)	R–S–S–2-pyridyl, R–S–S–4-pyridyl	75
Alkylation	X—CH_2COOH (X = Br or I)	R—S—CH_2COOH	50
	CH_3I	R—S—CH_3	76
	aziridine (C_2H_4NH)	R—S—$CH_2CH_2NH_2$	77
	N-ethylmaleimide, NCH_2CH_3 (NEM)	R–S–succinimide–NCH_2CH_3	73
Cyanylation	O_2N–C_6H_3(COOH)–SCN (NTCB)	R—S—CN	78
Oxidation	Many Oxidants	R—S—S—R, R—SOH, R—SO_2H, R—SO_3H	

4.3.1 *Reduction and S-carboxymethylation of disulphide-containing proteins*

The original procedure for the preparation of completely reduced (by mercapto-ethanol) and *S*-carboxymethylated (by iodoacetic acid) proteins was described by Crestfield *et al.* (31). The following procedure is a slightly modified one and is often used in our laboratory for the reduction and *S*-carboxymethylation of hen egg lysozyme.

(i) Prepare an 8 M urea solution (0.575 M Tris–HCl buffer at pH 8.6 containing 8 M urea and 5.25 mM EDTA) using de-ionized urea.
(ii) Dissolve the protein (1–50 mg) in 2 ml of the 8 M urea solution in a 20-ml egg-shaped flask equipped with a stopper.
(iii) Flush N_2 into the flask to remove the air and add 25 μl of 2-mercaptoethanol; flush with N_2 again, close the flask with the stopper, and then incubate the flask at 40°C for 0.5–2 h.
(iv) Add a freshly prepared solution of 67 mg of iodoacetic acid in 250 μl of 1 M NaOH and keep the mixture at room temperature for 15 min in the dark.
(v) Remove the excess reagents by gel filtration of the mixture through a 1.5 × 60-cm column of Sephadex G-25 equilibrated with 10–50% acetic acid or by exhaustive dialysis against distilled water in the dark. The reduced and *S*-carboxymethylated protein is recovered by lyophilization.

During the storage of the 8 M urea solution, cyanic acid may be formed owing to the decomposition of urea. Cyanate ion reacts with amino groups (*Table 1*) as well as with thiol groups (65) to form carbamylated products. However, upon reduction and *S*-carboxymethylation of lysozyme using the above 8 M urea solution that has been stored in a refrigerator for more than 1 month, we have never observed such side-reactions. The cyanate ion formed must be removed by reaction with the high concentrations of the amino group of Tris (0.575 M) in the 8 M urea solution.

Since thiol groups react with iodoacetic acid very rapidly under the conditions described above, the carboxymethylation is highly specific to thiol groups. Exclusion of light after the addition of iodoacetic acid is important to prevent the formation of iodine which may react with tyrosine, tryptophan and histidine residues.

The number of *S*-carboxymethylcysteine residues can be determined by amino acid analysis (31).

4.3.2 *Introduction of thiol groups (thiolation)*

Introduction of artificial thiol groups into proteins may be useful because the resultant thiol groups can be variously modified for a variety of purposes. In the following, the procedure for thiolation of proteins by modification of amino groups with *N*-succinimidyl 3-(2-pyridyldithio)propionate (SPDP) (80) is described.

SPDP was synthesized by Carlsson *et al.* in 1978 (80) as a bifunctional reagent to introduce thiol groups into proteins. This reagent contains one *N*-hydroxy-

Scheme IV

$$RNH_2 + \text{(succinimidyl)}NOC(=O)CH_2CH_2SS\text{-(2-pyridyl)} \longrightarrow RNHC(=O)CH_2CH_2SS\text{-(2-pyridyl)} + \text{(succinimidyl)}NOH$$

$$RNHC(=O)CH_2CH_2SS\text{-(2-pyridyl)} \xrightarrow{2R^1SH} RNHC(=O)CH_2CH_2SH + R^1SSR^1 + S\text{=(2-thiopyridone)}$$

$$\downarrow R^1SH$$

$$RNHC(=O)CH_2CH_2SSR^1 + S\text{=(2-thiopyridone)}$$

succinimide ester moiety, which reacts with amino groups to give stable amide bonds, and one 2-pyridyl disulphide moiety, which is reduced to the thiol when it is needed or reacts with aliphatic thiols to form aliphatic disulphides (*Scheme IV*). The reagent is now commercialy available from Pharmacia.

The modification of amino groups of a protein with SPDP is carried out as follows (80).

(i) Dissolve the protein (0.05–2 μmol) in 2 ml of 0.1 M sodium phosphate buffer containing 0.1 M NaCl, pH 7.5, in a 10-ml vial containing a magnetic stirring bar.
(ii) Depending on the degree of modification desired, add 0.01–0.3 ml of SPDP (20 mM in ethanol) dropwise to the stirred protein solution.
(iii) Leave the mixture at room temperature (23°C) for 30 min with stirring.
(iv) Remove the excess reagent by gel filtration on Sephadex G-25 in 0.1 M sodium phosphate buffer containing 0.1 M NaCl, pH 7.5.
(v) Store the modified protein at 4°C.

The modified protein contains 2-pyridyl disulphide groups. The content of these disulphide groups in the protein is determined by adding 0.1 ml of 50 mM dithiothreitol (DTT) in distilled water to a properly diluted protein solution (2.0 ml). The amount of 2-thiopyridone (or 2-mercaptopyridine) released, which is equivalent to the content of 2-pyridyl disulphide groups in the protein is calculated using its molar absorbance of $8.08 \times 10^3\ M^{-1}\ cm^{-1}$ at 343 nm.

The pyridyl disulphide groups in the modified protein can be converted to thiol groups by reduction with DTT, usually without concomitant reduction of native protein disulphide bonds. The procedure is as follows (80).

(i) Change the buffer of the modified protein to sodium acetate buffer, pH 4.5, containing 0.1 M NaCl, by dialysis or gel filtration.
(ii) Add DTT dissolved in the same buffer to the protein solution to a final concentration of 25 mM; leave the mixture at room temperature for 30–40 min.

(iii) Remove the excess DTT and 2-thiopyridone by gel filtration.

The thiol groups thus produced are very reactive and can take part in unwanted reactions. Therefore, store the modified protein in the pyridyl disulphide form and reduce it just before the protein thiol is used. The pyridyl disulphide groups are very stable and no decrease of the pyridyl disulphide content in the modified protein may be observed after storage at pH 7.5 and 4°C for 3 months.

Besides SPDP, methyl 4-mercaptoalkanimidates (81), 2-iminothiolane (82) and *N*-acetylhomocysteine thiolactone (83) are also employed for thiolation of proteins by the modification of amino groups (*Scheme V* and Chapter 5). Recently, a method for thiolation of proteins by the modification of tryptophan residues with 2,4-dinitrobenzenesulphenyl chloride followed by reduction (*Scheme V*) has been developed (84).

Scheme V

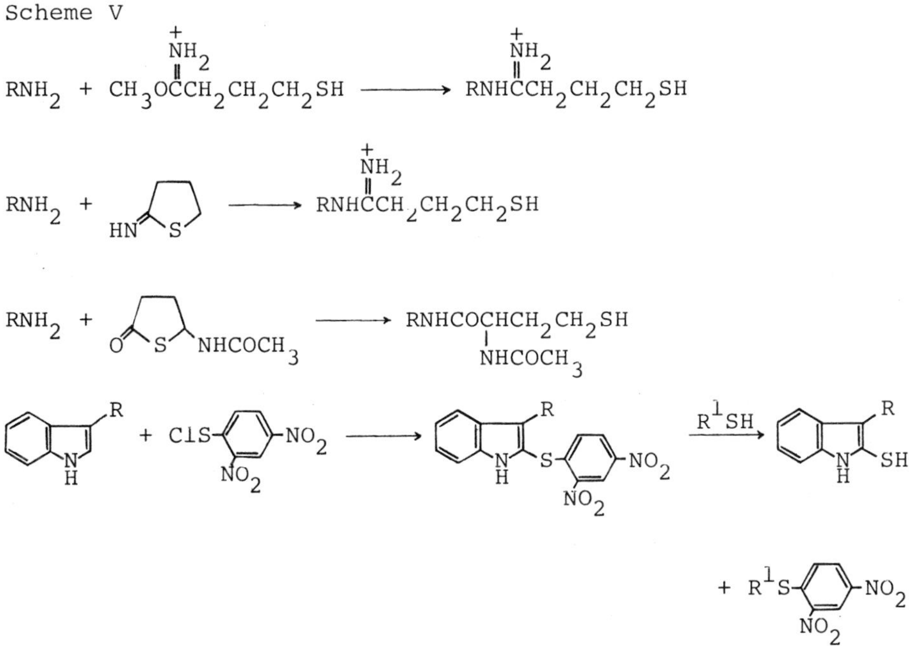

$RNH_2 + CH_3OC(=\overset{+}{N}H_2)CH_2CH_2CH_2SH \longrightarrow RNHC(=\overset{+}{N}H_2)CH_2CH_2CH_2SH$

RNH_2 + (2-iminothiolane: HN, S) $\longrightarrow RNHC(=\overset{+}{N}H_2)CH_2CH_2CH_2SH$

RNH_2 + (N-acetylhomocysteine thiolactone: O, S, $NHCOCH_3$) $\longrightarrow RNHCOCH(NHCOCH_3)CH_2CH_2SH$

(indole, R, N–H) + ClS–C₆H₃(NO₂)–NO₂ $\longrightarrow$ (2-(2,4-dinitrophenylthio)indole, R, N–H) $\xrightarrow{R^1SH}$ (2-mercaptoindole, R, N–H, SH) + R^1S–C₆H₃(NO₂)–NO₂

4.3.3 *Disulphide bond formation*

2-Pyridyl disulphide groups introduced into proteins can react with thiol-reagents or thiol-proteins (for example, native thiols or chemically introduced thiols of proteins) via thiol–disulphide exchange to form disulphide-linked protein–reagent or protein–protein conjugates. Since some protein–protein conjugates may be useful for the radio- and enzyme immunoassays of biologically important materials, a representative procedure for preparation of protein–protein conjugates by disulphide bond formation (80) is described.

(i) Dissolve 64 nmol of the 2-pyridyl disulphide form of the protein in 1.2 ml of 0.1 M sodium phosphate buffer, pH 7.5, containing 0.1 M NaCl.
(ii) Mix the solution with a solution of 68 nmol of the freshly-prepared thiol-containing protein in 0.9 ml of the same buffer and leave the mixture for 20 h at room temperature (23°C).
(iii) Isolate the protein–protein conjugate from non-conjugated proteins by gel filtration.

When a protein that contains only one 2-pyridyl disulphide group reacts with a protein that contains only one thiol group, usually only one thiol–disulphide exchange will occur, leading to the disulphide-linked conjugate between two proteins. More complex conjugates may be prepared by using proteins with more than one 2-pyridyl disulphide and thiol group, respectively.

4.4 **Tyrosine residues (nitration)**

Tyrosine contains a phenolic hydroxyl group, the p*K* of which is normally 10.1 (85). Since the ionized form, a phenoxide ion, is electron rich, aromatic rings of tyrosine residues of proteins can be modified with electrophiles (electrophilic substitution). Tetranitromethane (TNM) can nitrate tyrosine residues under very mild conditions to form 3-nitrotyrosine residues (86). 3-Nitrotyrosine is coloured, with a p*K* of about 7.2, and the absorption spectrum is pH-dependent (86). The absorption maximum at 428 nm at alkaline pH with molar extinction coefficient (ε_{428}) of 4200 M^{-1} cm^{-1} shifts to 360 nm in acid with isosbestic point at 381 nm and ε_{381} = 2200 M^{-1} cm^{-1}. Since the generation in a protein of a chromophore that absorbs in the visible region of the spectrum, such as 3-nitrotyrosine, provides a number of convenient experimental approaches, the general procedure for the nitration of tyrosine residues of proteins with TNM is described here.

(i) Wash TNM (Aldrich Chemical Co.) with water to remove impurities and dilute it with 10 volumes of 95% ethanol (to 0.84 M).
(ii) Dissolve the protein (to $\sim 10^{-4}$ M) in 0.05 M Tris–HCl buffer, pH 8.0, in a vial or small beaker containing a magnetic stirring bar.
(iii) Add the TNM solution (1–10 equivalents of tyrosine residues) to the vigorously stirred protein solution, and continue stirring for 1 h at room temperature.
(iv) Terminate the reaction by passing the mixture through a column of Bio-Gel P-4 in 0.05 M Tris–HCl buffer at pH 8.

The degree of nitration is determined by the absorption at 381 nm (ε = 2200 M^{-1} cm^{-1}), the isosbestic point of ionized and non-ionized 3-nitrotyrosine residues. Amino acid analysis can be employed as another method to quantify the nitration of tyrosine residues. 3-Nitrotyrosine elutes from the standard amino acid analyser as a discrete peak just after phenylalanine.

Under the conditions described above, TNM also oxidizes thiol groups (87). For their protection, thiols are converted to *S*-sulphenylsulphonates by the treatment of the proteins with sodium tetrathionate. After nitration, thiols are regenerated by thiol reagents such as mercaptoethanol and DTT (88–90).

In some cases, TNM induces polymerization of proteins due to inter-molecular coupling of tyrosine residues (91). Therefore, check the formation of polymerized proteins by gel filtration or by SDS–polyacrylamide gel electrophoresis.

Nitration of tyrosine residues with TNM is dependent upon the pH, with little or no modification at pH 6 and very rapid modification at pH 9. Nitration and thiol oxidation can often be distinguished by performing the reaction at pH 6, where only oxidation takes place (86).

Detailed investigations of the nitration of phenol model compounds by TNM suggested the following mechanism (91). TNM forms a charge-transfer complex with phenoxide ion in aqueous solution, thus accounting for the pH dependence of the reaction. This is followed by a rate-determining electron transfer to generate the phenoxide (XPhO·) and nitrite ($NO_2^{\cdot}$) radicals plus the nitroformate anion [$C(NO_2)_3^-$]. Nitration results from the coupling of $NO_2^{\cdot}$ and XPhO·, but an additional reaction is made possible by the coupling of phenoxide radicals with each other, accounting for the formation of the polymerized proteins.

Since a protein with fully exposed tyrosine residues would tend to polymerize, this would be reduced by carrying out the reaction on dilute solutions of the protein. When the nitration leads to de-stabilization of the protein, the unfolded protein would also tend to polymerize. In this case, addition of the substrate or an analogue to the reaction medium to stabilize the native conformation should reduce the polymerization. The following provides an example.

Hen egg lysozyme contains three tyrosine residues, Tyr20, 23 and 53. These tyrosine residues were sequentially nitrated in the order Tyr23, Tyr20 and Tyr53 (92). The mono-nitrotyrosyl (Tyr23), di-nitrotyrosyl (Tyr20,23) and tri-nitrotyrosyl (Tyr20,23,53) lysozymes could be separated by ion-exchange chromatography on Bio-Rex 70 at pH 10. When lysozyme (100 mg, 7 μmol) was nitrated with 1.9 times the molar amount of TNM (2.6 mg, 13.3 μmol) in 10 ml of 0.05 M Tris–HCl buffer at pH 8 for 2.5 h, 36% of mono-nitrotyrosyl lysozyme was obtained with formation of a considerable amount of polymerized lysozyme which was precipitated during the chromatography. When 215 mg of lysozyme (15 μmol) was reacted with 13.2 times the molar quantity of TNM (39 mg, 199 μmol) in 20 ml of 0.05 M Tris–HCl buffer at pH 8 for 7 h, di-nitrotyrosyl and tri-nitrotyrosyl lysozymes were obtained in yields of 11 and 5%, respectively, and most of the remaining protein was found to be polymerized. When the latter reaction was carried out in the presence of 10 mg of the mixture of the dimer and trimer of *N*-acetylglucosamine [$(GlcNAc)_n$, $n = 2$ and 3; $(GlcNAc)_2/(GlcNAc)_3 = 3/7$], substrate analogues of lysozyme, the formation of polymerized lysozyme was reduced, accompanied by increased formation of di-nitrotyrosyl lysozyme (32%) and tri-nitrotyrosyl lysozyme (9%) (unpublished results).

The nitotyrosyl residue is a potential site for derivatization because it can be reduced to an amino group using sodium hydrosulphite ($Na_2S_2O_4$) (93). When a 5- to 6-fold molar excess of $Na_2S_2O_4$ is added to nitrotyrosyl derivatives at pH values between 6 and 9, the yellow colour of nitrotyrosine disappears within a few minutes to form the aminotyrosyl derivative as a major product. The introduction of aminophenols into the primary sequences of proteins provides the opportunity for further specific chemical modification at the newly formed amino groups. 3-Aminotyrosine can be quantified by amino acid analysis (93).

4.5 **Tryptophan residues (oxidation)**

The indole ring of the tryptophan residue is one of the reactive functional groups in proteins and is modified with various electrophilic reagents and oxidants. While tryptophan possesses a large hydrophobic side-chain and its content in proteins is generally low, tryptophan residues are often found on the surface of proteins. This suggests that tryptophan residues are often involved in protein function. In consequence, selective modification of particular tryptophan residues in a protein is generally desired and cases of complete modification are rare. *Table 3* provides some examples of the modification of tryptophan residues in hen egg lysozyme, which contains six tryptophan residues. The reactions listed in *Table 3* may indicate the nature of tryptophan residues.

This section describes the oxidation of tryptophan residues by several reagents.

4.5.1 *N-Bromosuccinimide* (NBS)

N-Bromosuccinimide (NBS), which was introduced in 1958 by Patchornik *et al.* (101), in acidic medium oxidizes the indole groups of tryptophan residues in proteins to oxindole groups. In the case of hen egg lysozyme, Trp62 was selectively oxidized to oxindolealanine62 by NBS (97). The following procedure is that slightly modified for the preparation of oxindolealanine62 lysozyme.

(i) Dissolve the protein (10 μmol) in 20 ml of water in a small beaker containing a magnetic stirring bar, and adjust the pH to 4.5 with dilute HCl.
(ii) Cool the solution in an ice-water bath with stirring.
(iii) Add 1 ml of an aqueous solution containing 11 μmol of NBS dropwise to the vigorously stirred protein solution, and continue stirring for 15 min.
(iv) Dialyse the mixture against distilled water.

In many cases, the oxidation of tryptophan residues with NBS is carried out in acetate buffer at pH 3–6 (102). When acetate buffer is employed, however, it may produce δ_1-acetoxytryptophan residues, which must be hydrolysed to oxindole-alanine residues by prolonged dialysis or incubation at neutral pH and room temperature after the reaction. In the case of lysozyme, the δ_1-acetoxytrypto-phan62 derivative was obtained as an intermediate when 0.1 M acetate buffer (pH 4.5) was used, and it gradually decomposed to oxindolealanine62 lysozyme at pH 3.9 and room temperature, with a half-time of 44 h (103).

When excess NBS is used, it may cause the cleavage of peptide bonds (104) as well as bromination or oxidation of methionine, tyrosine, histidine and lysine residues. Therefore, the modification with NBS is most suitable for limited oxidization of the exposed tryptophan residues of proteins.

Table 3. Modification of tryptophan residue (examples in hen egg lysozyme).

Reagent	*Condition*	*Product*	*Modified residue and activity*	*Reference*
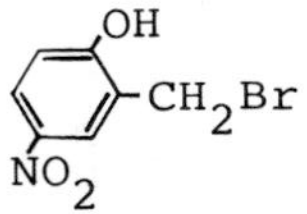 OH, CH_2Br, NO_2	2.5 mol equivalent, pH 2.8	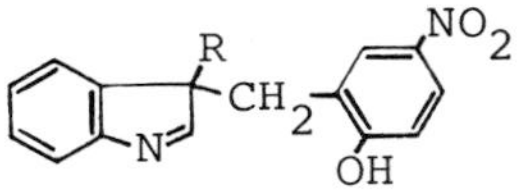 R, CH_2, NO_2, OH, N	Trp62, inactivation, reversible	94
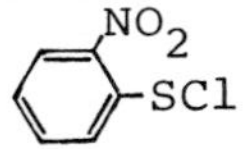 NO_2, SCl	40 mol equivalent, pH 3.5	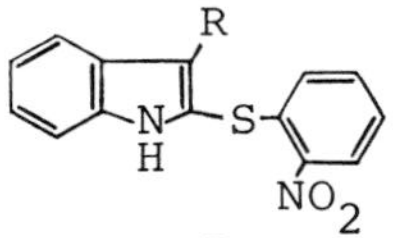 R, N, H, S, NO_2	Trp62, inactivation	95
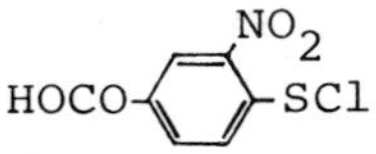 HOCO, NO_2, SCl	1 mol equivalent, 25% acetic acid	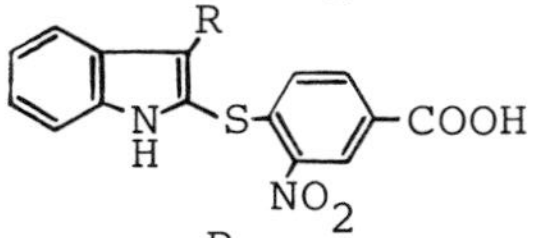 R, N, H, S, COOH, NO_2	Trp108, inactivation	96
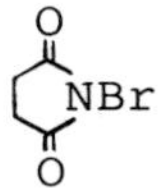 O, NBr, O	1 mol equivalent, pH 4.5	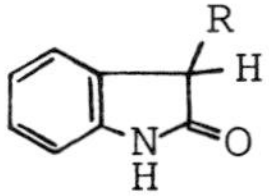 R, H, N, H, O	Trp62, inactivation	97
I_3^-	1/2 mol equivalent, pH 5.5	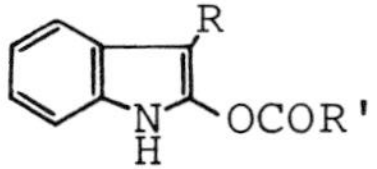 R, N, H, OCOR'	Trp108, inactivation R'COOH = Glu53	98
O_3	H_2O (pH 5.5)	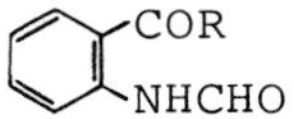 COR, NHCHO	Trp62, inactivation	99
		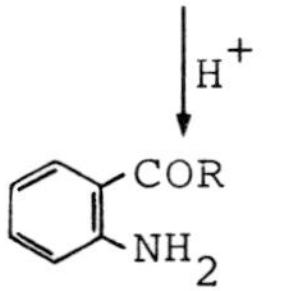 H^+ ↓ COR, NH_2	reactivation (60–70%)	100

Oxindolealanine62 lysozyme, as prepared in the procedure described above, was relatively pure, but it may contain a trace amount of unreacted native lysozyme. Native lysozyme was not removed by ion-exchange chromatography but it could be removed by passing the mixture through a column of chitin-coated Celite, an affinity adsorbent for lysozyme (3). Trp62 is involved in the substrate binding and its oxidation to the oxindolealanine residue greatly reduces the substrate-binding ability (7, 105).

4.5.2 *Ozone*

Ozone oxidizes tryptophan residues to *N'*-formylkynurenine residues. Although ozone also oxidizes cysteine, cystine, methionine, tyrosine and histidine residues, low concentrations of ozone reduce the side-reactions except for the oxidation of cysteine and methionine residues. In the case of hen egg lysozyme, the reaction was carried out in either water (99) or in formic acid (106) leading to the respective modifications of Trp62 (99) or Trp108 and Trp111 (106). *N'*-Formylkynurenine62 lysozyme was converted to kynurenine62 lysozyme by hydrolysis of the formamide bond in dilute HCl at −8 to −10°C (100).

4.5.3 *Iodine*

The oxidation of indole to oxindole by iodine was catalysed by a carboxylic acid, via formation of an ester intermediate (107). Glu35 and Trp108 are the sole vicinal carboxyl–indole pair in hen egg lysozyme, and the derivative cross-linked between Glu35 and oxidized Trp108 (oxindolealanine108) to an ester [(35–108)CL lysozyme] was prepared by iodine oxidation (98). The procedure employed for this conversion is described below.

(i) Prepare a tri-iodide solution (0.04 M I_2–0.48 M KI in water).
(ii) Dissolve hen egg lysozyme in water (10–20 mg/ml) and place the solution on a pH-stat.
(iii) Add the tri-iodide solution (0.5–0.6 mol I_2/mol of protein) in five portions to the lysozyme solution with stirring at room temperature; maintain the pH at 5.5 by the addition of 1 M NaOH. Allow the reaction of each tri-iodide addition to go to completion before adding the next, as judged by base uptake and disappearance of iodine colour.
(iv) After the reaction (3–4 h), dialyse the mixture against distilled water and recover the protein by lyophilization.

(35–108)CL lysozyme was isolated by ion-exchange chromatography of the products on Bio-Rex 70 at pH 10 in 25% yield based on lysozyme (or in 50% yield based on iodine). Iodinated lysozyme (20% of total lysozyme) and unreacted native lysozyme (55%) were also isolated by the ion-exchange chromatography.

The ester intermediate isolated in this way can be converted to oxindolealanine108 lysozyme as follows.

(i) Dissolve 50 mg of (35–108)CL lysozyme in 5 ml of 3.7 M GdmCl and lower the pH to 1.0 with 6 M HCl.

(ii) Incubate the solution at 50°C for 20 min, and then dialyse it against distilled water exhaustively.

4.6 **Intra-molecular cross-linking**

Introduction of inter-molecular or intra-molecular cross-links are useful tools to study the mutual orientations of subunits in oligomeric proteins (Chapter 5) or the distances between the particular residues in folded proteins. Intra-molecular cross-links are also useful to stabilize folded proteins. In this respect, many bifunctional reagents have been prepared thus far and are reviewed by Wold (108).

Selective introduction of an intra-molecular cross-link into a protein is generally very difficult. To do so, we must first know the nature and the reactivities of the functional groups of the protein involved, and then we must design or select the reaction and/or the bifunctional reagent suitable to the individual protein. *Table 4* lists some successful examples of the selective introduction of the intra-molecular cross-links into proteins. In this section, we show how we designed the reaction and reagents to introduce intra-molecular cross-links into hen egg lysozyme as examples. The cross-linking between Glu35 and Trp108 has already been described in Section 4.5.3.

4.6.1 *Cross-linking of lysozyme between Lys13 (ε-amino) and Leu129 (α-carboxyl) by imidazole-catalysed carbodiimide reaction* (4, 29, 34)

As was shown in Section 4.1.1 (i), the water-soluble carbodiimide EDC catalyses the formation of amide bonds between protein carboxyl and amino groups. Since hen egg lysozyme contains a salt bridge between the ε-amino group of Lys13 and the α-carboxyl group of Leu129 (114), we tried to cross-link this salt bridge in an amide bond by reaction with EDC. In the absence of amine nucleophiles, EDC catalyses the undesirable polymerization of lysozyme due to the formation of the inter-molecular amide bonds (28). Therefore, the strategy was to activate the carboxyl group of Leu129 and simultaneously to protect the other carboxyl groups by EDC reaction.

We found first that this salt bridge could be cross-linked by the reaction with EDC in the presence of 1 M imidazole at pH 5, followed by dialysis at pH 10 (29). The presence of imidazole prevented the formation of lysozyme polymer and increased the yield of the intra-molecularly cross-linked lysozyme. This suggests the following role of imidazole (29).

EDC activates some of the carboxyl groups of lysozyme as shown before [Section 4.1.1 (i)]. In the absence of imidazole, the activated carboxyls often react with amino groups of another lysozyme molecule to form polymeric lysozyme. In the presence of imidazole, the activated carboxyls are converted to acylimidazole groups. The reactivity of the acylimidazole group is lower than that of the EDC-activated carboxyl group, so lysozyme is protected from the polymerization by imidazole. Acylimidazole groups were hydrolysed during the incubation at pH 10. Further, a high concentration of imidazole increases the ionic strength of the

Table 4. Intra-molecular cross-linking of proteins.

Protein	*Reagent*	*Cross-linked residues and structure*	*Reference*
Papain	$ClCH_2COCH_2Cl$	Cys25—S—CH_2COCH_2—His159	109
RNase A	1,5-difluoro-2,4-dinitrobenzene (F, F, O_2N, NO_2)	Lys7-NH-[$C_6H_2(NO_2)_2$]-NH-Lys41 (O_2N, NO_2)	110
Muscle glyceraldehyde-3-phosphate dehydrogenase	1,5-difluoro-2,4-dinitrobenzene (F, F, O_2N, NO_2)	Cys149-S-[$C_6H_2(NO_2)_2$]-NH-Lys183 (O_2N, NO_2)	111
α-Subunit of *Escherichia coli* tryptophan synthetase	1,5-difluoro-2,4-dinitrobenzene (F, F, O_2N, NO_2)	Cys80-S-[$C_6H_2(NO_2)_2$]-NH-Lys108 (O_2N, NO_2)	112
Bovine pancreatic trypsin inhibitor	EDC	Ala58(α)CONH(α)Arg1	113
Lysozyme	I_3^-	indole (N–H) with Trp108 and OCO-Glu35	114
	EDC + imidazole	Lys13—NHCO—Leu129(α)	4, 29, 34
	$BrCH_2CONH$-$(CH_2)_n$-$NHCOCH_2Br$ (n = 2, 4, 6)	Lys1(ε)-NH-CH_2CONH-$(CH_2)_n$-$NHCOCH_2$-N(imidazole)-His15	5

reaction medium, which weakens the salt bridge and facilitates activation of the α-carboxyl group of Leu129 by EDC to increase the formation of the amide bond between Lys13 and Leu129.

In the course of study of this cross-linking reaction, we found a serious side-reaction where the α-peptide bond between Asp101 and Gly102 was partly inverted to a β-peptide bond (4). That is, about two thirds of the protein (both the non-cross-linked and cross-linked lysozymes) contained the β-aspartylglycyl sequence at Asp101-Gly102. The hydrolysis of the acylimidazole group at Asp101 was considered to accompany this peptide bond inversion because the next residue is glycine (Gly102). Therefore, in this reaction, three products were formed, cross-linked lysozyme [(13–129)CL lysozyme], cross-linked lysozyme with the β-aspartyl sequence at Asp101 [101-β-(13–129)CL lysozyme] and non-cross-linked lysozyme with the β-aspartyl sequence at Asp101 (101-β lysozyme). The β derivatives were indistinguishable from the α by routine separation methods, and we could separate them only by affinity column chromatography (*Figure 2*) (4). Since we knew that $(GlcNAc)_3$ hindered the activation of Asp101 with EDC (115), the reaction was finally carried out in the presence of $(GlcNAc)_3$ to reduce the side-reaction (34).

The following procedure is that used for the preparation of (13–129)CL lysozyme after the trials described above (4, 29, 34).

(i) Dissolve 100 mg of lysozyme and 15 mg of $(GlcNAc)_3$ in 10 ml of 1 M imidazole–HCl buffer, pH 5, in a vial containing a magnetic stirring bar.
(ii) Add 32 mg of EDC to the protein solution with stirring at room temperature, to initiate the reaction.
(iii) After stirring for 1 day dialyse the reaction mixture against 0.02 M borate buffer, pH 10, at room temperature for 4–5 h, and then against distilled water exhaustively.
(iv) Separate the cross-linked derivatives from the non-cross-linked by ion-exchange chromatography on Bio-Rex 70 at pH 10 and room temperature (a 1×200-cm column eluted with a gradient of 1 litre of 0.02 M borate buffer and 1 litre of the same buffer containing 0.16 M NaCl; the cross-linked derivatives appear as a single peak just after the peak of those non-cross-linked).
(v) After dialysis and concentration of the cross-linked derivatives, load the sample onto a 6.5×10-cm column of chitin-coated Celite (3) which is pre-equilibrated with 0.1 M sodium acetate buffer containing 0.5 M NaCl, pH 5.5, at 4°C. Elute the column with a gradient from 1 litre of 0.1 M sodium acetate buffer containing 0.5 M NaCl to 270 ml of 2.5 M acetic acid. The 101-β-derivative will be eluted first and then (13–129)CL lysozyme.

4.6.2 *Cross-linking of lysozyme between Lys1 (ε-amino) and His15 with bis-bromoacetamide derivatives* (5)

Hen egg lysozyme contains a single histidine residue, His15, which is selectively alkylated with haloacetic acids and their derivatives at pH 5.0–5.5 (44, 116–118). At pH 7–9, the ε-amino groups of lysozyme are also alkylated with iodoacetic acid

(117). Therefore, we tried to introduce a cross-link between His 15 and one of the ε-amino groups of lysine residues with bisbromoacetamide derivatives. The strategy was that His15 was alkylated monofunctionally first with one end of the bisbromoacetamide molecule at pH 5.5, excess bifunctional reagent was removed, and then the pH was raised to 9.0 to allow the other end of the reagent molecule to alkylate one of the ε-amino groups.

Although the ε-amino group of Lys96 is the closest amino group to His15 (7.8 Å), detailed study on the kinetics of the alkylation of His15 with bromoacetamide derivatives suggested that lysozyme has a small hydrophobic pocket that binds reagents in the direction from His15 to Lys1, the opposite direction to Lys96 (119). Thus, we prepared three bisbromoacetamide derivatives, $BrCH_2CONH(CH_2)_nNHCOCH_2Br$ (n = 2, 4 and 6), as bifunctional reagents for cross-linking two residues with a distance greater than 11 Å because the distance between His15 (nitrogen at ε_2) and Lys1 (ε-amino) is 11.8 Å (105). Employing these bifunctional reagents, we succeeded in the introduction of the cross-links into lysozyme between Lys1 and His15 (5).

The following procedure was employed for this cross-linking.

(i) Dissolve 300 mg of hen egg lysozyme in 10 ml of water in a jacketed titration vessel and adjust the pH to 5.5 with dilute acetic acid.
(ii) Place the vessel on a pH-stat and warm the solution to 40°C by circulating water.
(iii) Add $BrCH_2CONH(CH_2)_nNHCOCH_2Br$ (114 mg of the n = 2 reagent as a solid, 126 mg of the n = 4 reagent in 3 ml of ethanol or 75 mg of the n = 6 reagent in 4 ml of ethanol) to the protein solution with stirring, and maintain the pH at 5.5 by the addition of 0.2 M NaOH with the aid of the pH-stat for 6 (n = 2), 30 (n = 4) or 40 h (n = 6).
(iv) Remove the excess reagent by gel filtration through a Sephadex G-25 column (3 × 70 cm) eluted with 0.05 M KH_2PO_4.
(v) Dilute the protein fraction to 300 ml with 2 mM sodium tetraborate and adjust the pH to 9.0 with NaOH.
(vi) After incubation at room temperature for 1 day, preciptate the protein by adding 180 g of $(NH_4)_2SO_4$, collect the protein by centrifugation, re-dissolve it in about 5 ml of water and then pass the solution through a Sephadex G-100 column (1.5 × 150 cm) with 0.05 M phosphate buffer at pH 7.0 to remove any polymerized lysozyme.
(vii) Isolate the intra-molecularly cross-linked lysozyme by ion-exchange chromatography on Bio-Rex 70 (1.4 × 65-cm column) eluted with a gradient of 1 litre of 0.02 M borate buffer and 1 litre of the same buffer containing 0.15 M NaCl at pH 10.

5. REFERENCES

1. Walsh, C. T. (1984) *Annu. Rev. Biochem.*, **53**, 493.
2. Martinek, K. and Mozhaev, V. V. (1985) *Annu. Rev. Biochem.*, **57**, 179.
3. Yamada, H., Fukumura, T., Ito, Y. and Imoto, T. (1985) *Anal. Biochem.*, **146**, 71.
4. Yamada, H., Ueda, T., Kuroki, R., Fukumura, T., Yasukochi, T., Hirabayashi, T., Fujita, K. and Imoto, T. (1985) *Biochemistry*, **24**, 7953.

5. Ueda, T., Yamada, H., Hirata, M. and Imoto, T. (1985) *Biochemistry,* **24**, 6316.
6. Yamasaki, N., Hayashi, K. and Funatsu, M. (1968) *Agric. Biol. Chem.* (*Tokyo*), **32**, 55.
7. Imoto, T., Fujimoto, M. and Yagishita, K. (1974) *J. Biochem.,* **76**, 745.
8. Clark, P. I. and Lowe, G. (1978) *Eur. J. Biochem.,* **84**, 293.
9. Neet, K. E. and Koshland, D. E., Jr (1966) *Proc. Natl. Acad. Sci. USA,* **56**, 1606.
10. Polgar, L. and Bender, M. L. (1966) *J. Am. Chem. Soc.,* **88**, 3153.
11. Yokosawa, H., Ojima, S. and Ishii, S. (1977) *J. Biochem.,* **82**, 869.
12. Kuroki, R., Yamada, H., Moriyama, T. and Imoto, T. (1986) *J. Biol. Chem.,* **261**, 13571.
13. Kamisaki, Y., Wada, H., Yagura, T. and Matsushima, A. (1981) *J. Pharmacol. Exp. Ther.,* **216**, 410.
14. Lee, W. Y. and Sehon, A. H. (1977) *Nature,* **267**, 618.
15. Takahashi, K., Ajima, A., Yoshimoto, T. and Inada, Y. (1984) *Biochem. Biophys. Res. Commun.,* **125**, 761.
16. Inada, Y., Nishimura, H., Takahashi, K., Yoshimoto, T., Ranjan, A. and Saito, Y. (1984) *Biochem. Biophys. Res. Commun.,* **122**, 845.
17. Matsushima, A., Okada, M. and Inada, Y. (1984) *FEBS Lett,* **178**, 275.
18. Hirs, C. H. W. (ed) (1967) *Methods in Enzymology.* Vol. 11, Academic Press, New York.
19. Hirs, C. H. W. and Timasheff, S. N. (eds) (1972) *Methods in Enzymology.* Vol. 25, Academic Press, New York.
20. Hirs, C. H. W. and Timasheff, S. N. (eds) (1977) *Methods in Enzymology.* Vol. 47, Academic Press, New York.
21. Hirs, C. H. W. and Timasheff, S. N. (eds) (1983) *Methods in Enzymology.* Vol. 91, Academic Press, New York.
22. Hoare, D. G. and Koshland, D. E., Jr (1966) *J. Am. Chem. Soc.,* **88**, 2057.
23. Hoare, D. G. and Koshland, D. E., Jr (1967) *J. Biol. Chem.,* **242**, 2447.
24. Lin, T.-Y. and Koshland, D. E., Jr (1969) *J. Biol. Chem.,* **244**, 505.
25. Carraway, K. L. and Triplett, R. B. (1970) *Biochim. Biophys. Acta,* **200**, 564.
26. Perfetti, R. B., Anderson, C. D. and Hall, P. L. (1976) *Biochemistry,* **15**, 1735.
27. Carraway, K. L. and Koshland, D. E., Jr (1968) *Biochim. Biophys. Acta,* **160**, 272.
28. Timkovich, R. (1977) *Biochem. Biophys. Res. Commun.,* **74**, 1463.
29. Yamada, H., Kuroki, R., Hirata, M. and Imoto, T. (1983) *Biochemistry,* **22**, 4551.
30. Yamada, H., Imoto, T., Fujita, K., Okazaki, K. and Motomura, M. (1981) *Biochemistry,* **20**, 4836.
31. Crestfield, A. M., Moore, S. and Stein, W. H. (1963) *J. Biol. Chem.,* **238**, 622.
32. Canfield, R. E. (1963) *J. Biol. Chem.,* **238**, 2691.
33. Okazaki, K., Imoto, T., Yamada, H., Kuroki, R. and Fujita, K. (1982) *J. Biol. Chem.,* **257**, 12559.
34. Ueda, T., Yamada, H. and Imoto, T. (1987) *Protein Eng.,* **1**, 189.
35. Wang, R., Ozaki, T., Hayashi, K. and Funatsu, M. (1972) *Sci. Bull. Fac. Agric. Kyushu Univ.,* **27**, 1.
36. Imoto, T., Yamada, H., Yasukochi, T., Yamada, E., Ito, Y., Ueda, T., Nagatani, H., Miki, T. and Horiuchi, T. (1987) *Protein Eng.,* **1**, 333.
37. Kooistra, C. and Sluyterman, L. A. AE. (1987) *Intl. J. Peptide Protein Res.,* **29**, 357.
38. Bodlaender, P., Feinstein, G. and Shaw, E. (1969) *Biochemistry,* **8**, 4941.
39. Feinstein, G., Bodlaender, P. and Shaw, E. (1969) *Biochemistry,* **8**, 4949.
40. Frankel-Conrat, H. and Olcott, H. S. (1945) *J. Biol. Chem.,* **161**, 259.
41. Meerwein, H. (1973) In *Organic Syntheses*. Baumgarten, H. E. (ed. in chief), John Wiley and Sons, New York, Coll. Vol. 5, p. 1080.
42. Meerwein, H. (1973) In *Organic Syntheses*. Baumgarten, H. E. (ed. in chief), John Wiley and Sons, New York, Coll. Vol. 5, p. 1096.
43. Rawn, J. D. and Lienhard, G. E. (1974) *Biochem. Biophys. Res. Commun,* **56**, 654.
44. Parsons, S. M., Jao, L., Dahlquist, F. W., Borders, C. L., Jr, Groff, T., Racs, J. and Raftery, M. A. (1969). *Biochemistry,* **8**, 700.
45. Parsons, S. M. and Raftery, M. A. (1969) *Biochemistry,* **8**, 4199.
46. Wilcox, P. E. (1967) In *Methods in Enzymology*. Hirs, C. H. W. (ed.), Academic Press, New York, Vol. 11, p. 605.
47. Wilcox, P. E. (1972) In *Methods in Enzymology*. Hirs, C. H. W. and Timasheff, S. N. (eds), Academic Press, New York, Vol. 25, p. 596.
48. Reihm, J. P. and Scheraga, H. A. (1965) *Biochemistry,* **4**, 772.
49. Delpierre, G. R. and Fruton, J. S. (1965) *Proc. Natl. Acad. Sci. USA,* **54**, 1161.

50. Gurd, F. R. N. (1967) In *Methods in Enzymology*. Hirs, C. H. W. and Timasheff, S. N. (eds), Academic Press, New York, Vol. 25, p. 242.
51. Takahashi, K., Stein, W. H. and Moore, S. (1967) *J. Biol. Chem.*, **242**, 4682.
52. Yamada, H., Imoto, T. and Noshita, S. (1982) *Biochemistry*, **21**, 2187.
53. Kimmel, J. R. (1967) In *Methods in Enzymology*. Hirs, C. H. W. (ed.), Academic Press, New York, Vol. 11, p. 584.
54. Habeeb, A. F. S. A. (1972) In *Methods in Enzymology*. Hirs, C. H. W. and Timasheff, S. N. (eds), Academic Press, New York, Vol. 25, p. 558.
55. Ludwig, M. L. and Hunter, M. J. (1967) In *Methods in Enzymology*. Hirs, C. H. W. (ed.), Academic Press, New York, Vol. 11, p. 595.
56. Hunter, M. J. and Ludwig, M. L. (1972) In *Methods in Enzymology*. Hirs, C. H. W. and Timasheff, S. N. (eds), Academic Press, New York, Vol. 25, p. 585.
57. Inman, J. K., Perham, R. N., DuBois, G. C. and Appella, E. (1983) In *Methods in Enzymology*. Hirs, C. H. W. and Timasheff, S. N. (eds), Academic Press, New York, Vol. 91, p. 559.
58. Riordan, J. F. and Vallee, B. L. (1972) In *Methods in Enzymology*. Hirs, C. H. W. and Timasheff, S. N. (eds), Academic Press, New York, Vol. 25, p. 494.
59. Boyd, H., Leach, S. J. and Milligan, B. (1972) *Intl. J. Peptide Protein Res.*, **4**, 117.
60. Lindsay, D. G. and Shall, S. (1971) *Biochem. J.*, **121**, 737.
61. Riordan, J. F. and Vallee, B. L. (1967) In *Methods in Enzymology*. Hirs, C. H. W. (ed.), Academic Press, New York, Vol. 11, p. 565.
62. Klotz, I. M. (1967) In *Methods in Enzymology*. Hirs, C. H. W. (ed.), Academic Press, New York, Vol. 11, p. 570.
63. Butler, P. J. G. and Hartley, B. S. (1972) In *Methods in Enzymology*. Hirs, C. H. W. and Timasheff, S. N. (eds), Academic Press, New York, Vol. 25, p. 191.
64. Atassi, M. Z. and Habeeb, A. F. S. A. (1972) In *Methods in Enzymology*. Hirs, C. H. W. and Timasheff, S. N. (eds), Academic Press, New York, Vol. 25, p. 546.
65. Stark, G. R. (1967) In *Methods in Enzymology*. Hirs, C. W. H. (ed.), Academic Press, New York, Vol. 11, p. 590.
66. Hirs, C. H. W. (1967) In *Methods in Enzymology*. Hirs, C. H. W. (ed.), Academic Press, New York, Vol. 11, p. 548.
67. Yamada, H., Matsunaga, N., Domoto, H. and Imoto, T. (1986) *J. Biochem.*, **100**, 233.
68. Fields, R. (1972) In *Methods in Enzymology*. Hirs, C. H. W. and Timasheff, S. N. (eds), Academic Press, New York, Vol. 25, p. 464.
69. Means, G. E. (1977) In *Methods in Enzymology*. Hirs, C. H. W. and Timasheff, S. N. (eds), Academic Press, New York, Vol. 47, p. 469.
70. Bernad, A., Nieto, M. A., Vioque, A. and Palcian, E. (1986) *Biochim. Biophys. Acta*, **873**, 350.
71. Balls, A. K. and Wood, H. N. (1956) *J. Biol. Chem.*, **219**, 245.
72. Bornstein, P. and Balian, G. (1977) In *Methods in Enzymology*. Hirs, C. H. W. and Timasheff, S. N. (eds), Academic Press, New York, Vol. 47, p. 132.
73. Riordan, J. F. and Vallee, B. L. (1972) In *Methods in Enzymology*. Hirs, C. H. W. and Timasheff, S. N. (eds), Academic Press, New York, Vol. 25, p. 449.
74. Habeeb, A. F. S. A. (1972) In *Methods in Enzymology*. Hirs, C. H. W. and Timasheff, S. N. (eds), Academic Press, New York, Vol. 25, p. 457.
75. Grassetti, D. R. and Murray, J. F. (1967) *Arch. Biochem. Biophys.*, **119**, 41.
76. Jacobson, G. R. and Stark, G. R. (1973) *J. Biol. Chem.*, **248**, 8003.
77. Ruegg, U. T. and Rudinger, J. (1977) In *Methods in Enzymology*. Hirs, C. H. W. and Timasheff, S. N. (eds), Academic Press, New York, Vol. 47, p. 111.
78. Stark, G. R. (1977) In *Methods in Enzymology*. Hirs, C. H. W. and Timasheff, S. N. (eds), Academic Press, New York, Vol. 47, p. 129.
79. Kenyon, G. L. and Bruice, T. W. (1977) In *Methods in Enzymology*. Hirs, C. H. W. and Timasheff, S. N. (eds), Academic Press, New York, Vol. 47, p. 407.
80. Carlsson, J., Drevin, H. and Axen, R. (1978) *Biochem. J.*, **173**, 723.
81. Perham, N. R. and Thomas, J. O. (1971) *J. Mol. Biol.*, **62**, 415.
82. King, T. P. K., Li, Y. and Kochoumian, L. (1978) *Biochemistry*, **17**, 1499.
83. White, F. H., Jr (1972) In *Methods in Enzymology*. Hirs, C. H. W. and Timasheff, S. N. (eds), Academic Press, New York, Vol. 25, p. 541.
84. Wilchek, M. and Miron, T. (1972) *Biochem. Biophys. Res. Commun.*, **47**, 1015.
85. Edelhoch, H. (1962) *J. Biol. Chem.*, **237**, 2778.
86. Riordan, J. F. and Vallee, B. L. (1972) In *Methods in Enzymology*. Hirs, C. H. W. and Timasheff, S. N. (eds), Academic Press, New York, Vol. 25, p. 515.

87. Sokolovsky, M., Harell, D. and Riordan, J. F. (1969) *Biochemistry,* **8**, 4740.
88. Parker, D. J. and Allison, W. S. (1969) *J. Biol. Chem.,* **244**, 180.
89. Kassab, R., Roustan, C. and Praael, L.-A. (1968) *Biochim. Biophys. Acta,* **167**, 308.
90. Tsukamoto, S. and Ohno, M. (1978) *J. Biochem.* **84**, 1625.
91. Bruice, T. C., Gregory, J. J. and Walters, S. L. (1968) *J. Am. Chem. Soc.,* **90**, 1612.
92. Strosberg, A. D., Van Hoeck, B. and Kanarek, L. (1971) *Eur. J. Biochem.,* **19**, 36.
93. Sokolovsky, M., Riordan, J. F. and Vallee, B. L. (1967) *Biochem. Biophy. Res. Commun.,* **27**, 20.
94. Reiss, A. and Lukton, A. (1975) *Bioorg. Chem.,* **4**, 1.
95. Shechter, Y., Burstein, Y. and Patchornic, A. (1972) *Biochemistry,* **11**, 653.
96. Veronese, F. M., Boccu, E. and Fontana, A. (1972) *FEBS Lett,* **21**, 227.
97. Hayashi, K., Imoto, T., Funatsu, G. and Funatsu, M. (1965) *J. Biochem.* **58**, 227.
98. Imoto, T., Hartdegen, F. J. and Rupley, J. A. (1973) *J. Mol. Biol.,* **80**, 637.
99. Kuroda, M., Sakiyama, F. and Narita, K. (1975) *J. Biochem.,* **78**, 641.
100. Yamasaki, N., Tsujita, T., Sakiyama, F. and Narita, K. (1976) *J. Biochem.,* **80**, 409.
101. Patchornik, A., Lawson, W. B. and Witkop, B. (1958) *J. Am. Chem. Soc.,* **80**, 4747.
102. Spande, T. F. and Witkop, B. (1967) In *Methods in Enzymology.* Hirs, C. H. W. (ed.), Academic Press, New York, Vol. 11, p. 498.
103. North, R. S. and Allerhand, A. (1976) *Biochemistry,* **15**, 3438.
104. Ramachandran, L. K. and Witkop, B. (1976) In *Methods in Enzymology.* Hirs, C. H. W. (ed.), Academic Press, New York, Vol. 11, p. 283.
105. Imoto, T., Johnson, L. N., North, A. C. T. Phillips, D. C. and Rupley, J. A. (1972) In *The Enzymes.* Boyer, P. D. (ed.), Academic Press, New York, 3rd edn, Vol. 7, p. 665.
106. Previero, A., Coletti-Previero, M. A. and Jolles, P. (1967) *J. Mol. Biol.,* **24**, 261.
107. Imoto, T. and Rupley, J. A. (1973) *J. Mol. Biol.,* **80**, 657.
108. Wold, F. (1972) In *Methods in Enzymology.* Hirs, C. H. W. and Timasheff, S. N. (eds), Academic Press, New York, Vol. 25, p. 623.
109. Husain, S. S. and Lowe, G. (1968) *Chem. Commun.,* 310.
110. Marfey, S. P., Nowak, H., Uziel, M. and Yphantis, D. A. (1965) *J. Biol. Chem.,* **240**, 3264.
111. Shaltiel, S. and Tauber-Finkelstein, M. (1971) *Biochem. Biophys. Res. Commun.,* **44**, 484.
112. Hardman, J. K. and Hardman, D. F. (1971) *J. Biol. Chem.,* **246**, 6489.
113. Goldenberg, D. P. and Creighton, T. E. (1983) *J. Mol. Biol.,* **165**, 407.
114. Blake, C. C. F., Mair, G. A., North, A. C. T., Phillips, D. C. and Sarma, V. A. (1967) *Proc. R. Soc. Lond.,* Ser. B, **167**, 365.
115. Kuroki, R., Yamada, H. and Imoto, T. (1986) *J. Biochem.,* **99**, 1493.
116. Piszkiewicz, D. and Bruice, T. C. (1968) *Biochemistry,* **7**, 3037.
117. Kravchenko, N. A., Kleopina, G. V. and Kaverzneva, C. L. (1964) *Biochim. Biophys. Acta,* **92**, 412.
118. Goux, W. G. and Allerhand, A. (1979) *J. Biol. Chem.,* **254**, 2210.
119. Yamada, H., Uozumi, F., Ishikawa, A. and Imoto, T. (1984) *J. Biochem.,* **95**, 503.

CHAPTER 11

Site-directed mutagenesis, based on the phosphorothioate approach

JON R. SAYERS and FRITZ ECKSTEIN

1. INTRODUCTION

The most specific method of altering the covalent structure of a protein for structure–function studies is by mutagenesis of the gene coding for it (1, 2). In this way, any changes may be made to the protein primary structure that involve the 20 natural amino acids. Amino acid residues can be interchanged, deleted, or inserted, either individually or numerously. This approach requires that there be available a gene coding for the protein, either natural or synthetic, and an expression system for producing protein from the gene. The technology for cloning and expressing genes is outside the scope of this volume, but both are now standard procedures in molecular biology that are described in detail in very many places and are even available commercially in kits (e.g. Amersham International, Pharmacia LKB). Consequently, it is appropriate to include here a description of the application of these technologies toward studying protein structure and function using site-directed mutagenesis.

The basic principles of oligonucleotide-directed mutagenesis may be summarized as follows. The DNA of interest is cloned into the double-stranded form of a single-stranded bacteriophage such as one of the M13 series of phages (3). The DNA carrying the insert is then transfected into a bacterial cell line that amplifies the DNA, producing many copies of the viral (+) strand. These single-stranded copies are then packaged into phage and excreted from the cell. Such vectors are routinely used in DNA sequencing, as the single-stranded nature of the DNA lends itself to the Sanger dideoxy sequencing method (4).

A mismatch primer, a short oligonucleotide, is used to introduce the desired mutation. The primer is complementary to the target sequence except for the mutated, or mismatch, position. The single-stranded recombinant DNA is then annealed to the mismatch primer. The primer-annealed template is converted to a double-stranded closed circular DNA (RF IV) through the action of a suitable polymerase in the presence of the four deoxynucleoside triphosphates and a ligase. This DNA is a heteroduplex as it contains the desired mutant sequence on the viral (−) strand opposite the original wild-type sequence on the (+) strand, that is, it contains a mismatch. On transfection into a suitable cell line, mutant progeny with the mutation on both DNA strands may be obtained (5).

Unfortunately the percentage of phage progeny containing the desired mutation is low. As there are two DNA strands in the heteroduplex, it may seem reasonable to assume that 50% of the phage progeny should carry the desired mutation. The bacterial cell lines used for the transformation process, however, possess repair mechanisms capable of recognizing the (+) strand and are able to discriminate against the mutant (−) strand. This is due to *in vivo* methylation of the viral (+) strand. The *in vitro* generated (−) strand carries no such methylated bases. This reduces the mutational efficiency to a low level, typically less than 10% (6). Low mutational efficiency means that large numbers of putative mutant clones must be screened in order to obtain a clone with the required sequence. This involves a lot of effort in the form of hybridization screening or the sequencing of large numbers of clones. Thus, high efficiency site-directed mutagenesis is a desirable technique.

The phosphorothioate-based oligonucleotide-directed mutagenesis method (now also available as a kit from Amersham International), overcomes the problems associated with transfection of a heteroduplex species (7–11). This is achieved by destroying the wild-type sequence opposite the mismatch primer and repairing the DNA *in vitro*. Thus, the resultant mutant DNA is transfected into competent cells as a fully complementary homoduplex species ensuring high mutational efficiencies.

2. DEVELOPMENT OF THE PHOSPHOROTHIOATE METHODOLOGY

The phosphorothioate-based oligonucleotide-directed mutagenesis method is based on the observation that certain restriction endonucleases are incapable of hydrolysing phosphorothioate inter-nucleotidic linkages (12, 13). Thus, double-stranded DNA containing phosphorothioate linkages in one strand only may be nicked in the non-substituted strand. In our mutagenesis procedure, the mismatch oligonucleotide primer is annealed to the (+) strand of a single-stranded circular phage DNA. The primer is extended by a polymerization reaction in which one of the natural deoxynucleoside triphosphates is replaced by the corresponding deoxynucleoside 5′-*O*-(1-thiotriphosphate), dNTPαS (*Figure 1*). Thus, phosphorothioate groups are incorporated exclusively in the (−) strand of the newly synthesized RF IV DNA. This results in a strand asymmetry that may be exploited.

Reaction of such DNA with one of several restriction enzymes (e.g. *Nci*I) produces a nick in the (+) strand, as this enzyme is unable to cleave the (−) strand carrying a phosphorothioate group at the position of cleavage. The nick in the (+)

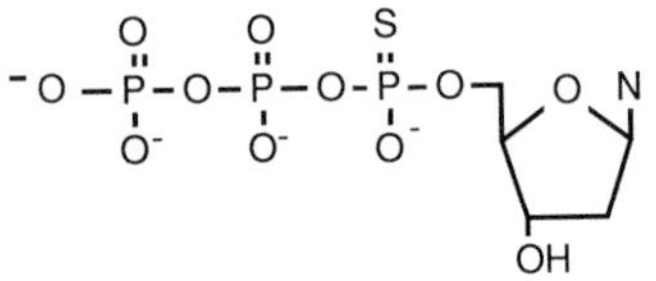

Figure 1. Structure of a deoxynucleoside 5′-*O*-(1-thiotriphosphate), dNTPαS. The sulphur atom replaces a non-bridging oxygen.

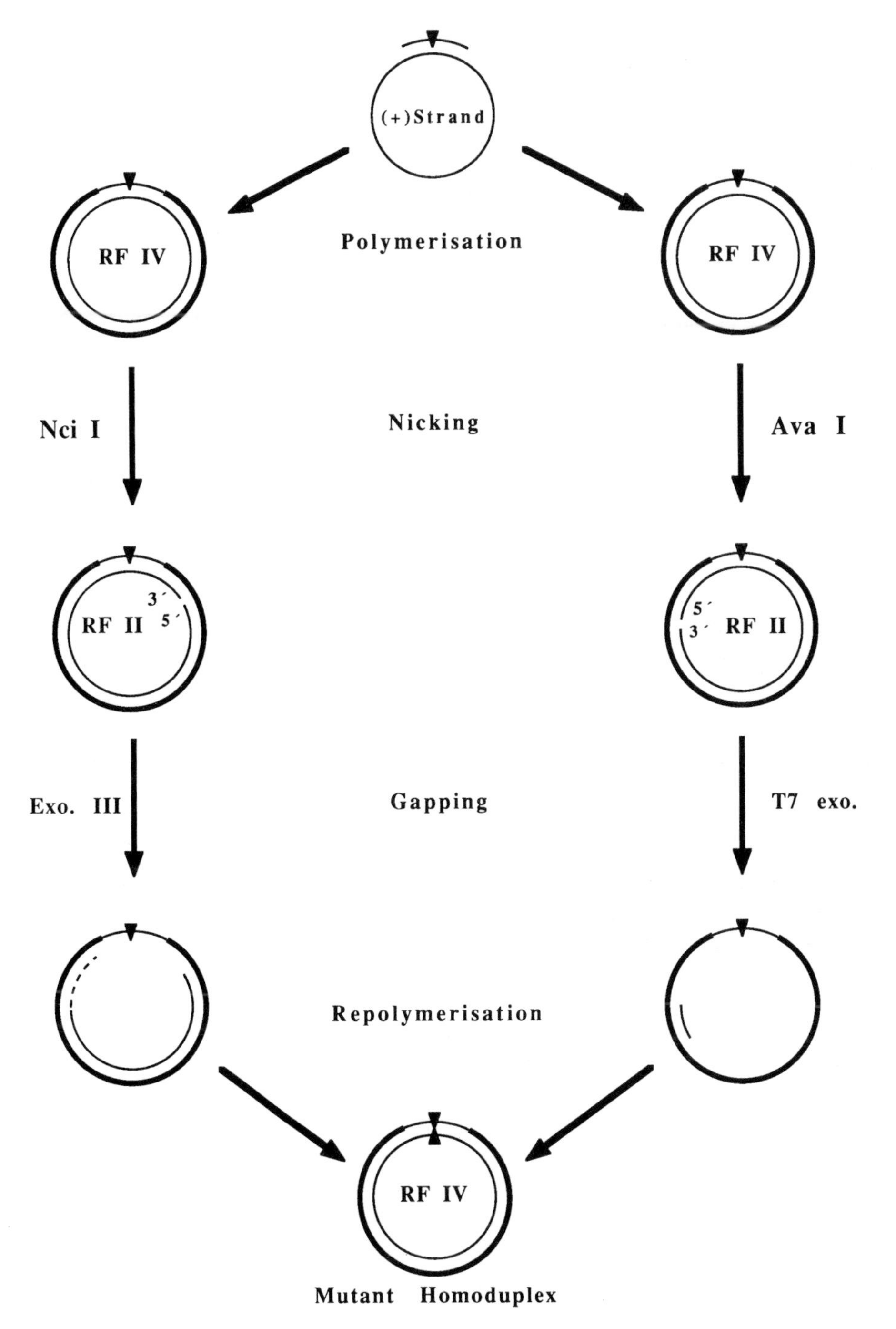

Figure 2. Schematic representation of the phosphorothioate-based mutagenesis method. Single-stranded DNA, annealed with a mismatch primer, is converted to RF IV DNA using Klenow fragment, T4 DNA ligase, and dCTPαS. The newly synthesized (−) strand is shown by the heavy lines. The (+) strand of the phosphorothioate-containing DNA is then nicked by reaction with a restriction endonuclease, such as *Nci*I or *Ava*I. The wild-type sequence is then digested away by either a 3′–5′ or a 5′–3′ exonuclease. A fully complementary homoduplex RF IV molecule is generated on re-polymerization.

strand is extended into a gap by reaction with a suitable exonuclease. Thus, the exonuclease digests away the wild-type sequence opposite the mismatch introduced by the primer. On re-polymerization, the gapped DNA is repaired using the mutant-carrying (−) strand as the template. The mutant sequence is now present in both strands of the DNA as a fully complementary homoduplex species. Transfection with this DNA produces mutational frequencies of the order of 85% (8–11). The polymerization, nicking, gapping and re-polymerization reactions must be performed carefully to obtain the highest possible mutational efficiencies (*Figure 2*).

2.1 **Preparation of single-stranded template DNA**

The protocol shown in *Table 1* produces M13 single-stranded DNA suitable for use in the mutagenesis procedure without the need to perform DNase treatments or caesium chloride gradient purifications (9). This protocol provides approximately 200–300 μg of single-stranded DNA template and may be conveniently scaled down by a factor of three as required. It is important that the template is free from small pieces of RNA or DNA, as such fragments may be capable of acting as primers in the polymerization reaction. A badly contaminated template may ultimately lead to the formation of wild-type double-stranded DNA and thus to reduced mutational efficiency. We routinely perform a self-priming test on our single-stranded DNA (*Table 2*). No significant amounts of polymerized material should be evident from this test.

During all work with phage it is extremely important that all re-usable equipment is sterilized. Phage-contaminated equipment should never be allowed to come into contact with any of the solutions used for transformation of the mutated DNA. Any contaminating wild-type phage will effectively render the procedure useless.

2.2 **The mismatch oligonucleotide**

The size of the mismatch primer obviously depends on the number of mismatches, or the size of deletions or insertions, that are required in the mutant sequence. The oligonucleotide used should have at least three bases to its 3′ side to protect it from the 3′–5′ exonuclease proofreading activity of the DNA polymerase Klenow fragment (14). The number of nucleotides to the 5′ side of the mismatch depends on the quality of the Klenow enzyme used, which should be free of any residual 5′–3′ exonuclease activity. For single or double base mismatches, we routinely use oligomers of 18–22 nucleotides in length, with the mismatch positioned toward the centre; such oligonucleotides bind quite specifically to their target sequence.

A deletion of 16 bases was achieved by using a 24-mer (12 bases either side of the deleted region). Similarly an insertion of 16 bases was created by the use of a 40-mer as the mutagenic primer (12 bases either side of the inserted region). In both cases the mutational efficiency obtained was greater than 80% (10). It is possible to make quite large deletions (180 bp) using relatively short primers (T. Pope and M. Hetherington, personal communication). Such primers are readily

Table 1. Preparation of template DNA.

This protocol assumes that the phage carrying the insert of interest has already been prepared and its sequence determined. Plate out the phage to give single plaques.

1. Prepare an overnight culture of, for example, SMH50 or TG1 cells in 3 ml of 2YT medium[a].
2. Prepare fresh cells by adding one drop of overnight culture from step 1 into 3 ml of fresh 2YT and incubate at 37°C for 3 h in a shaker.
3. Pick a single plaque into 100 μl of fresh cells and incubate overnight at 37°C.
4. At the same time set up another overnight cell culture (3 ml).
5. Inoculate 100 ml of 2YT media (in a 250 ml Erlenmeyer flask) with 1 ml of fresh cells from step 4 and grow, with shaking at 37°C, to an absorbance of 0.3 at 600 nm.
6. Add the phage solution prepared in step 3. Continue incubation for 5 h.
7. Transfer the solution to centrifuge tubes and pellet the cells by centrifugation (e.g. 20 min at 23 000 *g* in a Sorvall centrifuge using a GSA rotor).
8. Immediately decant the supernatant and add $\frac{1}{5}$ the volume of 20% polyethylene glycol (PEG) 6000 in 2.5 M NaCl. Allow the phage to precipitate for 30 min (or overnight) at 4°C.
9. Centrifuge at 3500 *g* for 20 min at 4°C. Discard the supernatant, remove traces of liquid with a screwed up tissue or drawn out pipette.
10. Add 1 ml of TE buffer[b] (pH 7.4), 9 ml of water to re-suspend the phage pellet.
11. Centrifuge at approximately 3500 *g* for 20 min at 4°C. Transfer the phage-containing supernatant to a clean centrifuge tube.
12. Add 2.2 ml of 20% (w/v) PEG in 2.5 M NaCl. Precipitate at 4°C for 30 min. Centrifuge at ~3500 *g* for 20 min at 4°C.
13. Discard the supernatant, dissolve the phage pellet in 500 μl of NTE buffer[c] and transfer to a sterile microcentrifuge tube.
14. Add 200 μl of buffer-equilibrated phenol, vortex for 30 sec, spin briefly in a microcentrifuge. Transfer the aqueous (upper) layer to a fresh microcentrifuge tube.
15. Repeat step 14.
16. Add 500 μl of water-saturated diethyl ether, vortex for 30 sec and spin briefly in a microcentrifuge. Discard the upper (ether) layer. Repeat the process three times.
17. Add 50 μl of 3 M sodium acetate (pH 6), mix and divide the solution into two microcentrifuge tubes.
18. Add 700 μl of absolute ethanol to each tube, cool to −78°C in a dry ice/propanol bath for 60 min. Centrifuge for 15 min in a microcentrifuge.
19. Discard the supernatant, add 700 μl of 70% ethanol, invert the tube to drain off the wash. Take great care not to dislodge the pellet.
20. Mark the tubes A and B, add 50 μl of DNA buffer[d] to tube A. Re-suspend the pellet by repeated vortexing. Transfer the buffer from tube A to tube B.
21. Add another 50 μl of DNA buffer to tube A, vortex, spin briefly and transfer the contents to tube B. Tube B now contains 100 μl of buffer and DNA. Vortex to re-suspend the pellet.
22. Take a 10 μl sample, dilute to 1000 μl and read the optical density on a UV spectrometer at 260 nm and 280 nm in a 1 ml quartz cuvette. The ratio of OD 260/280 should be 1.8 or higher; if not repeat the phenol extraction and following steps. 1 OD_{260} corresponds to ~37 μg of single-stranded DNA. Keep this sample as a standard for gel analysis, use 14 μl diluted with stop mix.

[a] 2YT contains 16 g of tryptone, 10 g of yeast extract and 5 g of NaCl per litre, sterilized by autoclaving.
[b] TE buffer contains 10 mM Tris–HCl pH 8 and 1 mM EDTA.
[c] NTE buffer contains 100 mM NaCl, 10 mM Tris–HCl pH 8 and 1 mM EDTA.
[d] DNA buffer contains 20 mM NaCl, 20 mM Tris–HCl pH 8 and 1 mM EDTA.

Table 2. Self priming test for single-stranded DNA.

This test should be performed routinely.
1. 5 μl of 500 mM Tris–HCl, pH 8.
2. 5 μl of 500 mM NaCl.
3. 5 μl of single-stranded DNA template.
4. Adjust the final reaction volume to 23 μl with water.
5. Incubate at 70°C for 5 min in a water bath. Transfer immediately to a heating block at 37°C and leave for 20 min. Place on ice.
6. Add 2.5 μl of 4 × dNTP nucleotide mix[a].
7. 2 μl of 500 mM Tris–HCl, pH 8.
8. 5 μl of 10 mM ATP.
9. 2 μl of 100 mM $MgCl_2$.
10. 5 units each of DNA polymerase I and T4 DNA ligase. Incubate at 16°C overnight. Remove a 2 μl sample for gel analysis.

[a] The 4 × dNTP nucleotide mix contains dATP, dCTP, dGTP and dTTP at 10 mM in each nucleotide, sterilized by filtration.

available through automated DNA synthesis. One final consideration is that the primer must not contain a recognition site for the restriction enzyme that is to be used in the nicking reaction. Such a site, if present, would lead to the linearization of the DNA during the nicking reaction, as the primer does not contain any phosphorothioate groups.

2.3 Preparation of RF IV DNA

The first step is the phosphorylation of the mismatch primer. The primer must be phosphorylated so that it can function as a substrate in the ligation reaction. This phosphorylation is catalysed by the enzyme polynucleotide kinase and requires ATP. A portion of the phosphorylated primer is then combined with the template DNA. Although the phosphorylated primer may be stored for some time at −20°C, we prefer to use freshly phosphorylated oligomers (*Table 3*). The annealing step is carried out by heating approximately two molar equivalents of primer with the single-stranded template DNA in a high salt buffer. The denatured mixture is then placed immediately at, for example, 37°C, allowing the primer to anneal to the target sequence.

A high quality Klenow fragment is required in the polymerization reaction (*Table 4*). It should be free of all detectable 5′–3′ exonuclease activity. Any such residual activity may serve to digest the mismatch primer from the 5′ end and could ultimately remove the mismatch. The polymerization reaction is usually carried out at 16°C to reduce strand displacement synthesis to a minimum. We usually obtain satisfactory yields of RF IV DNA from such reactions after 16 h. Longer reaction times of up to 40 h do appear to increase the ratio of RF IV to RF II. Indeed, a polymerization started late on a Friday afternoon may safely be left until Monday morning when performed at 16°C. However, reasonable, although lower, yields of RF IV DNA can also be obtained by performing the polymerization reaction at 37°C for 2–3 h.

Table 3. Phosphorylation of mismatch oligonucleotide.

1. Prepare the following mixture of reagents in a sterile 1.5 ml microcentrifuge tube. It is not necessary to use siliconized tubes.
   14 μl of H_2O;
   6 μl of 500 mM Tris–HCl, pH 8;
   2 μl of oligonucleotide primer, (stock solution with an absorbance at 260 nm of 5);
   3 μl of 10 mM ATP;
   3 μl of 100 mM $MgCl_2$;
   2 μl of 100 mM dithiothreitol (DTT);
   5 units of polynucleotide kinase[a].
2. Vortex the tube to mix the contents and centrifuge briefly to collect the solution.
3. Incubate at 37°C for 15 min in a heating block then at 70°C for 10 min in a water bath.
4. Store on ice.

[a] Polynucleotide kinase was obtained from United States Biochemicals.

Table 4. Preparation of RF IV DNA.

Annealing of primer to template DNA
The annealing reaction is normally performed as follows for an oligomer of ~18–24 nucleotides.
1. Prepare the mixture in a sterile microcentrifuge tube:
   10 μl of 500 mM Tris–HCl, pH 8;
   10 μl of 500 mM NaCl;
   6 μl of phosphorylated primer solution prepared as in *Table 3*;
   10 μg of single-stranded DNA template (typically 2–5 μg/μl).
2. Adjust the final reaction volume to 36 μl with water.
3. Incubate at 70°C for 5 min in a water bath.
4. Transfer immediately to a heating block at 37°C and leave for 20 min.
5. Place on ice.

Polymerization
1. Add the following reagents to the template/primer mixture after the annealing mixture has been cooled on ice:
   2.5 μl of 10 mM dATP;
   2.5 μl of 10 mM dGTP;
   2.5 μl of 10 mM dTTP;
   2.5 μl of 10 mM dCTPαS[a];
   10 μl of 10 mM ATP;
   10 μl of 100 mM $MgCl_2$;
   3 μl of 500 mM Tris–HCl, pH 8;
   10 units of Klenow enzyme[b];
   15 units of T4 DNA ligase.
2. Adjust the volume of the reaction to approximately 80 μl with water, mix, spin briefly and then place in water bath at 16°C.
3. Incubate for 16–40 h.
4. Heat-inactivate at 70°C for 10 min.
5. Remove a 2 μl sample for gel analysis. 20 μg of double-stranded DNA is thus prepared from 10 μg of template.

[a] The choice of phosphorothioate depends on which restriction enzyme is to be used in the nicking reaction. They are available from New England Nuclear.
[b] The Klenow enzyme supplied by New England Nuclear is adequate for this polymerization step. Alternatively, Amersham International supply a kit comprising the enzymes and protocols required to perform the entire mutagenesis procedure.

The ratio of Klenow fragment to T4 DNA ligase is an important factor. Too much Klenow favours strand displacement synthesis and is to be avoided. We have observed that a ratio of about 10 units Klenow to 15 units ligase produces the best results. Which of the deoxynucleoside phosphorothioates is used in the RF IV preparation is dictated by the choice of restriction enzyme to be used in the nicking reaction. The restriction enzymes *Nci*I and *Ava*I have been used extensively and both require dCTPαS in the polymerization reaction in order to yield nicked DNA.

2.3.1 *Filtration through nitrocellulose*

The correct performance of this step is extremely important in terms of getting the highest possible mutational efficiency. The RF IV preparation usually contains small amounts of single-stranded DNA. This is, of course, wild-type DNA and is capable of producing wild-type plaques if transfected along with the mutated DNA. Any single-stranded DNA present at the end of the polymerization reaction is efficiently removed by a simple filtration through two nitrocellulose filters. Under the conditions described in *Table 5* single-stranded DNA binds to the filter while polymerized material passes through. After filtration the DNA is subjected to a standard ethanol precipitation. It is extremely important that the

Table 5. Nitrocellulose filtration and ethanol precipitation.

1. Using forceps, place first the rubber seal and then *two* nitrocellulose filters (*do not use autoclaved filters*) in the female part of the filter housing[a].
2. Carefully wet the filter discs with 40 μl of 500 mM NaCl and assemble the unit.
3. Attach a 2ml disposable syringe to the outlet side of the filter unit using a short length of silicon tubing.
4. Add 6 μl of 5 M NaCl to the polymerization reaction. Mix and apply to the inlet side.
5. Slowly draw the sample through the filter unit using the syringe plunger. You may need to tap the top of the housing gently in order to collect the filtrate. Not all of the solution volume is drawn through in the first stage.
6. Apply 50 μl of 500 mM NaCl to the top of the filter unit and draw the wash through.
7. Carefully remove the filter unit and place the filtrate in a fresh sterile microcentrifuge tube. Transfer any remaining droplets with a micropipette.
8. Rinse the syringe with 50 μl of 500 mM NaCl and combine with the filtrate.
9. Add 400 μl of cold absolute ethanol, mix and chill at -78°C for 15 min.
10. Spin in a microcentrifuge for 15 min at ~14 000 r.p.m.
11. Discard the supernatant; a small pellet of salt/DNA should be visible.
12. Carefully add 400 μl of 70% ethanol (ethanol/water 7:3 v/v), invert the tube and check that the pellet is still stuck to the tube. Open the cap so that the liquid drains away. Be very careful not to disturb the pellet.
13. Carefully remove any traces of liquid with a drawn out pipette or a screwed up tissue. Alternatively, dry on a Speed-vac concentrator for 2–3 min.
14. Re-suspend the pellet in a total volume of 250 μl of nicking buffer (see *Table 6*).

[a] Filters are 13 mm in diameter, 0.45 μM pore size (SM11336) supplied by Sartorius. Filter units were from Millipore. The kit as supplied by Amersham International contains filter units that are operated by centrifugal force instead of a syringe system. This gives greater efficiency of DNA recovery.

filtration step be carried out regardless of whether or not any trace of single-stranded DNA is apparent in the gel analysis.

2.4 The nicking reaction

Several restriction endonucleases may be successfully employed to produce a nick in the wild-type (+) strand, although some appear to work better than others (8–11). *Nci*I and *Ava*I have been used most frequently in this laboratory. They both require that dCTPαS be used in the polymerization reaction. Such DNA then contains phosphorothioate linkages to the 5′ side of each dC residue in the (−) strand, except for the region covered by the mismatch primer. It is important that the nicking reaction proceed to completion so that all the RF IV DNA is converted to RF II (nicked) DNA. Any RF IV DNA left after the nicking step cannot be gapped by the exonuclease and lowers mutational efficiency. Protocols for *Ava*I and *Nci*I are given in *Table 6*.

Table 6. Strand-specific nicking of RF IV DNA.

RF IV DNA containing dCMPS may be nicked by the enzyme *Nci*I[a] as follows.

1. Re-suspend the pellet from *Table 5* in 190 μl of water and add:
   - 6 μl 500 mM Tris–HCl, pH 8;
   - 25 μl of 100 mM DTT;
   - 15 μl of 100 mM $MgCl_2$;
   - 15 μl of 500 mM NaCl;
   - *Nci*I (120 units).
2. Incubate at 37°C for 90 min.
3. Heat-inactivate the enzyme at 70°C for 10 min.
4. Keep a 6 μl sample for gel analysis.

RF IV DNA containing dCMPS in the (−) strand may also be nicked with the enzyme *Ava*I.

1. Re-suspend the DNA pellet (*Table 5*, ~20 μg) in 160 μl of water.
2. Vortex to dissolve the pellet.
3. Add the following solutions:
   - 13 μl of 500 mM Tris–HCl, pH 8;
   - 13 μl of 100 mM DTT;
   - 25 μl of 100 mM $MgCl_2$;
   - 37 μl of 500 mM NaCl;
   - *Ava*I (70 units).
4. Incubate at 37°C for 180 min.
5. Heat-inactivate at 70°C for 10 min. Keep a 6 μl sample for gel analysis. The same protocol may also be followed for nicking with the enzymes *Ava*II[b] and *Ban*II[c].

[a] The double site in the polylinker region (CCCGGG 6247/8) of M13mp18 is not nicked at all by *Nci*I when dCMPS is present in the (−) strand. The nearest 'nickable' *Nci*I site to the polylinker is the one at 6838.

[b] *Ava*II may be used if dGTPαS was used in the polymerization reaction.

[c] Note that the *Ban*II recognition sequence GAGCTC present in the polylinker of M13mp18 is linearized by this enzyme. *Ban*II is thus unsuitable for a vector containing this site. However, the site GGGCTC in, for example, M13mp2, mp7, mp8, mp9 or mp10 may be nicked with DNA containing dCMPS. Restriction enzymes were obtained from Amersham International, Boehringer Mannheim or New England Biolabs.

2.5 The gapping reaction

Exonuclease III was originally used in the method. It digests double-stranded DNA containing a free 3′ terminus in the 3′–5′ direction. Thus, a restriction site to the 3′ side of the target mutation relative to the (+) strand is required. Work with the M13mp18 vector has frequently used exonuclease III in conjunction with *Nci*I to degrade the (+) strand. *Nci*I sites are at positions 1924, 6247, 6248 and 6838. Interestingly enough the double site in the polylinker region (CCCGGG 6247/8) is not nicked at all when dCMPS is present in the (−) strand (9). The nearest nickable *Nci*I site to the polylinker is the one at position 6838. This is important to note, as exonuclease III gaps nicked DNA at a rate of about 100 nucleotides per minute under the conditions shown in *Table 7*. A mismatch site (e.g. in the middle of the polylinker) in an insert of 1000 bases would therefore be approximately 1100 bases distant from the *Nci*I site at position 6838. To ensure that the mismatch (+) strand is completely digested, a reaction time of about 20 min (allowing a safety margin) would be required. Exonuclease III may, of course, be used in conjunction with restriction endonucleases other than *Nci*I provided that the buffer conditions are adjusted accordingly.

The exonuclease III reaction is very sensitive to buffer conditions and the protocol should be followed exactly. The procedure described assumes that the nicked DNA is in the *Nci*I reaction buffer. This is first converted to a higher salt concentration, which is essential for reproducible gapping of the DNA with exonuclease III. Obviously the enzyme may be used in conjunction with other restriction enzymes capable of producing nicked DNA if required. We have found that exonuclease III gaps best in a buffer containing approximately 100 mM NaCl, 50 mM Tris–HCl (pH 8), 6 mM $MgCl_2$ and 10 mM dithiothreitol (DTT), using 15 units of exonuclease per μg of nicked DNA. Gel analysis of the exonuclease III gapping reaction should reveal a distinct band whose electrophoretic mobility increases progressively with longer incubation time. Extended reaction is possible because exonuclease III can remove most of the (+) strand (10), resulting in a discrete band migrating very close to a marker of single-stranded DNA. However, we have observed that large amounts of DNA may be destroyed by contaminating endonucleases if the reaction is allowed to continue for too long.

The 5′–3′ exonucleases T7 and λ exonuclease gap nicked double-stranded DNA

Table 7. The gapping reaction using exonuclease III.

The gapping reaction with exonuclease III is usually performed on *Nci*I-nicked DNA. This protocol assumes that the nicked DNA is initially present in the *Nci*I nicking buffer.

1. Add 4 μl of 5 M NaCl to the *Nci*I-nicked DNA (*Table 6*).
2. Add ~200 units of exonuclease III[a], vortex, spin briefly and incubate at 37°C for the time period required to gap through the mismatch (assume 100 bases per min with a safety margin of +50%).
3. Heat-inactivate the exonuclease at 70°C for 15 min. Place on ice.
4. Remove an 8 μl sample for gel analysis.

[a] Exonuclease III was obtained from New England Biolabs.

Table 8. The gapping reaction using T7 or λ exonuclease.

Both of these exonucleases appear to function in the buffers used for performing the nicking reaction.

1. Add 5 units of T7 exonuclease per μg double-stranded DNA present, i.e. 100 units for the *Ava*I-nicked sample from *Table 6*.
2. Incubate at 37°C for 30 min, heat-activate at 70°C for 10 min. Place at 37°C for 20 min, then on ice.

Alternatively, nicked DNA may be gapped by using λ exonuclease.

1. Add 4 units of λ exonuclease per μg double-stranded DNA present, i.e. 80 units for the *Ava*I-nicked sample from *Table 6*.
2. Incubate at 37°C for 120 min, heat-inactivate at 70°C for 10 min. Place at 37°C for 20 min, then on ice.
3. Remove a 14 μl sample for gel analysis.

T7 Exonuclease may be obtained from United States Biochemicals, λ exonuclease from New England Biolabs.

in the opposite direction to exonuclease III. Both are capable of removing almost all the nicked (+) strand under normal conditions, but unlike exonuclease III, discrete partially-gapped DNA species cannot be detected by gel analysis. Commercial samples of these enzymes appear to be endonuclease free, and prolonged incubation with either is usually possible. These exonucleases have been used in conjunction with restriction enzymes possessing a recognition site to the 5′ side of the insert. For most purposes the protocol given in *Table 8* produces a high degree of gapping; T7 exonuclease removes most of the (+) strand in less than 15 min with nicked M13mp18 as substrate. This latter exonuclease has emerged as the reagent of choice for the gapping reaction because it is available in concentrated form, is tolerant of variations in buffer conditions, and does not destroy the (−) strand. The choice of exonuclease will depend on position of the nick relative to the insert of interest. Whichever gapping system is chosen, it must be heat-inactivated when complete.

The behaviour of the three exonucleases toward nicked, mismatch primed DNA is shown in *Figure 3*. A partial exonuclease III digest, allowed to continue for approximately 10 min, is shown together with T7 and λ exonuclease digests of the same DNA.

2.6 Preparation of the mutant homoduplex: the re-polymerization step

The gapped DNA must be re-polymerized before transfection in order to obtain the best results. Transfection of gapped DNA seems to result both in low plaque yields and very low mutational efficiency. The re-polymerization reaction is carried out with DNA polymerase I (not Klenow fragment) in the presence of T4 DNA ligase (*Table 9*). The polymerase should be of high quality (i.e. endonuclease free). The four natural deoxynucleoside triphosphates are used in this reaction. It may be carried out at 16°C or 37°C as is convenient. Note that the re-polymerization reaction must contain ATP as well as dATP, as was the case with the original RF IV preparation.

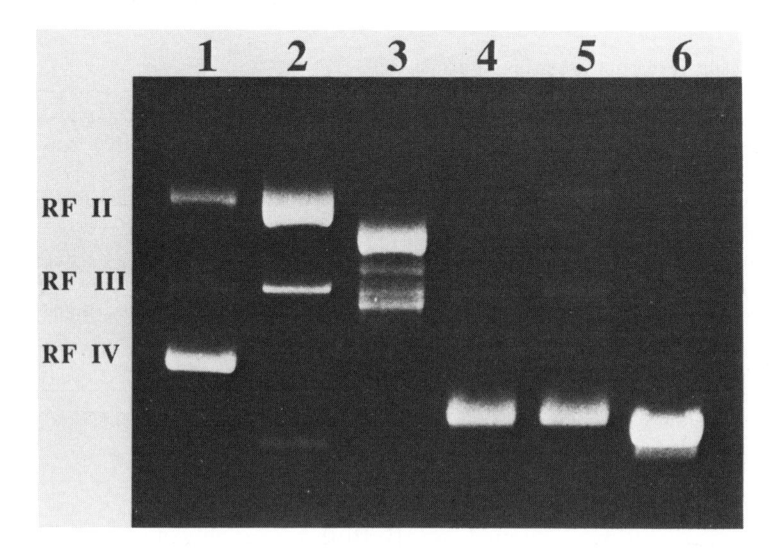

Figure 3. Gapping of nicked DNA with different exonucleases analysed by agarose gel electrophoresis. **Lane 1**, RF IV DNA prepared from a mismatch primer; **lane 2**, RF II DNA prepared by reaction with *Ava*I or *Nci*I; **lane 3**, partial gapping of RF II DNA with exonuclease III, reaction terminated after 10 min; **lanes 4** and **5**, gapping of RF II DNA with T7 and λ exonucleases, respectively; **lane 6**, marker of single-stranded circular M13 DNA.

Table 9. Re-polymerization.

1. To the gapped DNA solution prepared in *Table 7* or *8* add:
   20 μl of 10 mM ATP;
   5 μl of 4 × dNTP mix, 10 mM in each nucleotide (*Table 2*);
   10 units of DNA polymerase I;
   10 units of T4 DNA ligase.
2. Incubate at 16°C overnight or at 37°C for 180 min.
3. Remove a 14 μl sample for gel analysis.

2.7 Transfection

One advantage of the phosphorothioate-based mutagenesis method is that specialized cell lines are not required for transfection or for growth of the template DNA. Any cell line suitable for the growth of M13 may be used. We have found cell lines such as SMH50 (15) and TG1(16) are particularly useful in that they give consistently high transfection efficiencies (i.e. highest number of plaques per μg DNA transfected). A transfection protocol is described for competent cells prepared by the $CaCl_2$ method (*Table 10*). Obviously other transfection protocols may be used if preferred (17). Low plaque yield may be countered by precipitating the mutant DNA in 4 M ammonium acetate (18), or by passing the sample through

Table 10. Preparation of competent cells.

1. Add 3 ml of an overnight culture of, for example, SMH50 or TG1 cells to 100 ml of sterile 2YT media in a 250 ml Erlenmeyer flask.
2. Incubate in a shaker at 37°C, allow to grow to an absorbance of 0.6 at 660 nm.
3. Transfer the cells to suitable sterile centrifuge tubes, cap and spin at ~3000 *g* for 15 min at 4°C.
4. Discard the supernatant and re-suspend the cells in a total volume of 50 ml pre-chilled 50 mM $CaCl_2$ solution (sterile).
5. Leave on ice for 30 min. Centrifuge as in step 3.
6. Discard the supernatant and re-suspend the cells in a total volume of 20 ml of pre-chilled 50 mM $CaCl_2$ solution (sterile). Store at 4°C.

Cells prepared in this manner may be used for up to 1 week but produce the best results ~24–48 h after this treatment.

Table 11. Transfection.

1. Chill five sterile 10 ml polypropylene tubes on ice.
2. Add 300 μl of competent cells (*Table 10*) to each tube.
3. Dilute 20 μl of re-polymerized DNA solution to 50 μl with sterile water.
4. Add 2 μl, 5μl, 10μl and 20μl portions of diluted DNA (step 3) to the competent cells.
5. To the 5th tube make a mock transfection with 20 μl of sterile water used in diluting the DNA.
6. Swirl the tubes gently to mix the contents and place on ice for ~40 min.
7. Place 3 ml portions of molten top agar at 55°C in a water bath (in sterile polypropylene tubes).
8. Combine 1400 μl of fresh cells with 280 μl of aqueous IPTG and 280 μl of X-gal in dimethylformamide[a].
9. To each aliquot of transformed competent cells (steps 4 and 5), add 270 μl of fresh cell mix from step 8.
10. Add 3 ml of top agar to each mixture and plate out immediately on plates pre-warmed to 37°C. Allow to set and invert.
11. Incubate overnight at 37°C.
12. Pick 2–5 (blue) plaques for sequencing.

[a] Prepare the IPTG and X-gal solutions immediately before use. Dissolve 30 mg IPTG in 1 ml water; sterilize by filtration. 20 mg of X-gal should be suspended in 1 ml of dimethylformamide; do not attempt to sterilize!

G-50 Sephadex in a spun column (18). Both of these methods remove nucleoside triphosphates that may be responsible for low plaque yield (19). Even if only little RF IV can be detected by gel analysis after the re-polymerization, it is worth transfecting a sample of the DNA anyway.

3. MONITORING THE PROCEDURE AND DE-BUGGING

Each stage of the phosphorothioate-based mutagenesis procedure is readily monitored by gel electrophoresis. This simple analytical technique is performed in horizontal 1% (w/v) agarose slab gels, containing 0.4 μg/ml ethidium bromide (18). We recommend that the novice checks each stage of the mutagenesis procedure before carrying out the next step. This saves time in the long run. Once

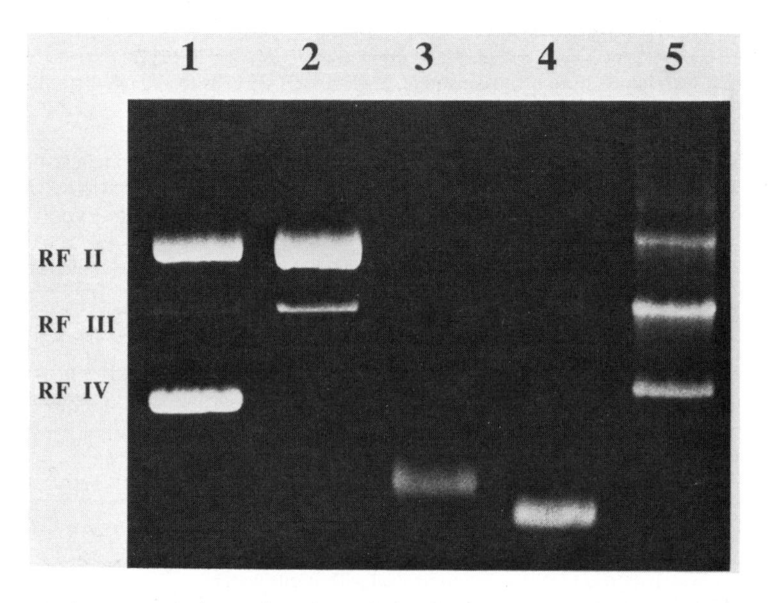

Figure 4. Agarose gel electrophoresis analysis of a complete mutagenesis procedure. **Lane 1**, polymerization reaction with dCTPαS; **lane 2**, *Ava*I nicking reaction; **lane 3**, gapping of RF II DNA with T7 exonuclease; **lane 4**, marker of single-stranded circular M13 DNA; **lane 5**, re-polymerization of gapped DNA.

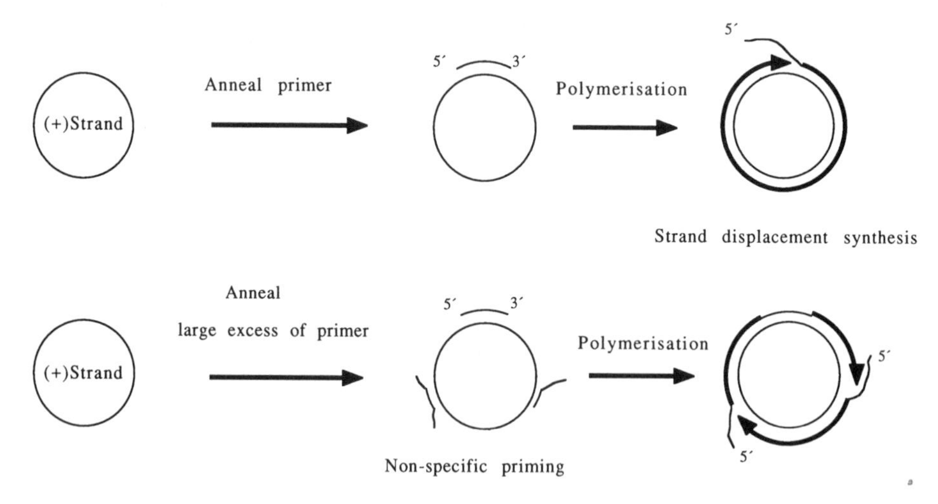

Figure 5. Two possible causes of unligatable RF II-like species. In the upper diagram, the correct amount of primer has been annealed, but strand displacement results due to, for example, the use of excess of Klenow enzyme over ligase. The growing (−) strand eventually displaces the 5′ end of the mismatch primer. In the lower diagram unligatable species arise due to annealing too much primer. Non-specific priming may occur, again leading to strand displacement synthesis.

the techniques have been mastered, the analysis can be performed at the end of the sequence to save time. It is seldom worth transfecting a sample of hopefully mutant DNA if, for example, the nicking reaction failed! A gel electrophoresis analysis is shown for the complete reaction sequence from polymerization to re-polymerization (*Figure 4*). Note that it is advisable to run the samples along with a single-stranded DNA standard.

Gel analysis is very useful in fault diagnosis. A polymerization reaction that produces little or no RF IV is readily detected. Such a result may indicate problems in the phosphorylation step or it may throw suspicion on the ligation reaction. Both of these reactions require ATP. A broad band in the vicinity of RF II, with little or no RF IV DNA present, may also indicate that the polymerization reaction has not reached completion; a longer incubation period may help.

The use of a large excess of primer may result in the primer initiating polymerization at sites other than that desired, resulting in unligatable DNA, as the Klenow fragment has no 5′–3′ exonuclease activity to digest away non-complementary ends (*Figure 5*). Such non-specific priming is evidenced by the polymerization reaction resulting in mostly RF II-like material with bands of higher molecular weight DNA visible. The use of too little primer is easily identified as large amounts of single-stranded DNA will be detected in the gel analysis if performed prior to the nitrocellulose filtration step.

The nicking reaction should convert all RF IV to RF II DNA, with the minimum of linearization. Should linearization result (i.e. RF III DNA), either the primer contains a recognition site for the restriction enzyme used or the wrong dNTPαS was employed in the polymerization reaction! It must be realized that not all the recognition sites of a restriction enzyme get cleaved at the same rate. The protocols given for *Ava*I, *Ava*II, *Ban*II and *Nci*I (*Table 6*) normally result in complete conversion of RF IV to RF II in the vectors we have used. Should the nicking reaction leave some RF IV DNA, it must be repeated or extra enzyme added to ensure complete conversion of RF IV to RF II DNA.

The gapping reaction is easily monitored in the case of the 5′–3′ exonucleases because they usually gap the DNA almost completely, although faint traces of RF II may remain. DNA gapped with either of these exonucleases will often run very close to a marker of single-stranded DNA, and intermediately gapped species are not usually observed. The mobility of DNA that has been gapped with exonuclease III will depend on the duration of the gapping reaction. Gaps of less than about 400 bases may be very difficult to visualize. The re-polymerization reaction usually yields reasonable amounts of RF IV DNA, although an increased amount of linearized material is observed. Should the only products be RF II and RF III DNA, then the ligase or ATP has probably been left out of the reaction mixture.

4. MODIFICATIONS OF THE BASIC SYSTEM

As with any general scheme there will be occasions when the procedure needs to be varied. The protocols given are based on primers in the 18–24 base size range. The use of primers designed to introduce deletions or insertions may require slightly different annealing conditions; for example, a 40-mer designed to introduce an insertion of 16 nucleotides was subjected to an intermediate heating phase in the standard annealing protocol. Thus, after the 70°C denaturation, the mixture was placed at 55°C for 20 min before being placed at 37°C for a further 30 min. This resulted in production of mostly RF IV DNA. Similarly, longer primers may require variations in the annealing step. The ratio of primer to template may also be varied over the range (e.g. 1:1 to 6:1). However, at higher primer to template ratios, the risk of non-specific priming increases, resulting in incomplete ligation (*Figure 5*).

Another interesting feature of insertion or deletion mutagenesis is that the gapping reactions are slowed down or arrested by the single-stranded regions created by such a mismatch primer. In the case of a deletion the exonuclease must initially hydrolyse duplex DNA. Once the looped out deletion is reached, the exonuclease has to gap a single-stranded region of DNA (i.e. the loopout). The effect of an insertion primer is to place the loopout structure on the (−) strand; again the exonuclease has to pass through a shorter, single-stranded region. The exonucleases mentioned gap through such single-stranded regions quite poorly. However, DNA polymerase I, which is used in the re-polymerization reaction, destroys the remaining wild-type strand, allowing high mutational efficiency to be maintained. Gel analysis of such reactions may seem somewhat unusual, in that the fully gapped species usually observed with T7 or λ exonuclease may not appear. Instead a band with intermediate electrophoretic mobility will be seen.

One of the restriction endonucleases mentioned will usually prove useful for most practical purposes, as we have a recognition site for *Nci*I to the 3′ side of the polylinker in M13mp18, for example. Whilst the restriction enzymes *Ava*I and *Ava*II have recognition sites to the 5′ side of the polylinker. *Ava*II requires dGTPαS in place of dGTP in the polymerization reaction, thus producing RF IV DNA containing dGMPS in the (−) strand; otherwise the procedure for nicking with this enzyme is the same as that for *Ava*I. *Ban*II may be used with several vectors such as M13mp2, mp8 or mp10. *Ban*II cannot be used with the vector M13mp18, because the GAGCTC site in the polylinker is linearized by this enzyme. However, these are not the only enzymes capable of giving nicked DNA. Other enzymes, such as *Hin*dII, which requires dGTPαS, and *Pvu*I, which requires dCTPαS during polymerization, may also be used provided that the restriction enzyme is removed from the reaction by phenol extraction after completion of nicking (8). We have recently extended the range of restriction endonucleases suitable for use in the mutagenesis protocol by performing the reaction in the presence of ethidium bromide. In this way it was possible to use *Bgl*I, *Hpa*II, *Hin*dIII and *Pvu*II in the mutagenesis procedure (11). However, in most instances either *Nci*I or *Ava*I should prove effective in the procedure.

5. SCOPE AND LIMITATIONS OF THE PROCEDURE

The DNA to be mutated must be present in a single-strand form, for example as an insert in one of the M13 derivatives. Alternatively, a vector containing a single-stranded phage origin of replication may be used [e.g. the pEMBL plasmids (20)]. The recombinant DNA must have a restriction site for one of the restriction enzymes capable of producing nicked DNA. This site may be in the vector itself or in the insert. The oligonucleotide used to introduce the mutation must not contain a site for the restriction enzyme, which would simply linearize the DNA.

Single and multiple base mismatches are equally accessible, as are deletions and insertions. The size of a possible insertion is obviously limited by the length of reasonably pure oligonucleotide that may be prepared. Small insertions could also be used to enable easier cloning of an insert into an expression plasmid (e.g. by creating different sticky ends compatible with the new vector). Frameshift mutations may be created simply by deleting or inserting one or two bases in the coding region of the insert, for example.

Specialized host cells are not required, either for amplification or transfection. Due to the high efficiency of mutation it is usually sufficient to sequence a small number of clones. Hybridization experiments are not required and the sequencing, or restriction analysis, of large numbers of clones should be a thing of the past.

6. REFERENCES

1. Knowles, J. R. (1987) *Science,* **236**, 1252.
2. Shaw, V. W. (1987) *Biochem. J.,* **246**, 1.
3. Messing, J. (1983) In *Methods in Enzymology.* Wu, R., Grossman, L. and Moldave, K. (eds), Academic Press, New York, Vol. 101, p. 20.
4. Biggin, M. D., Gibson, T. J. and Hong, G. F. (1983) *Proc. Natl. Acad. Sci. USA,* **80**, 3963.
5. Zoller, M. J. and Smith, M. (1982) *Nucleic Acids Res.,* **10**, 6487.
6. Friedberg, E. C. (1985) *DNA Repair*, W. H. Freeman, New York.
7. Sayers, J. R. and Eckstein, F. (1988) In *Genetic Engineering: Principles and Methods.* Setlow, J. K. (ed.), Plenum Press, New York and London, Vol. 10, p. 109.
8. Taylor, J. W., Ott, J. and Eckstein, F. (1985) *Nucleic Acids Res.,* **13**, 8765.
9. Nakamaye, K. L. and Eckstein, F. (1986) *Nucleic Acids Res.,* **14**, 9679.
10. Sayers, J. R., Schmidt, W. and Eckstein, F. (1988) *Nucleic Acids Res.,* **16**, 791.
11. Sayers, J. R., Wendler, A., Schmidt, W. and Eckstein, F. (1988) *Nucleic Acids Res.,* **16**, 803.
12. Potter, B. V. L. and Eckstein, F. (1984) *J. Biol. Chem.,* **259**, 14243.
13. Taylor, J. W., Schmidt, W., Cosstick, R., Okruszek, A. and Eckstein, F. (1985) *Nucleic Acids Res.,* **13**, 8749.
14. Gillam, S. and Smith, M. (1979) *Gene,* **8**, 81.
15. LeClerc, J. E., Istock, N. L., Saran, B. R. and Allan, R. (1984) *J. Mol. Biol.* **180**, 217.
16. Carter, P., Bedouelle, H. and Winter, G. (1985) *Nucleic Acids Res.,* **13**, 4431.
17. Hanahan, D. (1985) In *DNA Cloning: A Practical Approach.* Glover, D. M. (ed.), IRL Press, Oxford, Vol I, p. 109.
18. Maniatis, T., Fritsch, E. F. and Sambrook, J. (1982) *Molecular Cloning: A Laboratory Manual* Cold Spring Harbor Laboratory Press, Cold Spring Harbor, New York.
19. Taketo, A. (1974) *J. Biochem.,* **75**, 895.
20. Dente, L., Sollazzo, M., Baldari, C., Cesareni, G. and Cortese, R. (1985) In *DNA Cloning: A Practical Approach*, Glover, D. M. (ed.), IRL Press, Oxford, Vol. I, p. 101.

APPENDIX

Suppliers of specialist items

Aldrich Chemical Company, Milwaukee, WI, USA; The Old Brickyard, New Road, Gillingham, Dorset SP8 4JL, UK
Amersham International, Amersham Place, Little Chalfont, Buckinghamshire HP7 9NA, UK; 2636 S. Clearbrook Drive, Arlington Heights, IL 60005, USA
Analtech Inc, 75 Blue Hen Drive, PO Box 7557, Newark, DE 19711, USA
Anglian Biotec, Whitehall House, Whitehall Road, Colchester, Essex CO2 8HA, UK
BDH Ltd, Broom Road, Poole, Dorset BH12 4NN, UK
Beckman Instruments, 2500 Harbor Boulevard, PO Box 3100 Fullerton, CA 92634, USA
Bio-Rad, 1414 Harbour Way South, Richmond, CA 94804, USA; Caxton Way, Watford Business Park, Watford, Hertfordshire WD1 8RP, UK
Boehringer Mannheim Gmbh, Postfach 31 01 20, D-6800 Mannheim 31, West Germany; PO Box 50816, Indianapolis, IN 46250, USA
Brinkmann Instruments, Cantiague Road, Westbury, NY 11590, USA
BRL, see Gibco BRL
DuPont Company, Biotechnology Systems Division, BRML, G-50986, Wilmington, DE 19898, USA; Wedgwood Way, Stevenage, Hertfordshire SG1 4QN, UK
Eppendorf, PO Box 65 06 70, D-2000 Hamburg, West Germany; distributed in USA by Brinkmann Instruments.
Gibco BRL, Grand Island, NY, USA; PO Box 35, Trident House, Renfrew Road, Paisley PA3 4EF, UK
Hoefer Scientific Instruments, 654 Minnesota Street, PO Box 77387, San Francisco, CA 94107, USA
IBI Ltd, 36 Clifton Road, Cambridge CB1 4ZR, UK
Idea Scientific Co., Box 2078, Corvallis, OR 97339, USA
Janssen Life Sciences Products, Lammerdries 55, B-2430 Olen, Belgium
Joyce-Loebl, Marquisway, Team Valley, Gateshead, Tyne & Wear NE11 0QW, UK
LKB, see Pharmacia LKB Biotechology
E. Merck, Frankfurter Strasse 250, Postfach 4119, D-6100 Darmstadt 1, West Germany; EM Reagents distributed in USA by Brinkmann Instruments.
Millipore Corp., 80 Ashby Road, Bedford, MA 01730, USA; B.P. 307, Saint-Quentin, F-78054, France
New England Biolabs, 32 Tozer Road, Beverley, MA 01915-9990, USA; Postfach 2750, 6231 Schwalbach/Taunus, West Germany
New England Nuclear, see Dupont

Pharmacia LKB Biotechnology AB, S-75182 Uppsala, Sweden; 800 Centennial Avenue, Piscataway, NJ 08854, USA
Pierce Chemical Company, Rockford, IL 61105, USA; PO Box 1512, 3260 BA Oud-Beijerland, The Netherlands
Sartorius Gmbh, Weender Landstrasse 94/108, Postfach 3243, D-3400 Göttingen, West Germany; 30940 San Clemente Street, Hayward, CA 94544, USA
Schleicher and Schuell, 10 Optical Avenue, Keene, NH 03431, USA; Hahnestrasse 3, Postfach 4, D-3354 Dassel, West Germany
Schwarz Mann, Division of Becton Dickinson Corporation, Mountain View, CA 94043, USA
Sharples Pharmaceutical and Biotechnology Products, 955 Mearns Road, Warminster, PA 18974, USA
Sigma Chemical Co., PO Box 14508, St Louis, MS 63178, USA; Fancy Road, Poole, Dorset BH17 7NH, UK
Spectrum Medical Industries, 60916 Terminal Annex, Los Angeles, CA 90054, USA
Ultra-Violet Products Ltd, Science Park, Milton Road, Cambridge CB4 4FH, UK
Union Carbide, 6733 West 65th Street, Chicago, IL 60632, USA
United States Biochemical Corp., PO Box 22400, Cleveland, OH 44122, USA
UV Photoproducts, see Ultra-Violet Products
Wako Pure Chemical Industries, 12300 Ford Road, Suite 130, Dallas, TX 75234, USA; Nissanstrasse 2, D-4040 Neuss 1, West Germany
Yellow Springs Instrument Co., Box 279, Yellow Springs, OH 45387, USA

INDEX